Plants for Dry Climates
How to Select, Grow and Enjoy

Mary Rose Duffield & Warren D. Jones

Contents

Published by H.P. Books
P.O. Box 5367, Tucson, AZ 85703   602/888-2150
ISBN 0-89586-042-2 Library of Congress Catalog Card Number 80-82535
©1981 Fisher Publishing, Inc. Printed in U.S.A.

The arid environment is a perfect example of an overdrawn bank account. Evaporation of water exceeds the amount "deposited," preventing any "savings" of moisture. Dealing with this deficit is what this book is about.

The Arid Environment

Growing plants in a hot, dry, sunny environment is pioneering in a very real sense. Desert areas present special challenges: Extremes of heat, aridity and problem soils make gardening different, even difficult at times.

Experienced gardeners in these regions have been taught many lessons. Out of necessity they have learned which plants accept the rigors of the climate, or have devised ways to grow and protect plants unadapted to these extreme conditions.

This book is about growing plants in the desert and similar climates. It too, is a pioneering effort, dedicated to helping you create your own low-maintenance, energy-saving, drought-resistant landscape.

DEFINING AN ARID CLIMATE
What makes a locale arid is not how much rain falls, but how much *accumulates* in the soil. A desert is a perfect example of an overdrawn bank

Many think it's barren, but the natural desert has a mood all its own. Plants adapted to dry climates can create a special elegance, as above.

account. Evaporation of water exceeds the amount of rainfall "deposited" on the ground, preventing any penetration or "savings" of moisture. This imbalance between rainfall and soil moisture and evaporation is what all deserts have in common, whether it is southern coast or hot interior valleys of California, the high cool plateau of northern Mexico or the hot sands of the Sahara.

Low rainfall, 5 inches or less per year, also makes a region arid, but there are places that get much more precipitation that are classified as arid. *When* and *how* a region gets its rainfall are very important factors. An area can have as much as 15 inches annually and still be arid if rainfall is spread evenly over the year. Periods of windy, very dry weather between rains cause evaporation so moisture is unable to accumulate in the soil. If rain falls in a concentrated time span, especially if the air is cool and moist at that time, moisture will accumulate. The area is then able to support a solid cover of grass and occasionally, shrubs and trees.

Heat is *not* a standard ingredient of desert climates. For instance, Saltillo, the capital city of the Mexican state of Coahuila, is considered a prototype for the dry, cool, subtropic Chihuahuan Desert. Summers are pleasant, and it has become something of a resort—a place to escape the summer heat of the northeast Mexican lowlands.

Low humidity is not a universal indicator for a desert, either. Some of the most desolate are fog deserts. The bleak, northern Chilean Coast and the western coast of Baja California are examples. Cold ocean currents along shore produce a cool climate with little rain and no summer heat.

There are many kinds of arid climates, including deserts in cold, higher latitudes that merge into the Arctic Circle. This book, however, concentrates on subtropical to somewhat cool deserts, and the adjoining regions, the arid grasslands. Arid grasslands are regions that are dry the balance of the year, briefly supporting a cover of annual grass. They are as hot and windy as true deserts, and support many of the same landscape plants.

Arid Climates

Warm, arid climates can be grouped into three general zones, and the arid grasslands that surround these zones. The zones are based on how often the temperature drops below freezing, how far it goes below freezing and how long it stays there. No matter what you do to modify climate, cold is the deciding factor. All plants have a cold tolerance point below which they are severely damaged or killed.

Low elevation climates are the closest to tropical climates. Elevation ranges from sea level (some areas are even below sea level) to 2,000 feet. In the lowest desert latitudes this climate will extend to a slightly higher elevation. The growing season—days between killing frosts—ranges from year-round along the coast to 302 days in Phoenix, Arizona. Average winter minimum temperature is 36° to 37°F (2°C). Freezing temperatures can occur, even in coastal deserts, and drop to 20°F (−7°C) in some locations. Average summer maximum temperatures are near 102°F (39°C), and are much higher in drier interior locations. Highs of 120°F (49°C) are not unusual. Summer nights also remain very warm, often staying above 80°F (27°C). Annual rainfall is 10 inches or less; some areas receive less than 5 inches. Wide temperature variations exclude some tropical plants inland which thrive in milder coastal climates. Fall months signal the beginning of the planting year in this zone. Early fall planting of annual flowers permits a full life cycle before the high temperatures of late April and May.

Medium elevation climates generally have mild winters. Their average 2,500-foot elevation causes the growing season to be shorter than in low climates, ranging from 220 to 242 frost-free days. Occasionally, these zones can reach a minimum temperature of 15° to 18°F (−10° to −8°C). Summer maximum temperatures will also be lower by 5°F (−3°C) or more than low climates. Summer nights are more comfortable, often dropping to a cool 70°F (21°C) or lower. These moderate summer nights, along with an earlier cooling trend in the fall, allow many cool-weather plants to succeed. Precipitation is generally higher than the low zone, especially in regions subject to summer rains, such as Tucson, Arizona. Some areas have summer *and* winter rains, receiving as much as 10 to 15 inches annually. This relatively high rainfall does not take these areas out of the desert classification. Evaporation during the long, dry periods between rains prevents any real accumulation or penetration of moisture into the soil. Subtropicals and tender plants must be protected from the hard frosts that can occur here. September, October and November are ideal months for planting annuals, perennials and basic landscape plants. This allows time for plants to become well established through the winter and spring before summer heat.

High elevation climates between 3,300 and 5,000 feet are transitional regions and many temperate zone plants can be grown here. The growing season lasts between 200 to 220 days. Areas that are located at higher elevations in lower latitudes will have about the same expected annual low temperatures as other high elevation zones, but the growing season will be longer with frost ending earlier in the spring. Temperatures approaching 0°F (−20°C) have been experienced in most high desert regions but 15° to 18°F (−10° to −8°C) is the usual low for the season. Minimum winter temperatures average around freezing, providing the necessary chilling for such landscape plants as lilac and crab apple. Winters are mild enough and summers are hot enough in most of this zone to grow *Oleander*, *Eriobotrya* (loquat), *Magnolia*, *Ligustrum* (glossy privet) and *Punicia* (pomegranate) species. Rainfall is generally higher because most of these regions are close to mountain ranges.

Arid grasslands, or savannahs, have growing conditions similar to the three climate types described. In fact, they often border these regions. The only difference is, when rain falls, it comes at a time when it can accumulate in the soil. This enables it to support briefly a complete cover of grass, usually an annual grass. Bakersfield in California's interior valley, as well as the warm, dry valleys of southern California and regions of the southern Mediterranean, are examples of this climate. The low rainfall coastal strip along the southern California Coast is an arid grassland, except summers are cool. Plants that require high heat such as mesquite and palo verde, may not bloom and grow very slowly. But the majority of plants in this book perform well in this region. These areas have the same variations in temperature and modifying conditions as desert climates. Rainfall is generally as low as the bordering desert or slightly higher (7 to 15 inches). Most home landscapes need to be irrigated 9 to 10 months of the year—sometimes all year.

Yreka · Willow Creek · Eureka · Redding · Chico · Ukiah · Yuba City · Sacramento · Napa · Oakland · Modesto · San Francisco · San Jose · Fresno · Santa Cruz · Salinas · Monterey · San Luis Obispo · Bakersfield · Santa Barbara · Lancaster · Los Angeles · Long Beach · Banning · Palm Springs · San Diego · Tijuana · Ensenada · El Centro · Mexicali · Yuma · San Felipe · Winnemucca · Elko · Reno · Carson City · Ely · Goldfield · Bishop · Lone Pine · Las Vegas · China Lake · Searchlight · Baker · Kingman · Barstow · Needles · Victorville · San Bernardino · Twentynine Palms · Blythe

4

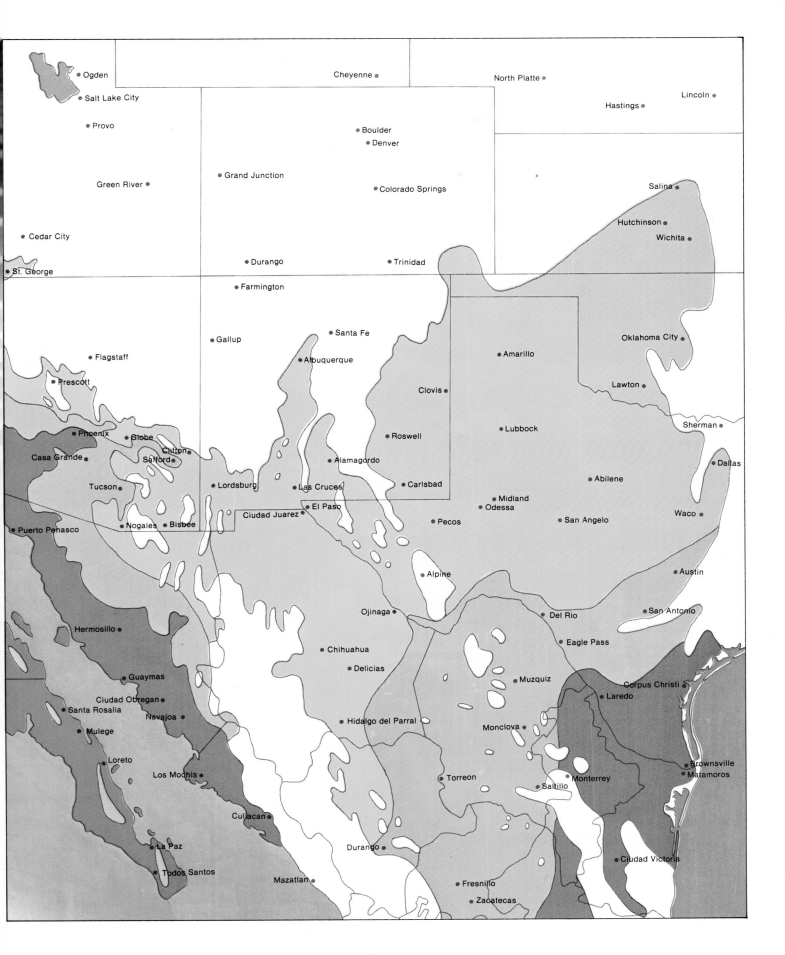

Your Climate Ingredients

Several major factors limit and influence plant growth in hot, arid regions. All these factors combine to make gardening different. Elements described in the following discussion, combined with your microclimates and the composition of your soil (see page 31), make up your plant's environment. Knowing these "ingredients" will take you a long way in learning how to select and care for landscape plants.

COLD AND FROST
Temperature is the key ingredient in selecting and using plants. As we mentioned on page 4, cold is *the* limiting factor in determining whether a plant will grow in a given area. In addition to the low point, the length of time the temperature remains below freezing determines the extent of damage that will be sustained by tender plants.

Freeze damage generally occurs when air temperature reaches 28°F (−2°C), but only if it remains there for several hours. For example, how often the 28°F (−2°C) temperature occurs and its duration establishes the lower limit of the citrus belt. This is because 28°F (−2°C) is the temperature at which citrus and most other sensitive plants sustain damage.

Heavy freezes—Day-to-night temperature changes in winter (see box) for many desert areas are not nearly as much of a threat to plants in the desert as occasional Arctic cold fronts. These fronts, or polar waves, sweep down from the north, bringing unusually cold temperatures. Unexpected heavy freezes periodically kill many plants that might otherwise do well in these climate zones. The cold is all-pervading—radiated heat from the sun collected on south walls or pavement during the day is not enough to compensate for the cold. Warm microclimates are also overcome, and get almost as cold as surrounding areas.

Covering your plants (see page 35) may give adequate protection during normal winter cold, but not during an Arctic freeze. Use heating devices such as electric lights to increase the temperature close to sensitive plants, even those in normally warm and protected spots.

Polar waves are particularly destructive if they follow a period of spring-like weather. Plants at that time have put out succulent new growth that has not had time to "harden off," mean-

At first glance, snow on cactus may seem unusual, but snowfall is not uncommon in the desert. Normally the amount that falls is slight and melts in a few hours.

Anatomy of a Frost
At sundown the earth begins to radiate skyward the heat it has accumulated during the day. Just before dawn, the earth reaches its coolest temperature. If the *dew point* is reached—the point where the ground and plants are cooler than the air—moisture will be deposited. However, the lower the humidity of the air, the closer to freezing the dew point will be. If the temperature drops below freezing, the moisture passes directly to the ground and plants as frost. This is most likely to happen on a still, clear, winter night when the earth is unprotected by a cloud cover.

A cold night with enough moisture to produce heavy frost may not be as destructive to tender plants as one (with the same minimum temperature) having humidity so low that *no* frost forms. This condition is sometimes referred to as a *black frost*. White frost forming on the foliage actually prevents the leaf from losing moisture. Low humidity on a very dry night causes foliage to lose moisture by evaporation. This lowers the leaf temperature below that of the air. Therefore, a wet surface can actually form ice or frost, even though the air temperature stays 4 to 5°F (−3°C) above freezing.

ing, become woody. This can be a serious problem even with a normal freeze. New succulent growth is naturally more susceptible to freezing than the rest of the plant. Even plants tolerant to cold, natives included, can suffer damage.

Temperature fluctuations—Plants must also be able to tolerate the alternating heat and cold between day and night, as well as winter lows. Inland deserts and interior valleys have greater temperature fluctuations than areas adjacent to large bodies of water, which tend to moderate temperatures. Annual variations may be 100°F (38°C) or more between a summer high and a winter low. Fluctuation between day and night temperatures is also pronounced, especially in spring and fall.

HIGH HEAT AND SUNSHINE
Except in arid regions by the sea, the most stressful period for plants is late May through September, when temperatures are very high and solar radiation is intense. During this period there may be 45 to 90 consecutive days of sunshine with temperatures reaching 100°F (38°C) or more. In regions where there are summer thunderstorms, the heat is moderated somewhat by afternoon cloud cover and humidity during stormy spells.

During this period of intense heat, the zone of greatest temperature variation is the ground surface where radiant heat from the sun is absorbed. Bare rocks, concrete and other dense objects can reach temperatures as high as 180°F (82°C) on a summer afternoon. These temperature extremes are especially tough on plants—dehydrating, bleaching, actually cooking young seedlings. With intense heat, broadleaf plants unadapted to the desert environment cannot *transpire* water to cool themselves rapidly enough to keep up with the sun. They wilt or suffer leafburn, a killing of all or part of the leaves. This happens most often if the soil is allowed to dry out and there is not enough water available. A moist soil surface is cooler, of course, and raises the humidity. This is why many plants appreciate shade on hot summer afternoons, especially in the low deserts. Mulches play an important role in modifying soil temperatures. They are discussed in detail on page 34.

Some plants cease to bloom when the daytime temperatures begin to hit the 100°F (38°C) mark. If they do bloom, their petals or fruiting parts are often damaged and they will not set fruit. Rose and gardenia blooms tend to dry out rapidly in the heat. With some plants, the continued warmth of night temperatures affects flower fertilization and fruit set. For instance, corn and tomatoes cease bearing when the minimum temperature stays above the mid-70's°F (21°C). Conversely, some plants, such as Bermudagrass, do not begin to grow well until the weather warms and night temperatures are in the 60's°F (18°C).

RAINFALL

Some arid regions have a winter and a summer rain season. The summer storms can be quite dramatic, coming on an area very quickly. They are often accompanied by strong winds that create a wall of dust and sand immediately preceding the storm. The rains last only a short time and are usually very localized, seldom getting more than three miles in diameter.

In areas that receive little rain, watering is especially important in early summer when plants are still trying to grow fast. Lots of water is needed to cope with the heat and for growth. As the hot season intensifies, plants may become semidormant and look poor until the cooler temperatures of fall, when they may again put out new growth. Irrigation practices are discussed further on page 33.

WINDS

Hot summer winds are often laden with dust and can be seen preceding a thundershower, covering all in their path with clouds of soil. The winds not only take their toll by drying your skin but suck moisture from tender succulent tissues of many plants.

Another desert occurrence during warm weather, especially when the ground is warmer than the air, is the *dust devil* or *whirlwind*. It is much like a small tornado, and may be started by something as small as the movement of a rabbit or by a car driving on a dirt road. Superheated air near the ground is set in motion, turning rapidly and spiraling upward, burdened with dirt and debris. It leaves a path of destruction or annoyance across the open spaces until it dissipates in a grove of trees or other barrier.

Dust devils are a common sight to travelers crossing the desert during the summer months.

Sometimes the only protection a desert dweller has against these great natural forces is a windbreak or

In many desert areas there are two periods of rainfall each year—summer and winter. Summer rains often produce torrents of water in a few hours, most of which runs off, as above. Winter rains are usually gentler, giving plants a better chance to absorb the moisture.

Dust devils are common in arid areas, especially in summer. Most are merely a nuisance, but large ones can be destructive.

windscreen. Cities benefit from the foliage of tall trees, breaking the force of the wind. Trees lessen the force of windflow through the "canyons" created by city buildings. Rows of trees planted at the edges of fields, along roads or close to houses or settlements in the open country filter the wind and make life in the desert environment more desirable. How to shelter your home with windbreaks is covered in detail on page 20.

HUMIDITY

Atmospheric humidity has a direct bearing on how much water plants must have to thrive, because much less ground moisture is needed where the air remains moist. The presence of humidity, either hot or cool, greatly reduces a plant's water demands. The amount of water required for transpiration—keeping the plant cool—far surpasses the need for plant growth. If the summer is cool and humid, a surprising amount of vegetation can exist even if annual rainfall is low.

Humidity also plays a part in determining which plant species will succeed in an area. It is also the climatic ingredient that is most difficult to adjust. Sprinkling the ground near plants raises humidity slightly, and the subsequent evaporation cools the area. Plants create a more humid environment as they transpire through their leaves. During summer rain periods or in irrigated farm areas, the humidity will often increase to an uncomfortable level. This is especially true in the hotter low deserts, coastal deserts and in wide low valleys which get little in the way of breezes.

Most regions with high humidity will support many of the plants traditionally associated with the tropics. As long as their irrigation needs are supplied, they perform as if in a truly tropical habitat. Landscape plantings in coastal desert cities such as San Diego, California, Guaymas, Mexico, and Jidda, Saudi Arabia, support many of the traditional tropicals.

Home Microclimates

Your own backyard has areas with different climatic conditions, called *microclimates*. Variables in sunlight, temperature, humidity and wind make each home landscape different, even from the site next door. Understanding your microclimates will help ensure success with your landscape plants. For example, cold-tender, sun-loving plants will perish if located in a cool, shaded part of your lot. Plants unadapted to high temperatures will die when exposed to the intense reflected heat from a south or west wall.

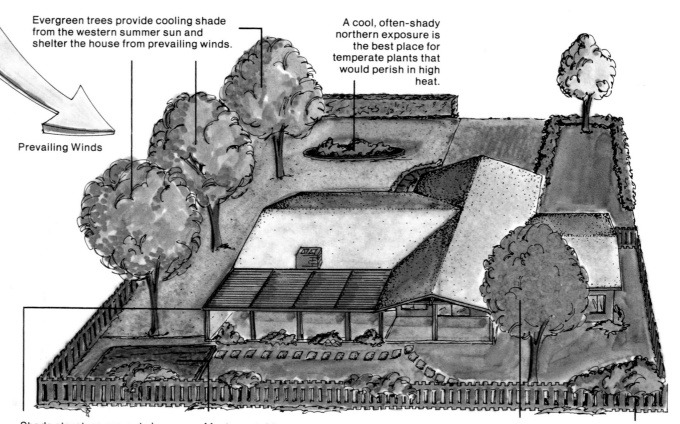

Evergreen trees provide cooling shade from the western summer sun and shelter the house from prevailing winds.

A cool, often-shady northern exposure is the best place for temperate plants that would perish in high heat.

Prevailing Winds

Shade structure prevents low winter sun from shining directly into windows and provides shelter for cold-tender plants beneath.

Most vegetables, especially cool-season crops, need plenty of sunlight. This raised garden is located in a sunny southern exposure.

Deciduous shade tree cools this part of house in summer. When leaves fall in winter, the sun adds warmth to the house.

Cold air can accumulate in low areas and will freeze tender plants. Spaces in fence allow cold air to drain away.

Microclimates

No absolute lines can be drawn to define climates. You cannot cross a given boundary and find yourself in the next zone. Instead, there is a gradual change as each merges with the next. Within each climate region you will also find many small climates, called *microclimates*. These places can be as large as a canyon or as small as a corner of your backyard. The slope of the ground, its surface composition, the direction and intensity of wind and sun, topography, nearby bodies of water and other factors combine to create these variations.

Large microclimates that are generally warmer than the surrounding climate are called *thermal belts*. They are usually areas of higher ground facing south or southwest. Because thermal belts receive maximum exposure to the sun's rays, their surfaces store up heat during the day and release it at night, while the cold air drains to lower levels. This keeps the minimum temperature higher, so a larger range of plants can be grown.

Microclimates generally cooler than the surrounding climate are called *cold air basins*. These are caused by cold air flowing down hillsides along natural drainage ways. Cold air is heavy and follows the same contours of a hillside as would a river of water. In the early morning hours it settles as "puddles" in low areas such as canyon or wash bottoms or the pockets of a valley. In a smaller sense, cold air can be contained by a solid wall or fence around your house. If there is an abrupt change in topography, the early morning temperature variations can be startling, even within short distances. You can check your own microclimates by walking around your property in early morning. In some instances a difference of 15°F (−9°C) can be recorded between a low, bottomland location and a nearby higher slope with good air drainage.

MODIFYING ELEMENTS

Large bodies of water also have an effect on the local climate. By absorbing heat, they keep temperatures warmer when it is cold. Through evaporation, they keep places cooler when it is warm. By the same token, a region adjacent to irrigated agriculture or a heavily watered landscape area such as a golf course or park has a lower nighttime minimum temperature because of evaporation at night.

Cities modify the minimum and maximum temperature ranges. Buildings and pavement act as heat sponges during the day much like rocks or gravel, and radiate heat at night. Heated buildings, factories and automobiles contribute additional heat and pollutants. The pollutants not only foul the air but put an atmospheric umbrella over the whole region, holding in the warmth at night like a giant greenhouse. This additional warmth is a benefit to the gardener. It is well known that the range of dependable plants increases to include more tender plants as the size and the density of the city increases.

The surface of a given area also affects the minimum and maximum temperatures. Rock outcrops, gravel and open sandy areas or bare earth will be warmer than areas covered with vegetation.

INDICATOR PLANTS

Certain plants growing in nature can be indicators of microclimates. In the Southwest, for example, you might notice *Vauquelinia californica* (Arizona rosewood) and *Quercus douglasii* (Mexican blue oak) growing on a north slope. This would indicate a cool location. A short distance away on a south slope, the presence of native plants such as *Simmondsia chinensis* (jojoba), *Encelia farinosa* (brittlebush) and *Olnea tesota* (ironwood) would indicate a warm spot. If you see ironwood growing in an area, for instance, this would also be a desirable location for citrus plants.

BACKYARD MICROCLIMATES

Microclimates in your own backyard follow the same principles as the larger areas just described. The direction of winds and air circulation cause modifications of temperatures. Paving, buildings, walls, trees, turf and other plantings affect the penetration and absorption of the sun's heat. Pavement and walls that face south store heat during the day and radiate it at night, moderating temperatures. Even after the sun has been down for some time you can feel the warmth coming from a south wall. Because this warmth is released gradually during the night, cold-sensitive plants such as bougainvillea may be grown here with success, yet fail just a few feet away in an open garden.

Likewise, the north sides of buildings are shaded and cooler than those of the hot south. This is the place for temperate plants such as viburnum, which would burn up if grown in the open western sun.

All of these factors operate within your general climate zone. It soon becomes apparent your microclimates have a great deal to do with the success or failure of your plants.

This bougainvillea is taking advantage of the microclimate created by a warm south wall and overhang. Heat accumulated by the wall during the day is released at night, which helps protect the plant from cold.

Adaptations of Desert Plants

The uniqueness of desert and arid land plants and their successful adaptations to their environment is important to consider. By understanding some of their adapted functions you can better understand how your landscape plants, desert and otherwise, cope with the arid environment. Even though you will probably select plants to create an integrated landscape, you should locate plants in ''zones'' according to their water needs.

For successful results, most non-desert plants require moderate to ample irrigation, while established desert natives and other plants from arid regions may require only occasional supplemental water. Some stay a lush green on modest amounts. Desert plants are especially adapted to arid landscaping, and have learned to live with dry conditions in a variety of ways, a few which sound like science fiction.

EVADERS AND ENDURERS

Plants with adaptations to periods of drought beat heat and deprivation of water in two ways: They *evade* it or they *endure* it, the latter by using water more efficiently.

Drought-evading plants simply disappear into a sort of time capsule until the temperature and moisture are right for them to grow again. The heart of the plant rests in seeds, bulbs or fleshy roots until moisture, temperature and the length of daylight signal them to reappear.

For the most part, arid land plants have found ways to endure periods of drought and high heat. The Mexican palo verde tree *(Parkinsonia aculeata)* is a prime example of an endurer. It has evolved many functions that combine to allow it to live in the desert. As a drought persists, it gradually sheds leaflets, then midribs, even twigs. By reducing tissues, a tree can maintain life in its core as long as possible until water returns. The palo verde's tiny leaves absorb little heat, and transpiration is reduced. Its open-branched form allows rain or dew to collect and flow down the tree to the roots. Green bark filled with chlorophyll continues *photosynthesis* (food production) even when leaves are absent. A shallow, fibrous root system extends outward near the soil surface, to absorb traces of rainfall. Simultaneously, a taproot or roots seek moisture at lower levels.

One of the ways the palo verde endures drought is by gradually dropping leaves, twigs and branches to conserve moisture. Green bark allows *photosynthesis* (food production) to continue even after leaves have fallen.

Arid land plants have evolved a number of other fascinating adaptations. Many have white or silvery leaf surfaces that reflect the sun. Some plants have the ability to turn away or fold up to protect them from the sun. Others, such as Texas ranger, have leaves that store water. Plant leaves become spongy and greenish during a rainy period, then turn small and reflective during a dry season. See photo below.

Cacti and other succulents also store water in their tissues. Their greenish surfaces produce food, and their tough, waxy skin prevents moisture loss. They often do not have leaves, or leaves may be temporary or undeveloped. Thorns growing at the bases of the undeveloped or temporary leaves protect them from animals.

Most plants continuously transpire water into the atmosphere and exchange carbon dioxide and oxygen through their leaves for their metabolic processes. A few desert plants have actually learned to ''hold their breath.'' They close their

The leaves of Texas ranger *(Leucophyllum frutescens)* are green and plump when water is plentiful. In dry periods the leaves shrink and turn a reflective silvery color.

Like many cacti and succulents, *Aloe arborescens* stores moisture in tissues when water is abundant. This moisture sustains the plant through dry periods.

stomata, tiny leaf pores, during hot days and open them only at night for necessary exchanges with the atmosphere. Others can store carbon dioxide and metabolize indefinitely within their bodies. Keeping their pores completely closed allows them to exist for indefinite periods during hostile climatic conditions, until moisture becomes available. Such plants are usually very slow growing and live for a long time. The giant saguaro cactus is an example.

WATER GATHERERS

Many plant forms have developed specialized water-gathering functions. They take any available moisture and guide it down the plant toward the root zone. The corrugated ribs of cacti and euphorbias accomplish this, as do the radiating leaf forms of the aloes and agaves. The open, branching form of desert trees, like the palo verde, does the same. Thorns on some plants are believed to collect dew, somehow absorbed by the plant. Foggy, coastal deserts, such as the Atacama in Chile, receive so little rain that plants have evolved which can survive solely on the humidity in the air.

The fleshy root parts developed by some plants, such as cat-claw vine, asparagus fern and queen's wreath are examples of more verdant plants that contend with extended dry periods through adaptations. Moisture is stored in the roots, and used when water becomes short.

Creosote bush *(Larrea tridentata)* is believed to produce a toxic substance, preventing other plants from growing in its vicinity. This eliminates competition for water. Only when the soil has been leached by rains will annual grasses or flowering plants grow for brief periods beneath creosote. Salty leaves that drop from the tamarisk tree (*Tamarix* species) are thought to perform the same function.

Arid land plants often have two kinds of roots. One kind is the wide-ranging *surface roots* that take up the slightest traces of moisture and nutrients. They function as soil holders and therefore make good erosion-control plants. The other kind is a *taproot,* which goes deep into the earth to seek out moisture stored below. They also anchor the plant against strong winds and the scouring action of fast-flowing water that accompanies desert storms. The taproot of the desert hackberry *(Celtis pallida)* will penetrate to 50 feet; mesquite *(Prosopis* species) roots penetrate to 150 feet or more. Taproots make it very hard to transplant or pull out tiny seedlings of mesquites, palo verdes and other desert plants. A 2 or 3-inch seedling may be anchored in the ground by a 10-inch root.

Some plants from arid regions, such as Bermudagrass, have developed the ability to penetrate the hard cement-like layers of caliche soil. The roots of the evergreen pistache are known to penetrate rock by secreting a dissolving acid.

RIPARIAN PLANTS

Not all plants you see in the desert and other arid lands are drought-resistant. Plants which grow along washes and other places where water collects are part of a *riparian,* or water-related community. Trees such as desert hackberry *(Celtis reticulata),* Arizona walnut *(Juglans major),* the sycamores *(Platanus* species), and the cottonwoods *(Populus* species) are riparian plants. They will grow only in locations where moisture is available all or most of the year.

The growth habit of mesquite also depends on how much water is available. It may never grow larger than a shrub in the dry mesa. But it becomes a large tree along flood plains or water courses, indicating ample moisture below. If the water table drops slowly, the mesquite adjusts by sending its roots deeper, following the moisture. If you should see a grove of dead mesquite trees, it could be that the water table dropped too fast or too deep for the roots to follow and the trees died of drought.

Rainfall is slight in this California desert; only a certain number of plants can survive on the available moisture. Wide, somewhat regular spacing between plants is a natural adaptation.

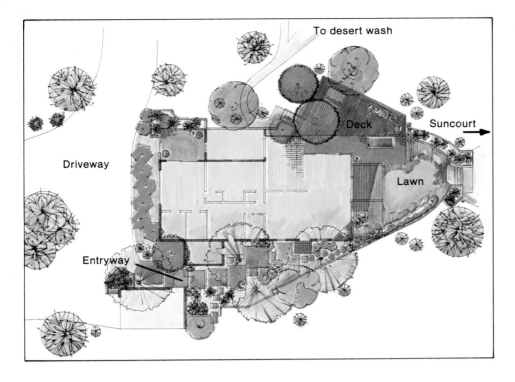

To desert wash

Deck
Suncourt →

Driveway

Lawn

Entryway

As you plan your outdoor area, keep the interests of you and your family foremost in your mind. The result will be a landscape you will use and enjoy.

A Landscape for You

Nearly everyone has some idea of what he wants his landscape to be, but not everyone feels he is getting the most from his outdoor space. Often, an undesirable landscape is the result of poor planning, caused by seasonal urges, nursery promotions or a natural desire to plant anything that will add color or greenery to a bare lot. The immediate effect is an improvement, but plants that are hastily selected and poorly located will develop into a hodgepodge. Established plants are difficult and expensive to move or replace if you change your mind. The final result is a landscape where most of your time is spent caring for it, rather than enjoying it.

THE OUTDOOR ROOM

Rather than going through this frustrating exercise, do some planning before you plant. One way to approach a landscape plan is to treat your outdoor area as an "outdoor room"—an extension of your home. Your room will have a "floor," "walls" and a "roof." The floor can be anything from lawn, to brick, to a redwood deck. Walls can be rows of trees, a hedge or an actual masonry wall. The roof can be made of wood, aluminum, shade cloth or canvas, a canopy-shaped tree, or the sky.

The landscape plan at the top of this page is the site pictured on page 12. Outdoor areas with different specific uses, according to the time of day or season, extend from the house in three directions.

Above all, keep two things in mind as you make *your* plan: First, and most important, your outdoor room should serve the interests of you and your family. Second, it should be an attractive, beautiful place, something you enjoy looking at. When these two features are skillfully blended together, the landscape is a success.

ASK YOURSELF SOME QUESTIONS

Planning begins with knowing what you want from your outdoor space. Before you rush out to the nursery, think about what you and your family like to do outdoors. The size and makeup of your family will of course be major factors. Small children, for example, have completely different needs from adults. Try to plan for multiple use later on as interests change. A sandbox can be easily transformed into a raised bed for vegetable gardening when your child's interest in sandcastles has waned.

☐ *Do you need space for recreation?* Backyards and family games are nearly synonymous. Lawns and paved areas support a number of activities for all ages. Perhaps you will want a special play yard for children, visible from indoors.

☐ *Do you want fruit, vegetable or flower gardens?* Gardening for food is increasing in popularity. Raised planter beds are neat, self-contained areas for growing vegetables or flowers. Improving the soil is also easier and more economical in a raised bed. Fruit trees, including citrus, should be considered. Dwarf varieties take up little space and grow well in containers. Espaliered fruit trees trained on a wall or fence produce fruit in a very small space.

☐ *Is there a need for privacy, or a private garden?*
A private place in your own backyard has a special appeal, offering retreat and relaxation. You will want to preserve scenic views, while the neighbor's *view of you* may need to be screened.

☐ *Will you do much entertaining?*
Patio areas adjacent to the house are desirable for large outdoor parties or small family gatherings. The size of your get-togethers naturally determines how large your patio should be. A paved area or plot of lawn nearby can serve as an "overflow" area for extra-large gatherings.

☐ *Does your climate require modification?*
The arid climate, unchecked, creates harsh living conditions indoors and out. Most sites have to be modified. Proper placement of trees, shrubs and ground covers will help make your home more livable than any other single factor. Using plants to temper difficult climates is discussed in detail on page 18.

☐ *What kind of gardener are you?*
This is an important question, because it deals with how much time you have, or want to spend caring for your landscape. For example, don't plan on having a large lawn if you really hate to water, mow and fertilize. Consider too, the total water requirements of your plants. The minioasis landscape, covered in detail on page 16, may be perfect for you.

☐ *Do you have any special needs?*
Areas for pets, storage, compost bin, work or hobbies should be considered. Open, multiple-use areas allow for changes as the family grows or interests change.

☐ *How much money do you have to spend?*
As a general rule, plan generously. Then, after you have roughed out costs, you may discover that you will have to start with smaller plants, reduce the size of the patio or deck area slightly or use less expensive building materials. Perhaps your landscape will have to be built a little at a time. Plant trees and shrubs first: They take the longest to develop. Watch for sales and nursery promotions. But don't compromise the basic quality of your plan; add to it gradually as your budget allows.

THE PLAN
The preceding questions have probably created a jigsaw puzzle of ideas in your mind. After you have had time to formulate your thoughts, sit down and put the pieces of your landscape together. One of the best methods, used by professionals and amateurs alike, begins with a sheet of graph paper. Use the graph squares to equal a unit of measurement (each square equals one foot for example) so it will be simple to plot distances. Draw in all existing features: house, property lines, power poles, walls, fences, views, existing trees and shrubs—anything that will affect the beauty and livability of the design. This would include notes on prevailing winds and how the sun shines on your lot in summer and in winter. Knowing these details will be important if you want to have a windscreen or trees for shade.

When you are satisfied that the major features of your site are represented on the graph paper, cover it with a sheet of tracing paper and sketch your ideas. Experiment. Modify. Erase. Analyze. Be bold. Changing a plan is simple; moving a landscape around is not. It might be helpful to follow these guidelines:

Begin with a simple, unified plan. Think of plants as groups and masses. Limit their forms, colors and textures. We've discussed some specifics on

A prime example of an *outdoor room*: Plants and materials combine to create a private, sheltered, relaxing retreat.

selecting plants on page 16. Likewise, limit the types of construction materials, and repeat the elements of the house if possible.

Work toward a basic concept. Don't concern yourself with minor details; they can be added later. It may help to have a landscape theme to follow. Several are described on page 15.

Keep plant size in proportion to your home. Those innocent-looking plants in the one-gallon containers can grow up to become giants. If large-growing plants are located near an entryway or sidewalk you will have to battle their natural growth habits, pruning and cutting them back continually. In addition, the appearance of the plants will suffer. It is important to know the *mature* size of your plants, and allow space for their growth.

Picture yourself in your completed garden. Having a good imagination is a real advantage in deciding if your proposed landscape is right for you. Spend some time outside during the day and night. Walk along paths you are considering, and envision how your proposed trees and shrubs will look after they have matured. Sit on the "deck" you have penciled on your plan and admire your view. Projecting yourself into your future landscape will give you a real feel of how you will like living in your outdoor room.

PROFESSIONAL HELP
If you just don't feel confident about your plan, or you wish to have the benefit of a professional opinion, you might want to consult with a landscape architect. The least expensive way is to have a landscape architect visit your homesite and offer his or her advice. This consultation will cost anywhere from $50 to $100. This isn't a lot if you are considering investing hundreds of dollars in plants and building materials—you want to be sure you are using them to the best advantage.

You may want a landscape architect to devise a plan from scratch. Some charge by the hour, others charge a percentage of the construction cost or charge a set fee. Fees will vary according to region. Cost for a typical base plan will be from $500 to $1,000, but this depends on many variables. If you have a small site or section of a landscape you want to have developed, some landscape architects will make a single trip to your home and draw a plan on site. This reduces their time, thus your cost. This type of plan ranges from $200 to $500.

Landscape Themes

A single landscape plan can be made to produce several different effects. It depends on which plants and building materials are used and how they are combined and maintained.

A theme gives your home and landscape a personality. It can be merely suggested by a few select plants, or carried out to the last detail. If you unconsciously tend to be a "one-of-this, one-of-that" kind of gardener, a theme can serve as an organizing and unifying device.

You'll find that plants of a similar nature grouped together are more pleasing to the eye than an unrelated mix of plants having opposing textures and growth habits. But don't force a style on your home; it will only appear contrived and unnatural. Get a feel for what will work best with your architecture, the layout of your site and your family's way of life.

Rustic landscapes can be produced by using rough, weathered materials such as railroad ties, used bricks, boulders and exposed-aggregate concrete. Construction is done in a loose, free manner. Plant forms should be natural and free in shape. Trimming or thinning is done to maintain an unobvious control of the plants, but there is no shearing or shaping. Maintenance is casual.

Formal landscapes are created by producing a feeling of geometry, precision and containment. Bricks or paving with sharply defined edges are appropriate elements. Plants that grow naturally into contained shapes or accept training well are at home here. Hedges, sheared screens and topiaries define the formal look.

Natural, wild and desert are similar themes. In its true form, a natural landscape includes plants native to an area, whether desert, chaparral or subtropical. It may or may not be enhanced by the addition of similar non-native plants.

A wild landscape may or may not include native plants, but gives the same feeling as a natural garden.

Desert landscapes may or may not include cacti, but most people consider cacti integral to the desert. Plants generally have wide soil and irrigation tolerances. The desert garden can be neatly planted like a collection or arranged more naturally.

All are informal in character. They have flowing lines, open spaces and rough textures. Stone, swept earth and gently contoured mounds of earth fit well here. Placement of plants is subtle and should be done as naturally as possible. One way to achieve an informal look and break the tradition of row planting is to set plants in a "lazy M" pattern.

Transitional landscapes include plants that look at home with the cultivated, tended garden, and the low-water landscape. The main function is to blend the two garden types together to create a unified whole.

Subtropical landscapes are somewhere between the somewhat sparse desert landscape and the verdant, tropical landscape. There is a festive feel to the plants in this type of garden—many have bold striking leaves, others have feathery leaves. Palms and citrus fit well here, as do flowering annuals and perennials.

Tropical landscapes generally work best in small areas in hot and dry climates, such as in a minioasis. Plants require protection from extreme cold, heat and wind. They are characterized by dense greenery and rich, vibrant color. They can be used to create a relaxing oasis and give a cooling effect if used with a pond.

Oriental gardens are landscapes of suggestion and illusion. They are seldom symmetrical. To Occidental eyes, Oriental landscapes may appear sparse and serene. Plants and materials are used as objects of art and sculptural beauty. Careful attention is given to creation of natural effects. Small, informal water elements, or even suggested streams in the form of dry streambeds work well here. Combining rocks, wood, gravel and other natural materials with picturesque plants makes the western version of the Oriental garden particularly attractive in arid regions. The style is well adapted to a small space, such as an entryway or townhouse patio. The effect is quite striking, yet requires little care.

Woodsy landscapes are informal and rustic with curving lines and natural materials. Wood, stone and tall trees are important features. The woodsy garden might include coniferous plants such as pines or junipers, but also requires deciduous trees. Vines and low ground covers in contrasting shades of green, along with colorful flowers add the finishing touches.

Railroad ties, concrete aggregate and casual groupings of plants blend to create a *rustic* look for this home entrance.

Cacti and other plants native to the area where this house is located are arranged to make this a *natural* landscape.

Drought-resistant plants are subtly combined with rocks and water elements to make an *Oriental* theme.

Selecting Plants

For many people, selecting plants at a nursery can be a bewildering experience. Row upon row of plants, most of them unfamiliar, await your selection. But a nursery is an excellent place to get to know plants. There you can study them and sort out your feelings about foliage texture, color and structure. You can also examine plants at various stages of growth. For example, you can look at a tree species in a 1-gallon can, 5-gallon can and a 15-gallon can noting changes that occur with age and size.

WATER NEEDS

An even more important part of plant selection is the amount of water they require. There are few regions in arid lands where there isn't a concern over the use of water. This dramatically modifies the approach to landscape planning, and critically affects plant selection. Plants naturally have varied water needs, from very little to ample. Before you actually purchase plants for your design, examine their total water needs. Are you willing to spend what it will take to keep them growing? Also, does your region have that much water to spend? All of these limitations have to be reckoned with before you make your selections.

THE MINIOASIS CONCEPT

For many, the ideal landscape is lush, green and inviting, filled with beautiful, healthy plants. At first this kind of landscape may seem difficult, expensive or even impossible in hot, arid regions. But even with very limited water there is a way to have your own lush, green place. It works like this: As you begin to select plants for your design, group them by their water needs. Those with high water requirements can be concentrated in one, small area close to your house—a minioasis. These plants should be your favorites—the ones you like to look at and touch. They should be attractive at close range, and look nice all year.

A minioasis is the perfect place for a small but lush green carpet of lawn.

Areas beyond this well-watered zone should be landscaped with plants having low water needs. The zoning of plants by their water requirements can be as extensive as you want to make it. These *transitional* areas gradually blend into a drier perimeter. Ideally, the plants should be similar in form and texture to those in the minioasis so that the transition is smooth. Or, if you choose, the transition can be immediate by placing a low wall or paved section between the high water and low water zones. A fine example of this is the full page photo at the beginning of this chapter.

The essence of the minioasis concept is *water budgeting*—spending water where it will be enjoyed the most. But the beauty of this plan is that none of the pleasure of living outdoors with plants is diminished. It is just done a little more thoughtfully. Improving the soil, usually necessary in desert areas, is also confined to a single area. It is important to realize the areas beyond the minioasis need not suffer in appearance. Attractive, drought-resistant plants are available to shade, screen or cover the ground just as efficiently as their water-demanding counterparts.

Plants and Pollen

Before you select plants for your landscape, you should be aware that some plants produce pollen that can cause allergic reactions. Generally, plants that have small, colorless flowers are the most troublesome. They rely on the wind for pollination and produce tremendous amounts of pollen. Some of the worst offenders are mulberry (*Morus alba*), willow (*Salix* species), African sumac (*Rhus lancea*), olive (*Olea europea*), tamarisk (*Tamarix* species) and common Bermudagrass. Plants with fragrant, colorful flowers also produce pollen but it is heavy and sticky. It is transferred by insects as they go from flower to flower. Some of these plants such as the acacias (*Acacia* species) produce a cloud of pollen near the plant, which may bother certain individuals, but it is not spread widely by the wind.

A minioasis planting adds just the right touch to this patio. Watering and caring for these moisture-loving plants is simple when they are grouped in one small area.

This *transitional* grouping of *Olea europaea*, bright green *Santolina virens* and flowering *Aloe arborescens* is lush in appearance, yet requires minimum water and maintenance.

Plant Selection Guide

The following plants are arranged by landscape use. Some are included in more than one category: Plants vary considerably in size a[nd] appearance, depending on the species and growing conditions. Also, many plants adapt well to training, which allows you to use them i[n] a variety of ways. The lists include some of the best plants for each use. They are not meant to be complete but serve as examples.

Ground Covers

Atriplex species .... Saltbush
Dichondra micrantha .... Dichondra
Euonymus fortunei radicans ...... Trailing euonymus
Phyla nodiflora .... Lippia
Gazania uniflora rigens .... Gazania
Hedera species .... Ivy
Juniperus species .... Juniper
Lantana montevidensis ... Trailing lantana
Osteospermum fruticosum Trailing African daisy
Rosmarinus officinalis 'Prostratus' .... Dwarf rosemary
Teucrium chamaedrys 'Prostratum' .... Germander
Trachelospermum species .... Asian, star jamine
Vinca species .... Periwinkle

Vines

Antigonon leptopus ....Queen's wreath
Bougainvillea species ....Bougainvillea
Ficus pumila ....Creeping fig
Gelsemium sempervirens .Carolina jasmine
Hedera species .... Ivy
Jasminum species ....Jasmine
Lonicera japonica 'Halliana' ....Hall's honeysuckle
Macfadyena unguis-cati .... Cat's claw
Passiflora alatocaerulea .... Passion vine
Rosa banksaie ....Lady Bank's rose
Trachelospermum jasminoides .... Star jasmine
Vitis vinifera .... Grape vine

Small Shrubs

Abelia grandiflora 'Prostrata' .... Abelia
Buxus microphylla japonica .... Japanese boxwood
Carissa grandiflora .... Natal plum
Euonymus fortunei .... Euonymus
Juniperus species .... Juniper
Justicia species ... Mexican honeysuckle
Myrtus communis 'Compacta' ... Compact myrtle
Nandina domestica 'Nana' ...... Heavenly bamboo
Pittosporum tobira 'Wheeler's Dwarf' .... Tobira
Plumbago auriculata ..... Cape plumbago
Punica granatum 'Nana' .... Dwarf pomegranate
Raphiolepis indica .... Indian hawthorn
Santolina chamaecyparissus..... Lavender cotton
Santolina virens .... Green santolina

Medium Shrubs

Abelia grandiflora .... Glossy abelia
Atriplex species .... Saltbush
Baccharis sarothroides ..... Coyote brush
Caesalpinia species ..... Bird of paradise
Callistemon citrinus .... Bottlebrush
Carissa grandiflora .... Natal plum
Cassia species .... Cassia
Cotoneaster species ..... Cotoneaster

Medium Shrubs continued

Dodonaea viscosa .... Hopbush
Elaeagnus ebbingei .... Ebbing silverberry
Euonymus japonica . Evergreen euonymus
Feijoa sellowiana .... Pineapple guava
Juniperus species .... Juniper
Leucophyllum frutescens ... Texas ranger
Ligustrum species .... Privet
Myrtus communis .... Myrtle
Nandina domestica .... Heavenly bamboo
Nerium oleander .... Oleander
Osmanthus fragrans .... Sweet olive
Photinia serrulata 'Nova' Chinese photinia
Pittosporum tobira .... Tobira
Pyracantha species .... Pyracantha
Raphiolepis indica .... Indian hawthorn
Simmondsia chinensis .... Jojoba
Viburnum species .... Viburnum
Xylosma congestum .... Xylosma

Large Shrubs

Cocculus laurifolius .... Cocculus
Cotoneaster species .... Cotoneaster
Dodonaea viscosa .... Hopbush
Elaeagnus ebbignei .... Ebbing silverberry
Euonymus japonica . Evergreen euonymus
Laurus nobilis .... Grecian laurel
Ligustrum lucidum .... Glossy privet
Nerium oleander .... Oleander
Olea europaea .... Olive
Photinia serrulata .... Chinese photinia
Platycladus orientalis .... Arborvitae
Punicia granatum .... Pomegranate
Pyracantha species .... Pyracantha
Rhus ovata .... Sugar-bush
Sophora secundiflora .... Mescal bean
Syringa species .... Lilac
Vauquelinia californica . Arizona rosewood
Viburnum species .... Viburnum
Xylosma congestum .... Xylosma

Small Trees

Acacia farnesiana .... Sweet acacia
Albizia julibrissin .... Silk tree
Bauhinia species .... Orchid tree
Butia capitata .... Pindo palm
Callistemon species .... Bottlebrush
Cedrus atlantica 'Glauca' ..... Atlas cedar
Cercidium microphyllum ...... Palo verde
Chilopsis linearis .... Desert willow
Chamaerops humilis ...... Mediterranean fan palm
Citrus species .... Citrus
Eriobotrya species .... Loquat
Feijoa sellowiana .... Pineapple guava
Ficus species .... Fig
Geijera parviflora .... Australian willow
Juniperus species .... Juniper
Lagerstroemia indica .... Crape myrtle
Laurus nobilis .... Grecian laurel
Ligustrum lucidum .... Glossy privet
Lysiloma thornberi .... Featherbush
Magnolia grandiflora .... Magnolia
Melia azedarach .... Chinaberry
Olea europaea .... Olive
Olneya tesota .... Ironwood
Parkinsonia aculeata . Mexican palo verde
Photinia species .... Photinia

Small Trees continued

Pithecellobium flexicaule .... Texas ebony
Pittosporum phillyraeoides .... Willow pittosporum
Podocarpus macrophyllus . Japanese yew
Prunus species ... Flowering plum, peach
Punica granatum .... Pomegranate
Pyrus kawakamii .... Evergreen pear
Rhus lancea .... African sumac
Sambucus mexicana . Mexican elderberry
Schinus species .... Pepper
Sophora secundiflora ... Mescal bean
Tamarix aphylla .... Tamarisk
Tecoma stans .... Yellow bells
Thevetia peruviana ...... Yellow oleander
Ulmus parvifolia 'Sempervirens' .. Chinese elm
Vitex agnus-castus .... Chaste tree
Xylosma congestum .... Xylosma
Zizyphus jujuba .... Chinese date

Medium Trees

Acacia abyssinica ..... Abyssinian acacia
Ailanthus altissima .... Ailanthus
Albizia julibrissin .... Silk tree
Arecastrum romanzoffianum . Queen palm
Casuarina stricta .... Beefwood
Cercidium floridum ..... Blue palo verde
Chorisia speciosa .... Floss-silk tree
Cercis canadensis ...... Eastern redbud
Cupressus species .... Cedar
Eucalyptus species .... Eucalyptus
Ficus species .... Fig
Fraxinus velutina .... Ash
Geijera parviflora .... Australian willow
Gleditsia triacanthos inermis Honey locust
Melia azedarach .... Chinaberry
Morus alba .... Mulberry
Parkinsonia aculeata . Mexican palo verde
Pinus thunbergiana .... Pine
Pistacia species .... Pistachio
Prosopis species .... Mesquite
Quercus species .... Oak
Rhus lancea .... African sumac
Robinia pseudoacacia ..... Black locust
Schinus species .... Pepper
Tamarix aphylla .... Tamarisk
Ulmus species .... Elm

Large Trees

Araucaria bidwillii .... Bunya-bunya
Carya illinoensis .... Pecan
Casuarina species .... Beefwood
Cedrus species .... Cedar
Cupressus species .... Cypress
Eucalyptus species .... Eucalyptus
Gleditsia triacanthos inermis Honey locust
Grevillea robusta .... Silk oak
Magnolia grandiflora .... Magnolia
Phoenix species .... Date palm
Pinus species .... Pine
Pistacia chinensis .... Pistache
Platanus species .... Sycamore
Populus species .... Poplar
Quercus species .... Oak
Robinia pseudoacacia .... Black locust
Washingtonia species ...... Fan palm

Modifying Your Climate

One of the most important services plants can do is increase human comfort. Plants are beautiful to look at, but in hot, arid regions they play an important part in tempering severe climatic conditions.

To be fully effective in modifying your climate requires careful placement of plants. You want to place the right plant where it will do the best possible job of modifying sun, wind and temperatures. Some homeowners merely guess at which plants they should grow and where they should be placed. But considering the amount of time it takes a tree to reach maturity and provide shade or act as a screen, it makes sense to choose and locate it carefully.

The first step to controlling your climate is knowing what needs to be modified. Study your site: Is your patio unprotected, facing the hot western sun? Do cold winter winds rip across the backyard, nearly always coming from the west? Are there windows in the house directly exposed to the summer sun? As you watch the weather on your site, you'll begin to notice particular patterns. Using this

knowledge, and being aware of your microclimates (see page 9) you can make the best plant selections and place them where they will do the most good for your indoor *and* outdoor environment.

THE SUN
In hot, arid regions, the most powerful climatic force is the sun. For example, the average maximum temperature in Palm Springs, California for July is 110°F (43°C), in Phoenix, Arizona it's 105°F (40°C). Modifying these kinds of temperature extremes should be one of your first landscaping priorities.

Due to the earth's rotation, the angle of the sun and its relationship to your home changes. During summer, the sun is high in the sky and the rays shine at a more direct angle and for a longer period. This increases temperatures. Conversely, the winter sun is lower in the sky. The less-direct angle and shorter days produce lower temperatures. The point is, you need to know these seasonal changes to choose the best location for your shade-producing plants. Keep mental

notes or better yet, sketch on paper the sun's path as it passes over your home. If you are new to your home and don't know how the sun will pass over in the respective seasons, ask your neighbor for help. For a further explanation, look at the illustrations opposite.

Because the sun is such a powerful force, it pays to use it to your best advantage. One of the best ways is with shade trees, properly placed to shade home or outdoor areas. *Deciduous* shade trees are those that lose their leaves each fall and produce new leaves in the spring. This works perfectly for climate control. When the summer sun is at its hottest, trees are in full leaf and block the hot rays. A house with a shaded roof remains 10°F (−6°C) to 20°F (−12°C) cooler inside. The same tree during the winter (with its lower temperatures) allows the sun to shine through its bare branches, warming the house. Thus deciduous trees save energy both seasons—reducing cooling bills in summer, heating bills in winter. Try to select trees with high arching branches that will create a canopy over

Deciduous trees are one of nature's ways of cooling and heating. In summer, this tree is in leaf and shades the house, reducing temperatures inside. A house that is shaded can be 10°F (−6°C) to 20°F (−12°C) cooler than one exposed to intense sunlight.

In winter, the same tree as at left has dropped its leaves. The sun is able to shine through the bare branches, adding warmth inside. Be sure to consider the seasonal patterns of sunshine (summer and winter) when you plant your shade trees.

A Dozen Trees for Fast Shade

The following list includes only a sampling of fast-growing trees for shade. It is best to select deciduous trees with high-arching canopies such as *Fraxinus* or *Prosopis* species for eastern and southern exposures. Low-branching trees or tall shrubs such as *Rhus lancea* or *Acacia* species will screen the low western sun.

E = evergreen D = deciduous

Acacia farnesiana (E) 15 to 20 feet
Eucalyptus microtheca (E) to 40 feet
Fraxinus velutina (D) 30 to 50 feet
Gleditsia triacanthos (D) to 75 feet
Morus alba (D) to 40 feet
Pinus halepensis (E) 30 to 50 feet
Populus fremontii (D) 50 to 100 feet
Prosopis chilensis (E) 25 to 30 feet
Prosopis glandulosa (D) to 30 feet
Rhus lancea (E) 15 to 20 feet
Shinus molle (E) 25 to 40 feet
Ulmus pumila (D) to 50 feet

your roof. Take a look at the list of shade trees on this page.

Solar energy installations require special consideration—they cannot be shaded from the sun. If you are considering placing a solar unit on your roof in the future, plan the placement of your trees accordingly.

Evergreen trees can also be used for shade, but because they are in leaf the entire year, getting the benefit of summer-shade and winter-sun is not as simple as with deciduous trees. One way is to grow evergreens that will be tall enough to shade the high summer sun, yet have a high canopy to allow the low winter sun to shine beneath, unimpeded. To do this, the tree has to be placed at just the right distance from your house.

Evergreens as well as deciduous trees can be used to block the low, yet hot rays of the setting summer sun. Because the sun sets in the northwest in midsummer compared to southwest in winter, there is no chance of blocking the desirable winter sun.

There are other ways of modifying the sun's heat and glare. Vines can be trained to grow on hot south and west walls, or trained on trellises to provide fast shade. Cat-claw vine *(Macfadeana unguis-cati)* climbs by itself, and is one of the best plants for a hot wall. For summer-shade winter-sun, deciduous grape vines *(Vitus vinifera)* are outstanding, fast-growing, and there is also the benefit of the fruit. Annual vines such as morning glory, thunbergia or perennials like queen's wreath are fast temporary covers.

Don't forget about manmade structures that provide shade. Overhangs above windows deflect the high summer sun, but allow the lower winter sun to enter. Wooden trellises, lathwork, shadecloth and other materials overhead will also reduce the sun's intensity, beneath them. One good idea is to use temporary vines or other covers until shade trees have grown.

REDUCING GLARE
When the sun hits light-colored pavement or buildings, bare earth or shiny objects, it reflects at an angle—partly as heat and partly as intense light. This reflection often heats the surroundings, enters buildings through windows, and is hard on the eyes. Tall

Manmade structures offer quick ways to provide shade. This protective arbor is covered with grape vines and *Rosa banksaie*, Lady Bank's rose.

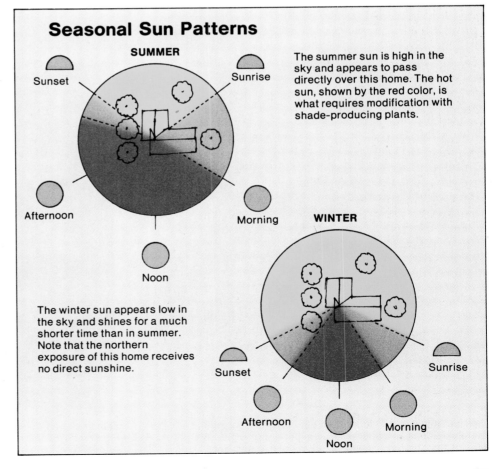

Seasonal Sun Patterns

SUMMER

Sunset / Sunrise / Afternoon / Morning / Noon

The summer sun is high in the sky and appears to pass directly over this home. The hot sun, shown by the red color, is what requires modification with shade-producing plants.

WINTER

Sunset / Sunrise / Afternoon / Morning / Noon

The winter sun appears low in the sky and shines for a much shorter time than in summer. Note that the northern exposure of this home receives no direct sunshine.

trees that shade the ground intercept the sunlight above building level, and shade the building. Ground covers such as lawn, junipers and rosemary accept the heat. These and other dark-colored plants with small leaves and needles are best at breaking down reflection. When placed on the south sides of buildings, they prevent the angled sun's rays from bouncing up into windows. Vines and espaliers on walls also cut down on heat reflected from light-colored surfaces.

Through transpiration, a living ground cover cools the air above it by about 10°F (−6°C). Paving, gravel and rock increase temperatures by storing heat during the day and releasing it at night.

CONTROLLING WIND AND AIR FLOW

Wind is a special problem in arid regions. It exaggerates the harsher elements. Wind-whipped sand and dust can be picked up and carried for miles, unimpeded by natural barriers. Hot, dry winds that occur in summer are more drying than still air, and cold winter winds are more penetrating.

Plants used as windscreens or windbreaks can play an important part in controlling and guiding air movement. Windscreens give shelter to homes and outdoor areas. They help alleviate dust problems along high-ways and keep topsoil from blowing away.

In urban areas, windbreaks shield playing fields, parks and school playgrounds. They help break up the wind-tunnel effect on city streets. Placement of trees around a structure can affect its ventilation and alter the climate by blocking or guiding the wind. A pocket of relatively still air can be created by placing a hedge 20 feet upwind of a building. A hedge placed 10 feet from a house will deflect the wind over the building and open windows will only get a low breeze backwash from the opposite side. Plants placed against a house can create a similar effect.

Windscreens set in the open are most effective when set perpendicular to the prevailing wind. A vertical, somewhat penetrable planting covering to the ground with a 50 to 60 percent density generally affords more protection than a wide, dense one. A screen that allows some wind flow is more effective for a greater distance because the wind slows as it filters through. A solid screen blocks the wind completely for a short distance, deflecting it upward. But after the wind passes over the screen, it comes back to earth with greater force. For this reason, a mix of plant sizes will do the best job at reducing wind flow.

Broadleaf evergreens and conifers (pines, junipers, cypress) with foliage that reaches to the ground are best as year-round windscreens. Deciduous trees lose their leaves in winter, so their effectiveness is reduced.

Even more important to the effectiveness of your windscreen is placement of plants. To figure out the prevailing wind patterns of your site, use a variation of the airport wind sock. Strips of cloth placed on fence-high stakes will show you at a glance which way the wind tends to blow. Chart the direction and intensity of the wind for a period of time, preferably both summer and winter. You'll notice patterns, and will then be able to place plants to filter any strong, annoying winds. As a general rule, optimum protection from the wind is obtained at a distance of three to seven times the height of the windbreak on the sheltered side. Ideally then, a row of 20-foot trees should be placed 60 to 100 feet away from the area you want to protect. Most home lots are not large enough to accommodate this kind of spacing, so placement will depend upon what space you have.

Not all winds are undesirable. Summer breezes have a cooling effect so you may want to direct them toward your site to create a breezeway. Again, this takes careful thought and research to be done effectively, but the end result is well worth the effort.

Evergreen plants that have foliage reaching to the ground are best for year-round windscreens. Junipers, shown above, provide shelter from the wind, and filter blowing dust and sand.

Tuna cacti (Opuntia ficus-indica) and privet (Ligustrum species) serve as a combination windbreak and barrier. To be most effective, rows of plants should be placed across the path of prevailing winds.

Vegetable Gardening

Many gardeners are surprised to discover that arid climates are among the best anywhere for growing vegetables. Irrigated desert valleys produce a huge percentage of the world's vegetable crops, especially in winter. The word *winter* emphasizes a very important aspect of vegetable gardening in the warm, arid regions: It's a two-season affair, and winter is often the most productive season. Planting at the right time is the key to success in these regions. The dates to plant your favorite crops are not as clearly defined as in temperate zones. In fact, you will be better off in most cases ignoring any "spring fever" planting urges.

An important part of knowing when to plant your garden requires a little understanding about vegetables as plants. Not all vegetables have the same cultural requirements. Some are better adapted to cool weather and can tolerate light freezes. Others require warm soil and temperatures to germinate and grow properly. Vegetables are generally divided into two climate categories: cool season and warm season. Because cool-season crops are hardy or frost tolerant and germinate in cool soil, they are usually planted from late summer to early spring, depending upon your climate.

Cool-season vegetables include beets, broccoli, cabbage, carrots, chard, lettuce, onions, parsnips, peas, potatoes, radishes, rutabagas, spinach and turnips.

Warm-season plants will not tolerate frost and need warm temperatures to set and mature fruit. Extreme heat prevents fruit set and reduces plant and fruit quality. Some common warm-season vegetables include beans, cucumbers, eggplant, melons, okra, peppers, pumpkins, soy beans, squash, sweet corn, sweet potatoes and tomatoes.

WHEN TO PLANT

The following are some planting guidelines for your climate zone, according to the map on pages 4 and 5. Keep in mind that climate varies greatly within a zone and from year to year. Your own experiences or those learned from your neighbor or nurseryman are invaluable when making planting decisions.

In the low and middle zones, two main planting periods are generally followed: late summer to winter for cool-season crops, early spring for warm-season crops. In the high zone there is one main planting period: spring to early summer.

Cool-season garden. The best time to garden in the low and middle zones is late summer to fall. Cooler temperatures, lower water needs and fewer pests make it easier to grow plants. This is the prime time to grow a cool-season garden: leaf crops like lettuce, cabbage and chard, and beets, carrots and turnips from seed. In the milder areas in these zones, early maturing summer crops like corn and tomato can sometimes be grown at this time. Many summer vegetables have trouble setting fruit when night temperatures stay up in the 70's and 80's°F (25° to 30°C). The cool but not cold night temperatures of early fall allow plants to come into bloom and set fruit.

In the higher zone, early frosts prevent fall planting in most areas. A late summer, cool-season crop planting with early-maturing varieties can sometimes produce before winter cold takes over.

Warm-season garden. Late winter through early spring is the time to plant warm-season vegetables in the low and middle zones. Generally, the lower the elevation, the sooner your plants can be set out. The idea is to get plants such as tomatoes and beans established as soon as possible so they will bloom and set fruit before the high heat of summer. The fruit will then mature during summer, even though blossoms cease setting fruit. And, if blooms start setting again when nights cool in fall, these same plants may bear a bonus crop before frost cuts them short.

Tender young plants require protection from frost. Cover them at night with protective devices. If your plants are not set out early, the midsummer bounty of vine-ripe vegetables is difficult to achieve. Some crops blossom and set fruit during midsummer heat: the melon and squash group and a few other heat lovers like peppers, eggplant, okra and sweet potatoes. They are usually planted in midspring.

In the high zone, because of late frosts, a late spring planting date is necessary. Again, put plants out as soon as possible and give them frost protection to stretch the growing season.

The basics for a successful vegetable garden in hot, arid regions are the same as for any other area. Soil preparation, a site with plenty of sunlight, and regular care and irrigation are necessary. The following chapter on Planting and Plant Care can help you with some of the fundamentals.

If you'd like information on the subject of desert vegetable gardening, send for the University of Arizona Bulletin Q 95, *Ten Steps to a Successful Vegetable Garden,* Agriculture Communications, College of Agriculture, University of Arizona, Tucson, AZ 85721.

Vegetable gardening in hot, dry areas is often a two-season affair: You can grow vegetables in summer and winter. Because of high heat in summer, winter is usually the best time. This Tucson garden produces crops throughout the year.

Lawns

There is nothing quite as green or inviting in hot, dry climates as an expanse of lawn. At first thought, it may seem that having a lawn in such an environment is difficult. But it isn't a question of whether a lawn will succeed; it's how much time, water and effort you are willing to spend. Actually, lawns are very successful in arid regions, if you plant the right kind of grass. All things considered, lawn is still the best ground cover for holding down large areas of dirt. It is also excellent for reducing glare and heat and softening the landscape.

SELECTING THE RIGHT GRASS
Summer heat is the limiting factor in growing a lawn in hot, dry regions. Many well-known temperate zone or cool-season grasses such as bluegrass and bentgrass will thrive here in winter, but are unable to survive a long, hot summer. Warm-season grasses such as Bermudagrass, St. Augustinegrass and occasionally zoysiagrass are the best choices.

Bermudagrass *(Cynodon dactylon)* is by far the most common lawn grass for these areas. Bermudagrass is very tolerant of alkaline water and caliche, and relatively free of diseases and insect pests. When given regular water, fertilizer and proper care, a first-class turf can be achieved, comparable to lawns anywhere. It can be invasive, however, so consider the planting site accordingly. Once established, it is difficult to get rid of.

The softer, finer-textured hybrid Bermudas—the 'Tif' series ('Tifdwarf', 'Tifgreen', 'Tifway', 'Tiflawn'), 'Santa Ana' and 'Midiron'—produce a lawn equal to the best bentgrass lawns. They require more water and fertilizer and more frequent mowing compared to common Bermuda, but they have advantages. They seldom go to seed and their fine texture is more pleasing to the eye and the touch. Pearl scale and Bermuda mite are occasional problems. Contact your cooperative extension service for the latest treatment.

St. Augustinegrass *(Stenotaphrum secundatum)* makes a very thick, bright green lawn in hot, wet climates such as along the Gulf Coast of the United States. It does reasonably well in hot and dry areas.

Vigorous and rather coarse bladed, it spreads rapidly, and is especially adapted to shade. St. Augustine needs more water than Bermudagrass to keep it looking prime. Problems sometimes occur with caliche subsoil that may cause it to become yellowish and chlorotic, requiring an application of iron chelates. Cold weather brings on dormancy: The turf turns brown when temperatures go below freezing. Overseeding with winter grass, a common practice with Bermudagrass, is not recommended. Spring recovery of St. Augustine is inhibited, and the combination of the two grasses, one fine and one coarse, creates an undesirable effect.

Zoysiagrass *(Zoysia tenuifolia)* is fine textured like Bermuda and very tough, standing up to heavy foot traffic. It is tolerant of shade but, like St. Augustine, is subject to chlorosis due to caliche and calcareous soils. It also has more insect and disease problems than Bermuda or St. Augustine. Zoysia goes dormant early in cold weather and seems to stay brown for a longer period than either Bermuda or St. Augustine. Overseeding is not recommended. If a green lawn is desired in winter, applying a lawn dye would be the best solution. See the top of page 24.

PUTTING IN A NEW LAWN
The most common way to start a new lawn is with seed. If you use this method, the most important rule to keep in mind is, *don't let the soil dry out.* Once germination begins, con-

stant moisture is necessary, or the seedlings will quickly die. This may mean watering lightly four or five times a day until the grass is up and growing. Pre-watering the area prior to seeding and top dressing with an organic material significantly increase the moisture-holding ability of the seed bed.

Other ways of starting a lawn include sodding, plugging, sprigging or stolonizing. These are considered *vegetative* means of propagating a lawn. Vegetative methods are best done in the early warm part of the year: May, June and July.

Sodding produces an instant lawn. Rolls of mature turf are laid down in strips like a carpet. This is more costly than a lawn from seed, but the effect is immediate.

Plugging is done by cutting small chunks or plugs of sod and planting them in a prepared seed bed. Spacing is as close as you can afford, or depending on how soon you want a complete cover. For example, 'Tifgreen' hybrid Bermuda placed at 8-inch intervals will cover in 3 to 5 months. Daily watering is necessary until new growth becomes established.

Stolonizing is the most often recommended method of vegetative propagation. Shredded turf segments are spread over a moist, prepared lawn bed and covered lightly with soil or organic mulch. This covering must be kept constantly moist until the individual segments have put down

Some gardeners in arid areas keep a lawn only in winter, when temperatures are cooler and rainfall is more plentiful. This annual ryegrass lawn grew from winter through spring before succumbing to high heat. The drifts of African daisies added color during the same period.

roots and have established themselves. Stolonizing is the fastest and smoothest way of establishing a hybrid Bermuda turf, short of sodding.

Sprigging involves hand planting small stolons into shallow trenches in the lawn bed. It is quite effective, but requires a lot of hand work.

The usual way to start a Bermudagrass lawn is with seed. In the low and middle zones, sow in the warmest part of the year—April through August. This allows the new grass to grow fast and establish before winter stops growth. If you live at a higher elevation where cool fall weather stops growth sooner, plant in late spring. Hybrid Bermudagrass varieties produce no viable seed, so they must be started by one of the vegetative methods described previously.

St. Augustine is sometimes sodded if sod is available, but is more often plugged or sprigged. Because of the coarseness of the individual runners, it is not stolonized.

Zoysia can be established by planting stolons or plugs, but not by seed. Perhaps one of its biggest drawbacks is that it is so infernally slow to cover, and takes a season or more before creating a lawn.

Keep in mind that no matter what method you use, it is very important that the soil be thoroughly prepared and moistened before you plant. Arid land soils are generally quite alkaline and lack organic matter. Adding organic material such as forest mulch or similar wood by-product will create a better growing medium for your lawn, and is one of the most important things you can do to ensure your lawn's success. Soils and soil amendments are discussed in detail in Planting & Plant Care, page 31.

THE DORMANT LAWN
Warm-season grasses such as Bermuda go dormant and turn a straw color in winter. Some people prefer an all-year green lawn so they *overseed* with cool-season winter grasses. Overseeding is essentially planting another quick-growing grass *over* existing dormant turf. The overseeded grass, adapted to cooler winter temperatures, will stay green and thriving through the dormant period of the warm-season grass. It then dies or is crowded out as hot weather comes on in spring as the Bermuda lawn really

begins to grow. Overseeding is generally done in the fall when there is still enough mild growing weather ahead, allowing the winter lawn to establish.

Annual ryegrass (*Lolium multiflorium*) is the grass most often used for overseeding. It is rather coarse and bunching, but it does have a bright, vibrant, lettuce green color and it grows very fast. However, because most winter grasses grow in clumps, the rate of seeding has to be fairly heavy to get an immediate effect: approximately 10 pounds per 1,000 square feet. Ryegrass has little tolerance to drying out and the soil should be kept moist at all times. It needs occasional light applications of fertilizer during the winter to keep it looking attractive. Damping off fungus can be a problem, especially if you tend to sow a lot of seed to create a quick, thick turf.

There are a number of other cool-season grasses occasionally used for overseeding Bermuda, and also for winter lawns. Among these are the new perennial ryegrasses, which are much finer in texture than older varieties. These grasses are well liked for permanent lawns in cooler climates. 'Manhattan,' 'Pennfine' and 'Denny' are recommended selections.

LOW-WATER LAWNS
In arid regions where water is at a premium, dormant lawn can be a blessing in disguise. It covers and protects the

ground with a vegetative carpet, without requiring nearly as much water as would a year-around, green lawn. In areas where dust and erosion control are problems, a Bermudagrass lawn can be established, then gradually allowed to dry out until it goes into a drought-induced dormancy. Afterwards, just enough moisture is supplied periodically to stimulate new growth and keep the turf intact. Bermuda retreats into clumps if kept dry too long.

This kind of lawn serves as a good walk-on ground cover. With irrigation, it becomes a green, vigorous lawn in a short time. Alternating watering and drying Bermuda turf works very well for playgrounds, parks and other large areas where water is costly or in short supply. In regions where summer rains occur, there is the welcome bonus of having these lawns suddenly turn green for a period after a good storm.

Another way to have a low-water lawn is to have lawn in winter only, allowing the area to lie fallow in the summer. This can be done by seeding each fall with cool-season annual or perennial grasses—those used to overseed dormant warm-season grasses. A winter lawn makes a lot of sense in many situations: It takes much less water to achieve a green effect in cool periods than it would to keep a lawn green all summer. And, even though the turf is dead during summer, dust and erosion are reduced.

Keeping your lawn area small is a simple and practical way to have a lawn without using exorbitant amounts of water. The cooling effect of a lush, green spot is retained, without the expense and effort required for a traditional, full-size, backyard lawn.

LAWN TINTS OR DYES

Dying or spraying dormant lawns with a special green paint may sound like a strange activity, but it is very common. It is one of the alternatives if you can't stand the tan, dormant color of your warm-season lawn. There is also an advantage compared to overseeding: Overseeded winter grass often delays the spring recovery of the Bermuda. Over the years, firms doing the spraying have developed some fairly natural-looking colors, but most everyone who lives in Bermuda country has seen dubious "dye jobs" on occasion. But quality dyes, when applied correctly by a reputable company, will not wash or rub off.

LAWN SUBSTITUTES

Dichondra (*Dichondra micrantha*), also known as *pony foot,* is perhaps the most attractive of the nongrass, walk-on ground covers occasionally used in arid landscapes. This creeping plant has small, round, bright green leaves, giving it a fresh, cool appearance. It will not stand excessive foot traffic, so use it in small areas or between stepping stones, or in places where mowing is difficult. Dichondra grows well in both sun and partial shade. It needs a rich, well prepared soil and ample moisture. Sow seed in the spring and fall, or plant from plugs or sod, if available.

Dichondra is sometimes attacked by flea beetles. This pest can be very destructive and can completely wipe out an entire lawn. Cutworms may also appear in spring. Their voracious appetites can wipe out a lawn if not sprayed. Rabbits love the tender leaves and can be a problem all year. Contact your local cooperative extension service for controls.

Lippia (*Phyla nodiflora*) is excellent on banks, holding soil with its deep-rooted runners. Lippia will withstand more foot traffic than dichondra but it has one drawback: It produces an abundance of cloverlike blossoms that attract bees. Avoid it if you like to run barefoot across the lawn. Lippia's small leaves are bluish green or almost grayish, especially if grown in full sun. It is quite drought-tolerant after it is established, but if you want a first-class cover, supply with regular irrigation. Lippia is available in flats at nurseries. Establish by setting out small plugs 12 to 18 inches on center in a prepared bed. Nematodes are sometimes a problem. See page 39.

White dutch clover (*Trifolium repens*) is often planted to fill in a turf during the dormant season, or as a temporary cover until a permanent lawn can mature. Clover is a fairly vigorous, spreading plant that rarely gets over 4 or 5 inches tall in these regions. It can achieve a solid cover when planted as a lawn, or as a bank cover. Clover also produces white blossoms that attract bees. It is tough and especially vigorous during the winter months, when some of the other walk-on ground covers are dormant. Clover is often combined with dichondra. Although the textures are different, a mix of the two does make a fairly dependable year-round green cover. Mowing helps improve growth and creates a tighter lawn. Sow seed in fall and water regularly to keep healthy.

LAWN CARE

A regular fertilization program is necessary to keep lawns growing well. Because desert lawns require so much water, nutrients are quickly leached from the soil. Rates for different grasses vary. Applications of fertilizers every 6 weeks during the growing season are recommended for hybrid Bermudagrass varieties. If your lawn shows signs of chlorosis (yellow color) it's probably because of a caliche underbase or general alkalinity. Iron must be added in the form of iron chelates. See page 35.

When it comes to lawn irrigation, keep one thing in mind: Never allow the soil to dry out completely if you want a healthy looking turf. Although Bermudagrass will withstand a certain amount of drought, it will look unattractive if kept too dry. If you have a deep, well prepared soil that retains water, irrigations during hot weather can be spaced three to four days apart to maintain an attractive appearance. Fast-draining sandy soils will have to be watered more often. Winds and high temperatures greatly increase water needs. The best advice is, get to know your lawn and water when it needs it. A first sign of water stress is a bluish cast or when footprints remain indented after you walk across the lawn.

Watering during cool weather requires a different set of guidelines. An actively growing winter lawn requires deep watering once or twice a week, depending on the weather and rainfall. Again, your own experiences and observations will determine the proper irrigation schedule.

Dormant lawns also need some irrigation. It is a common belief that they do not need water, but roots continue to develop even though surface growth has stopped. Many fine Bermuda lawns have been damaged when allowed to get bone dry. Soak dormant lawns at least once a month during the winter if no significant rainfall occurs.

'Tifgreen' hybrid Bermuda makes an attractive, fine-textured lawn. For best results, mow with a reel mower 1/2 to 1 inch high.

Dichondra is not a grass, but a broadleaf plant with small, round leaves. Given proper care, it makes a fine lawn for small areas.

Annual Color

A question often asked desert dwellers by visitors is, "When will the desert bloom?" This colorful and photogenic extravaganza has been so widely publicized that many assume it is an annual event, but the correct answer to this question is: "Sometime during February, March or April in the next five years." It takes properly spaced rains combined with other weather events to produce these great color shows. When the desert does bloom, it produces more flowers than anywhere else. Desert plants must flower abundantly in good years to maintain the seed supply that perpetuates the species. The desert gardener does not have to wait for just the right combination of weather to have his own flower show, because he controls the moisture that brings the flower garden to life.

PLANTING DATES

Many newcomers to desert regions are surprised to learn that the seasons are more or less reversed. October is the "spring" planting time. During this period, weather conditions are ideal for setting out new plants for a winter or spring garden. Plants require much less water during this cool period. And, flower color seems to be more magnificent if it occurs during the coolness of winter.

Cool-weather annuals should be planted when nights are cool but not cold, and days are warm but not hot. Plants are then able to reach a blooming phase before nights become too cold. If buds have formed, flowers will open throughout winter except during very cold spells. This show continues until summer heat comes in May. Many gardeners abandon flower gardening in summer, waiting until cool weather returns.

Many annuals thrive during hot weather. Year-round color is possible with a second planting season, usually beginning mid-May, for those who wish to garden through the summer. The list of summer-flowering annuals is shorter than cold-weather types and gardening problems are magnified by severe climatic conditions, but summer color can be very showy. Planting dates for summer flowers are not as critical as winter types because of the long, warm season ahead. Planting summer flowers early before nights have warmed doesn't create an advantage. Most will do nothing until it warms up.

PLANTS OR SEED?

Most annual plants can be grown easily from seed, but it is better to buy plants at the nursery for fall gardens. Unless you have a cooled greenhouse or a place indoors to start seedlings in late July or August, it is practically impossible to have plants ready for transplanting in time. A delay of a few weeks in planting can cause flowers to wait until spring before blooming, thus shortening the season drastically.

Summer-flowering annuals are also usually planted as nursery-grown plants, but many species can be sown directly in the flower bed. Because the air and soil temperatures are so warm, plants develop rapidly. Plant a little more seed than you need to assure adequate coverage, then thin seedlings if they become crowded.

SOIL AND WATER REQUIREMENTS

Annuals can be grown in nearly any garden soil. Poor, sandy or heavy soils can be made usable by adding humus and fertilizer. Most flower species require regular irrigation, and few are drought-resistant. Locate flower beds where they will be easy to water. Some drought-tolerant desert natives are occasionally grown as cut flowers but they are of limited value.

The following chart and the perennial chart on pages 28 and 29 were adapted from the University of Arizona Cooperative Extension Service. Use your elevation as a guide to the planting and flowering dates.

Two ways to design with color: Left, masses and borders of petunias, calendulas and sweet alyssum create a dramatic color showcase. At right, a simple planting of African daisies produces a casual, natural color display.

Annuals

Common Name / Botanical Name	Height in Inches	Planting Dates to 3,000 feet	Blooming Dates to 3,000 feet	Planting Dates 3,000 to 5,000 feet	Blooming Dates 3,000 to 5,000 feet	Comments
African daisy / *Dimorphotheca sinuata*	12-18	Sept. 15 to Nov. 30	Feb. to May	Sept. to Oct. 15	Feb. to May	Hardy and fast growing. Brilliant yellow and orange blooms, excellent massed or in beds. Blooms freely up to 6 months a year. Naturalizes and reseeds easily.
Aster / *Aster* species	12-24	Oct. to Nov. 15	April	April to May	July to Sept.	Beautiful bedding plants for higher elevations. Well-drained soil. Pompon type most hardy.
Begonia (waxleaf) / *Begonia* species	10-12	Sept. to Nov. 15	All year if frost free	April to July	March to Oct.	Partial to full shade. Often used as summer bedding plant. Pots, beds and borders. Red, pink and white flower colors. Bronze-foliage types take heat the best.
Calendula / *Calendula officinalis*	15-18	Sept. 15 to Dec.	Jan. to April	Aug. to Oct.	March to June	Hardy, early flowering, easy to grow from seed. Yellow and orange blossoms over a long season. Available in dwarf form. Not hardy in higher elevations.
Candytuft (hyacinth type) / *Iberis amara*	12-18	Sept. 15 to Nov. 30	Feb. to May	Sept. 15 to Oct. April to May	March to July	White fragrant flowers, also available in pink, lavender and purple. Longer season in partial shade, sometimes perennial. Good cut flower, beds and border.
Candytuft / *Iberis umbellata*	12	Sept. 15 to Nov. 30, Jan. to Feb.	Feb. to May	Sept. 15 to April	March to June	Flat flower clusters in assorted colors. Ground cover, rock garden and border.
Coleus / *Coleus* species and cultivars	6-18	April to June	June to Nov.	June 15 to July 15	July to Oct.	Grown for its multicolored foliage. Excellent container plant, indoors and out. Tender to frost. Needs some humidity.
Cosmos (yellow) / *Cosmos sulphureus*	36-72	April to June	July to Nov.	May 15 to June 15	July to frost	Brilliant, daisylike, orange to yellow-orange flowers. Use for cutting or screening. Tolerates heat. Late summer bloom period.
Dahlias (seed) / *Dahlia* species	12-36	March to April	April 15 to Nov.	June to July	July to frost	Tall and dwarf varieties. Excellent for cut flowers. Well-drained soil. Part shade in lower zone.
Delphinium / *Delphinium* species	30-60	Sept. 15 to Nov.	March to June	Sept. to Oct. 15	March 15 to June	True perennial used as an annual. Tall flower spikes, use as background. 'Pacific Giant' and 'Connecticut Yankee' do well.
Dianthus / *Dianthus barbatus*	10-15	Sept. 15 to Nov. 30	March to June	Sept. to March	April to July	Dense clusters of double or single flowers. Mixed colors. 'China Pink', 'Annual Carnation' and 'Sweet William' are popular. Often perennial in partial shade.
Globe amaranth / *Gomphrena globosa*	15-24	April to July 15	June to Nov. 15	May 15 to July 15	July to Oct.	Excellent border or low bedding plant. Cloverlike flowers are usually purple but are sometimes available in pink and white. Use in dried arrangements.
Impatiens (balsam) / *Impatiens balsamina*	6-24	April to May	May to Nov. 15.	May to June	June to Oct. 30	Outstanding shade plants, especially adapted to hanging baskets. Dwarf and semidwarfs available. Bright colors.
Larkspur / *Consolida ambigua*	24-48	Sept. 15 to Nov. 30	March to May	Aug. to Oct.	May to July	Tall, blue, red or pink flower spikes. Popular for cut flowers. Background plant for beds and screening. Naturalizes easily.
Lobelia / *Lobelia spacata*	6-10	Sept. 15 to Nov. 15	March 15 to May	April to May	April to May	Good for borders, edgings and rock gardens. Protect from temperatures below 28°F (-2°C). Summer annual in highest areas.
Marigold, African / *Tagetes erecta*	30-36	April 1 to June 30	June to frost	May 15 to June 30	July to frost	Tall flower for beds or cutting. Blossoms of many types and colors. Reseeds easily. Likes afternoon shade in low zone. Prone to leafhopper damage.
Marigold, French / *Tagetes patula*	12-18	April to June 30	May to frost	May to June 30	July to frost	Produces masses of orange and yellow blossoms. Best in fall and late spring gardens. Prone to leafhopper damage. Double or single types. Excellent color plant for higher elevations.

Common Name / Botanical Name	Height in Inches	Planting Dates to 3,000 feet	Blooming Dates to 3,000 feet	Planting Dates 3,000 to 5,000 feet	Blooming Dates 3,000 to 5,000 feet	Comments
Nasturtium / *Tropaeolum* species	10-18	Sept. to Nov.	April 15 to May 30	April 15 to May 30	June to Sept.	Protect from frost. Low beds, ground cover, borders and cut flowers. Single and double forms.
Pansy (garden) / *Viola* species	6-12	Oct. to Jan.	Sept. to Nov.	Sept. to Nov.	March to July	Multicolored flowers bloom over a long season. Plant in fertile soil. Tolerates partial shade.
Periwinkle (Madagascar) / *Catharanthus roseus* (*Vinca rosea*)	18-24	April 15 to May 30	June to Nov. (or to frost)	May 30 to June 30	July to frost	Hardy plant for beds or borders. One of the best summer bedding plants. Masses of white and lavender blossoms all season. Often perennial.
Petunia / *Petunia* hybrids	12-24	Oct. to April	Nov. to May	April to May	May to frost	Available in a variety of forms and colors. Adapted to many uses—pots, beds, ground covers. Best to plant transplants. Long blooming season. Requires well prepared garden soil.
Phlox (annual) / *Phlox drummondii*	6-12	Oct. to Jan. 1	March to June	Aug. to Sept., April to June	April to May, June to frost	Prolific color producer, good for borders, edging or rock gardens. Also as a cut flower.
Poppy (California) / *Eschscholzia californica*	10-12	Sept. to Nov. 30	March 15 to May 30	July to Sept.	April to August	Colorful border plant. Seeding is best. May be sown in open desert areas that get winter moisture. Naturalizes over large areas.
Poppy (Iceland) / *Papaver nudicaule*	12-18	Oct. to Nov. 30	Feb. 15 to May	Aug. to Oct.	April to June	Sow or set out plants early from flats in large beds for mass effect. 'Champagne Bubbles' is one of the best varieties; it blooms early.
Primrose (fairy) / *Primula malacoides*	6-12	Oct. to Nov.	Feb. to May	April to May	May 15 to June	Small pink, rose, red, lavender or white flowers. Blooms in shade. Needs regular care. In cold areas plant after major frost period is past.
Primrose (English) / *Primula polyantha*	6-12	Oct. to Nov.	Nov. to May	Oct. to March	March to May	Hybrid types in a variety of rich colors. Excellent for containers, beds and borders. Best winter color plant for shade.
Salvia / *Salvia patens*	15-30	April to June 15	June 15 to Nov.	June to frost	July to frost	Bedding plant for full sun to part shade. Tall, brilliant, red flower spikes.
Snapdragon / *Antirrhinum* species	6-36	Oct. to Nov. 30	March to June	March to June	May to frost	Excellent for cutting, borders or beds. Dwarf types provide earlier and longer period of bloom.
Stock / *Matthiola* species	18-30	Oct. to Nov. 30	Feb. 15 to June	July to Oct.	April to July	Excellent cutting, border or bedding flower. Plant early for best results. Single and double, white, pink, purple flowers are fragrant.
Sweet alyssum / *Lobularia maritima*	6-9	Sept. to March	Jan. 15 to June	Aug. to Sept., March to April	April to frost	Excellent for borders, ground cover, edging and rockeries. Blooms profusely and reseeds. White and purple colors. Hardy; grows all year with morning sun.
Sweet pea / *Lathyrus odoratus*	12-72	Oct. to Jan. 15	Feb. to May	March 15 to May 1	April to June	Early and late-flowering varieties. Excellent cut flowers. Vining plant: Train on fence or trellis. Available in bush form. Plant in spring in cold areas of high zone.
Verbena (common) / *Verbena* hybrids	8-12	Sept. 15 to Nov. 30	All year (best in spring)	Aug. to Sept., April to May	April to Nov.	Small colorful plants for edging, ground cover and window boxes. Very hardy; often perennial.
Viola / *Viola cornuta*	6-8	Oct. to Jan. 1	Dec. to May	Sept. to Nov.	April to July	Excellent edging or low bedding plant for winter and spring color. Light shade extends bloom.
Zinnia / *Zinnia elegans*	4-30	April 15 to July 15	June to frost	May 15 to July	July to frost	One of the best summer flowers. Loves sun, avoid shade. Tall, medium and dwarf types. Many forms and colors. Grows easily from seed.

Perennial Color

Perennials differ from annuals in that they live and flower for a number of years, depending on the species. You may note that a few perennials were listed previously as annuals, simply because that is how they are used in hot, dry regions. It is marvelous to have a plant that will put on a seasonal show year after year, and many gardeners use perennials for this reason. But they are not used quite as often in these regions as in temperate zones where they are a major source of garden color. This is because desert perennials produce color for maybe half the year or less, yet occupy garden space for the whole year. Many gardeners choose perennials that produce attractive foliage as well as flowers, so the unattractive "off-season" is much less noticeable.

Perennials require optimum growing conditions. Some will endure drought stress, but lack of water often puts them into dormancy. It is best to plant them in a well prepared soil and supply regular irrigation. Prune plants occasionally to initiate new growth.

Most perennials regrow from the same clump year after year, so plants should be dug up, divided and reset periodically to keep them growing and blooming well. Some plants will only last a few years, then need to be replaced. Changing the location of a perennial species every few years is also a good idea. This will prevent build up of diseases and insects particular to that plant or growing area.

Common Name *Botanical Name*	Height in Inches	Planting Dates S-Seed T-Transplant to 3,000 feet	Blooming Dates	Planting Dates S-Seed T-Transplant 3,000 to 5,000 feet	Blooming Dates	Comments
Asters, hardy (New England aster) *Aster novae-angliae*	24-36	S-April to Aug. T-Oct. to Dec.	Sept. to Dec.	S-June to Aug. T-Nov. to May	Sept. to Dec.	Purple, blue, pink and white daisy flowers in showy sprays. Full sun or partial shade. Very hardy; may crowd out other perennials. Clumps last indefinitely. Divide every 2 or 3 years.
Blue marguerite *Felicia amelloides*	18-24	T-Oct. to April	March to Dec. All winter if frost-free	T-April to May	May to frost	Azure blue daisies form on single stems above mounds of crisp, dark green foliage. Full sun—blooms almost continuously, but tender to frost. Excellent container plant. Replace plants every 3 to 4 years.
Canna *Canna* hybrids	14-36	T-May to June	May to frost	T-June	June to frost	Red, pink, orange, yellow and white orchid-textured flowers. Full sun. Dwarfs best in borders. Divide and reset every other year to prevent beds from crowding.
Carnation *Dianthus caryophyllus*	12-24	S-April to June T-Oct. to March	Sept. to Dec. and March to June	S-May to July T-Sept. to Jan.	April to frost	Red, pink, orange, yellow and white fringed flowers. Single and multiple stems on bluish gray foliage. Full sun. Very hardy. Cutting grown types are best.
Chrysanthemum (Mums) *Chrysanthemum morifolium*	18-36	S-March to May T-Dec. to March	Oct. to Dec.	S-May to July T-Jan. to April	Aug. to Nov.	Purple, red, pink, orange, yellow and white flowers. Sun or partial shade. Cushion and hardy types best for the garden. Use early-flowering varieties in upper elevations, pinch back early spring and late summer to encourage flowers and growth.
Columbine *Aquilegia* hybrids	24-36	S-March to Sept. T-Nov. to Feb.	April and May	S-April to Aug. T-Oct. to Jan.	April to July	Purple, blue, pink, yellow and white flowers are numerous, appear above fresh green foliage. Best in full or partial shade—good under trees. Old plants can be separated to reset but replacements are usually grown from seed.
Coreopsis *Coreopsis grandiflora*	24-36	S-April and Oct. T-Oct. to March	April to June	S-May to July T-Nov. to March	May to Oct.	Golden yellow flowers appear intermittently all summer. Medium green clumping foliage. Full sun. Very hardy. Plants last 2 to 4 years.
Daylily *Hemerocallis* hybrids	18-36	T-Oct. to Jan. All year if frost-free	March to June	T-Oct. to Jan.	May to July	Yellow, orange and maroon colors. Striking lily-shaped flowers borne above clumps of strap-shaped foliage. Sun or partial shade. Some leaf burn in low elevations, early types are best there. Blooms over a long season. Very hardy. Long-lasting clumps, but divide, thin and reset or they become "clump-bound."
Dianthus (Pinks) *Dianthus plumarius*	10-15	S-April to June T-Oct. to March	March to June	S-July to Sept. T-Oct. to Feb.	April to June	Dainty clusters of fragrant, carnationlike flowers in white, pink, salmon and rose. Attractive foliage all year. Very hardy. Full sun to partial shade. Needs well drained soil high in organic matter. Clumps are long lasting. Reset new rooted cuttings when old plants become woody.

Common Name / Botanical Name	Height in Inches	Planting Dates S-Seed T-Transplant to 3,000 feet	Blooming Dates	Planting Dates S-Seed T-Transplant 3,000 to 5,000 feet	Blooming Dates	Comments
Feverfew / *Chrysanthemum parthenium*	18-24	S-April to May T-Oct. to March	April to June	S-May to Sept. T-Nov. to March	April to frost	Full sun. Very hardy. Single and double, white, button-shaped flowers. Bright green foliage. Old plants can be divided but it is better to replant every 3 to 4 years.
Foxglove / *Digitalis purpurea*	24-72	S-April to May T-Oct. to Dec.	April to Sept.	S-May to July T-Oct. to Feb.	May to Sept.	Shade and partial sun. Tall spikes of bells create a bold vertical display. Large grayish leaves. Replace with new plants when old ones become woody and die out.
Gaillardia (blanket flower) / *Gaillardia grandiflora*	12-30	S-April to Sept. T-Dec. to Feb.	April to Dec.	S-May to July T-Oct. to Feb.	April to Nov.	Yellow and red daisy flowers with grayish green foliage. Best in full sun. Very hardy border plant that blooms all summer and off and on all year. Replace with new seedlings when old plants decline.
Geranium / *Pelargonium hortorum*	12-36	S-April and May T-Sept. to May	All year, but most stop in hot season	S-Start inside March to April T-April to June	May to frost	Sun or partial shade. Variety of double and single flowers in large flat clusters. Very tough but frost tender in upper elevations. Long lasting if not frosted.
Gerbera (Transvaal daisy) / *Gerbera jamesonii*	12-18	S-April to July T-Oct. to Feb.	March to June	March to April	June to Summer	Red, pink, orange, yellow and white fringy flowers appear on long stems above pointed green foliage. Full sun or partial shade. Drainage is important.
Hollyhock / *Alcea rosea*	48-72	S-March to Oct. T-Oct. to March	April to June	S- April to Aug. T-Oct. to March	May to July	Red, pink, yellow and white flowers on giant 4 to 6-foot spikes. Full sun. Very hardy and reseeds easily. Single and double flower forms.
Iris (bearded iris) / *Iris germanica*	12-48	T-July to Nov.	Feb. to May	T-Aug. to Dec.	April to June	Large, purple, blue, yellow, orange, pink and white flowers. Bold, gray-green, spiky, strap-shaped leaves. Full sun to partial shade. Very hardy—not too much water required after blooming. Divide in late summer or fall. Reset every 4 to 5 years.
Penstemon (bearded tongue) / *Penstemon gloxinioides*	14-30	S-April to Oct. T-Oct. to Jan.	April to June	S-May to Aug. T-Nov. to Feb.	May to Summer	Red, pink, white, blue and purple flowers. Vigorous upright plant, almost shrublike. Sun or partial shade. Very hardy. Good summer color.
Phlox / *Phlox paniculata*	18-36	T-Oct. to Jan.	July to Sept.	T-Sept. to Dec.	June to Aug.	Outstanding, purple, red, pink and white flowers. Plant in partial shade or in east or northeast exposure. Divide and reset every 3 or 4 years.
Physostegia (false dragon head) / *Physostegia virginiana*	24-36	S-April to June T-Oct. to March	Aug. to Oct.	S-June to July T-Nov. to March	Aug. to Oct.	Pink, lavender and white colors. Full sun to partial shade. Flowers borne in terminal spikes—good for background planting. Cut back plant in off season.
Rudbeckia (gloriosa daisy) / *Rudbeckia hirta*	30-48	S-April to Oct. T-Oct. to Dec.	Aug. to Oct.	S-Aug. to Oct. T-Oct. to Jan.	Aug. to Oct.	Yellow and mahogany colors. Bushy plants need space. Full sun. Very hardy—fall flowering. Divide every third year. Often better to reset with new plants when old clumps weaken.
Shasta daisy / *Chrysanthemum maximum*	18-30	S-April to Oct. T-Jan. to Feb.	April to June	S-May to Aug. T-Oct. to March	May to frost	Huge daisy flowers—semidouble types available. Full sun or partial shade. Divide every 3 or 4 years.
Sunflower (perennial) / *Helianthus multiflorus*	36-60	S-April to May T-Oct. to March	Aug. to Nov.	S-May to June T-Dec. to March	Aug. to Nov.	Spectacular clusters of orange, yellow, white and maroon flowers. Full sun. Very hardy. Tall background plant with mass of flowers—clean yellow color. Divide and reset or plant new plants every 3 or 4 years.
Sweet violet / *Viola odorata*	8-10	T-Oct. to Feb.	Jan. to March	T-Oct. to March	Jan. to April	Purple, blue and lavender colors. Sun to shade—excellent bedding plant for shady location. Very hardy. Violet-blue flowers in winter and in early spring. Heart-shaped evergreen leaves. Reset every 3 to 5 years.

It is important to give plants what they need to maintain health and growth. Knowing the basics of planting and plant care will take you a long way toward a successful landscape.

Planting & Plant Care

No matter what kind of climate you live in, hot, arid or otherwise, the key to success with plants depends on understanding their needs for growth. Plants have extremely varied cultural requirements, but the basics remain the same: They need a good soil that drains well, appropriate amounts of sunshine, a uniform supply of moisture and nutrients, and protection from the elements.

THE IMPORTANCE OF GOOD SOIL

In nearly every instance, a "good soil" is high in organic matter. Lack of organic matter and the high alkalinity of soils makes growing plants different in arid areas. Soils tend to be dense and dry. Whether rocky, clayey, gravelly or sandy, they usually have little *humus*—organic matter—often less than one percent by volume.

SOIL AMENDMENTS

When you *amend* the soil with organic matter, drainage is improved, and the

These healthy plants show the importance of basic plant care. Above: Regular irrigation is necessary with most plants, especially young plants.

soil is able to retain moisture and nutrients longer in the root zone. Organic amendments normally added to the soil include composted forest products, peat moss, grass clippings and homemade compost.

High summer heat causes many fine-textured organic materials such as peat moss to dissipate quickly—sometimes in less than a year. For that reason some of the best and cheapest amendments are the forest products: ground bark, sawdust and other by-products from the mill. The coarseness of these amendments is an advantage. They take longer to decompose and the soil "fluffs up," adding beneficial air spaces.

Commercially available bark products such as forest mulch are *stabilized* with nitrogen. This additional nitrogen is necessary because bacteria uses nitrogen in the soil to decompose organic materials. It is recommended that you use amendments that are already nitrogen stabilized.

Manures are not often used as amendments in arid areas because of their high salt content: Arid land soils are usually too salty to begin with. Manures are sometimes used as a top

dressing on a new lawn. If you do use manure, be certain it is well composted and salts are leached away before incorporating it into the soil.

When adding amendments, you need to add enough to change the structure of your soil. A little bit of material won't do any good. If you were to prepare a planting bed for a lawn, vegetable garden or ground cover, for instance, you need to spread at least 3 inches of material over the planting area. This should be mixed thoroughly with the existing soil to a depth of 8 to 12 inches.

PROBLEM SOILS

The term *alkaline* is used in this book to refer to soils of a high pH—7.0 or above. There are three basic types of alkalinity: *calcareous, saline* and *sodic*. These conditions can exist separately, but they are often found in combination. All of these conditions can cause nutrient deficiencies in plants. They "tie up" manganese, zinc and most importantly, iron, preventing them from dissolving so plants cannot absorb them. This results in unhealthy plants with reduced growth and vigor.

Calcareous soils usually contain calcium carbonate, which is only slightly soluble in water. This material is most troublesome when it forms an impervious layer known as *caliche* or *hardpan*. This layer can be up to 6 feet or more thick. It is formed when rain or surface water continually deposits calcium at a certain soil depth, usually 6 to 20 inches. The water either evaporates or is absorbed by plants, leaving the calcium to accumulate in a cementlike layer. Heavy accumulations may take centuries to build up. See photo below.

A cross-section of soil reveals a very thick layer of caliche—the lighter, cement-colored section. The darker layer at the top is topsoil. As you can see, plant roots are unable to penetrate the caliche.

Suggested Sizes for Planting Holes:

- 6x6x5 feet deep for 36 to 48-inch boxed tree
- 5x5x4 feet deep for 30 to 36-inch boxed tree
- 4x4x3 feet deep for 15-gallon can or 24-inch boxed tree
- 3x3x3 feet deep for 5-gallon can
- 2x2x2 feet deep for 1-gallon can
- 1-1/2 feet deep for a flower bed or ground cover
- 1 foot deep for a lawn

Saline soils contain soluble salts such as sodium chloride—common table salt. Soluble salts may be present naturally in the soil, but they can also come from irrigation water or sea spray. You can often see salts as a white crust on the soil surface, left by evaporated water. Similar crusts may also form in the bottom of a plant pit, which can crowd or burn roots. Most salts are easily *leached* (washed from the soil) below root depth by occasional deep soakings.

Sodic soils are sometimes referred to as *alkali, black alkali* or *slick spot* soils. They are caused by having too much sodium in the soil in relation to amounts of calcium and magnesium. Sodic soils usually have a pH higher than 8.5. When dampened, they are often dark and become slick and slimy on the surface. Their structure as a soil has broken down. They are impervious to water, have no air spaces and do not allow roots to grow. Sometimes sodic soils occur in combination with soluble salts, called *sodic-saline* soils. Few plants can grow under either condition. Adding large quantities of gypsum to the soil can help, but if you think you have one of these soils, it is best to consult an expert for treatment.

Plow pan is caused when soil has been tilled for a period of time to a certain depth. It is found in old orchards or areas farmed for commercial crops. A layer of salts may accumulate at the surface of the unbroken subsoil, or the subsoil may become packed. Plant roots sometimes find it difficult to break through. In both cases, water will not soak into the subsoil and forms standing pools. Before planting, the subsoil should be broken up and soaked to wash away salts.

If you have questions about soil quality, check with your county agent, nearby university extension service or a private soil lab. A soil analysis test can be obtained for a relatively small fee.

PLANTING TREES AND SHRUBS FROM CONTAINERS

In very hard-packed, low-humus soils, the planting hole may be the only space your plants will have for root development. In caliche soils, farmed areas with plow pan, or dense clay soils, planting holes should be deep and have adequate drainage. Plants will certainly fail if they are rootbound or waterlogged.

When gardening with difficult soils, the saying is, "the larger the planting hole, the better." But not everyone can afford to rent a backhoe to dig holes or fill large holes with planting soil. The rule, is to make the hole at least twice as large as the rootball.

Once you have decided on placement of your plants, dig the hole or bed using the guide at lower left. *Check for drainage.* Four inches of water in the bottom of a hole should drain in 3 to 4 hours. If water remains after that time, dig the hole deeper or make a 6-inch wide "chimney" 2 to 3 feet deep in center of hole. If the hole still won't drain, consider planting somewhere else. Or, select a plant tolerant of these conditions. Mesquite or palo verde trees, for example, will grow in caliche.

Well-prepared, amended soil is especially important in the top 8 to 12 inches, where plant feeder roots develop. Save topsoil dug from hole, but discard rocks, caliche and salty layers. Retain subsoil for use later as a retainer for a watering basin. See photos on page 33. Throughly mix one-third to one-half parts forest mulch or similar product with topsoil. If soil is dense and sticky, indicating a high clay content, add one part sand to improve drainage.

BARE-ROOT PLANTING

Deciduous plants such as roses, fruit trees, shade trees and shrubs can be planted bare root during their dormant season. They are less expensive than plants in containers. Buy fresh stock at the beginning of the bare-root season in your area, and be sure roots are not dried out.

To plant, make a damp mound of soil mix in center of hole. Set plant on the mound, spreading roots. Make sure the *bud union,* the place on grafted plants where the top joins roots, is above ground level. Allow at least 2 to 4 inches for settling. Fill hole to this level, and add an earth berm for a watering basin. As you fill with soil, gently press down around roots to remove air spaces. Soak planting hole, and adjust soil level if necessary.

After planting, thin plants back by one-third or more. This is necessary because roots were trimmed by the grower and cannot support a large top. If you don't cut back enough, the plant may decline and die.

Keep plant moist until leaves appear. Then follow a regular watering schedule for plant as given in the plant descriptions.

Planting Step-by-Step

1/Dig hole according to size of plant container using guide on page 32. Prepare planting mix by mixing one-third to one-half parts soil amendment with topsoil.

2/Dig some planting mix into existing soil in the hole to create a transition zone for plant roots. Partially fill hole with mix, leaving space for rootball.

3/Carefully remove plant from container, keeping rootball intact. Trim broken or matted roots. If rootbound, gently free roots at the rootball base.

4/Set plant in hole and carefully press soil around roots to remove pockets of air. Plant should be planted at same level as it was in its container.

5/Water to settle soil. Check again to see if plant is situated at proper level. There is often a soil-stained ring around trunk that will tell you this.

6/Make a temporary, inner water basin to make sure moisture gets to rootball. Also make a larger, permanent basin. Water. Keep soil damp for about 10 to 14 days.

STAKING TREES

Many young trees need to be staked for support, especially if strong winds occur in your area. Place stakes at right angles to the strongest wind. You can also train a tree to a desired height by staking it, or encourage a shrubby plant or a plant with a weaving trunk to become more erect in form. Often you will want to use more than one stake. Tying should be done loosely with a pliable material or flexible plastic tie so the tree can move in the wind and gain strength on its own. Ties should be loosened as the tree grows. Remove supports when tree will remain erect on its own.

WATERING BASICS

"When" and "how much" are the kind of questions many gardeners ask about watering plants. It would help if there were set rules, but there are simply too many variables. Different plants require different amounts of moisture, but other factors come into play, such as the plant's age and size, the composition of the soil, the time of year and the weather.

Actually, the factor that most determines the answers to watering questions is *you.* Your experience with the plants in your landscape is the best guideline to follow. As you get to know your plants' varied needs, a watering schedule will emerge.

Here are a few guidelines to follow that will help you get started: Try to water plants just before they begin to show signs of stress. Look for slightly wilted foliage, or a loss in luster or color change that is darker, grayer, or bluer than normal. Other signs are curling or falling leaves. If leaf edges turn dry or leaves yellow, this indicates a more advanced state of drought.

Deep, periodic waterings are usually best for plants in any region, but they are especially useful in arid lands because of the high percentage of soil salts. Long, slow waterings, up to four hours, will wash salts from the root zone. Letting the soil get somewhat dry between irrigations, allows air, necessary for plant life, to get to the roots. Never allow soil to become bone-dry between waterings, except for well-established desert plants.

Watering newly planted plants. Keep plants damp for 10 to 14 days (longer in warm weather) until they begin to grow. Then gradually reduce the watering schedule to match the plants' particular needs.

Natives and other drought-resistant plants are often neglected when they are first planted because of their reputation for toughness. But when *you* choose a plant's location, it is not necessarily getting all the advantages plants in the wild receive. Treat native plants with the care you would give any nursery plant, and give them regular irrigation for at least the first year.

SEASONAL WATERING GUIDELINES

A heavy irrigation should be given to most plants in early spring, slightly later in higher and colder zones. It can be done just prior to the first feeding of the year. This leaches salts which may have collected in the soil during the previous year or over winter.

Maintain steady soil dampness during spring to encourage growth, flowering and fruiting. Plants that are stressed do not perform as well.

In summer, especially if the weather is hot and dry, maintain a diligent watering schedule. This is the period of greatest stress and weak plants are more subject to pests and disease.

Taper off water as autumn nears to harden growth so that it becomes woody. Ample water encourages succulent growth that can be damaged by cold, even cold-hardy plants.

Don't forget to water plants periodically in winter. Roots continue to grow, but at a slower rate. Lack of water can damage plants, or ultimately cause death.

DRIP IRRIGATION

When you apply water to your plants, it is important to put it in the root zone where it will do the most good. The rising cost and decreasing availability of water has led to increased use of drip systems throughout the world.

Drip irrigation heads or tubing apply water very slowly to the soil in a pear-shaped area, from 3 to 6 feet in diameter and up to 8 feet or more deep, depending on soil composition. When a system is working properly, there is no loss of water due to runoff or puddling. Water basins are not required around plants with a drip system. And, because there is not as much surface water, there are fewer weeds and evaporation is reduced. Most drip systems operate on very low water pressure that can reduce water costs.

Drip systems are seldom used in lawn or ground cover areas, but are valuable for watering individual trees and shrubs. Plants or trees on slopes that seldom receive enough deep moisture are perfect candidates for drip irrigation. The result is better root growth and deeper water penetration. In patios or similar areas, containers and hanging plants can be watered simply and efficiently with individual tubing going to each plant.

Many types of drip systems are available. Your nurseryman or an irrigation contractor can help you design a system that is right for your landscape.

MULCHES

A mulch is any material used to cover the soil. Mulches can be organic or inorganic: Bark products, wood chips, leaf mold, grass clippings and manure are organic mulches. Inorganic mulches are gravel, decomposed granite, rock and plastic.

Mulches have many uses in the landscape. On a small scale, they can be used around plants to modify temperatures, reduce loss of moisture and reduce weeds. Organic mulches improve soil structure. The cooling effect they have on a plant's root zone is of great value in hot climates. For example, soil is approximately 10°F (−6°C) cooler when covered with a 3-inch layer of mulch.

Mulches, especially inorganic types, are inexpensive, low-maintenance ways to cover large areas. They can be a permanent, attractive part of a landscape design. This is a common practice in many arid areas where water is at a premium. But certain mulches cause more weed seeds to germinate than if the ground were bare. Heavy black plastic (at least 4 mils thick) can be sandwiched between layers of protective sand and overlaid with mulch to stop weeds. Holes are cut in the plastic for plants. By creating a depression around planting holes and sloping plastic toward them, natural runoff will water plants. One drawback with plastic is that it doesn't always last or stay concealed by the mulch covering. This can give the landscape a shabby

Drip irrigation allows you to put water where it's needed most—in the plant's root zone. Container plants which require a lot of water are prime candidates for a drip system.

Organic products such as forest mulch can be used to amend the soil for planting, or applied as a mulch around the root zone of plants to modify soil temperatures.

look. An alternative method involves mixing cement into the sand layer before topping with the mulch. This mixture forms a type of crust and reportedly does an excellent job of keeping weeds out permanently. A pre-emergent spray is another method of weed control. It is applied to the soil, usually in early spring, to kill weed seeds *before* they germinate.

FEEDING PLANTS
Like people, plants need proper nutrition to be healthy. Nutrients most often needed in arid climates are nitrogen and iron. Slow growth, small leaves and a pale to yellowish look, especially on old foliage, can indicate a need for nitrogen. New growth that is yellow often indicates a need for iron.

Labels on fertilizer packages identify contents of nutrients with three numbers, such as 24-4-4. The first number is the percentage of *nitrogen,* the second is the percentage of *phosphorus* and the third is *potassium* or *potash*. Potash is not usually needed by plants in soils with high pH and a high lime content. But container plants in acid potting mixes may need it. Look for mixes that are high in nitrogen.

There are fertilizers formulated for certain plants such as gardenias or citrus, or plants in containers. Many nurseries put out their own fertilizer, especially balanced for local soils and conditions.

Iron chlorosis. *Chlorosis* means a yellowing of plant leaves. Iron chlorosis is an iron deficiency in plants. It is caused by hot temperatures of summer, extremes of wet or dry soils, inadequate depth of watering, salty soils or caliche (high-calcium) soils. Symptoms are most often seen on new growth. See photo below right.

Improving soil gradually with additions of organic matter should correct the situation in time. Or iron can be added in the form of *chelates*, either directly to the soil or sprayed on plant leaves. Ferrous sulfate (iron) can be added by making *banks* in the soil. Punch some holes several inches deep in the root zone and add a tablespoon or two of iron. The only drawback with this method is that the iron is absorbed slowly.

Soil sulfur or another acidifying agent changes soil composition and makes iron and other minerals available to plants. Small amounts of sulfuric acid placed in a tiny area near the root zone can provide a bank of released minerals for roots to draw from.

PRUNING
Pruning or thinning is done to improve structure, form or size, or to increase blooming or fruiting. Heavy pruning is usually done at the end of winter or in early spring. Wait until the coldest weather is over, before new growth begins. This allows plants to take advantage of their greatest natural growth period. Severe pruning performed at any other time may subject plants to damage from cold or sunburn.

Thin plants to hold their natural form, rather than shear them. Thinning causes plants to fill out with new foliage. This not only improves their appearance, but the new growth shields the bark and roots from sunburn. Clipping and light trimming can be done any time to maintain a desired shape.

PROTECTING PLANTS
Cold protection is sometimes necessary for borderline plants or if an unusual cold spell occurs. Flood lights placed in or near plants can produce quite a bit of heat, especially if there is a covering over plants. Sometimes a covering is all a plant needs to see it through a light frost. Burlap works well. If you use plastic, keep it from touching the plant. Plastic can become so cold it may freeze the plant leaves. Make a wooden frame and place the cover over it. Flooding planting beds with water can make temperatures rise a few degrees. Burning petroleum blocks will also increase temperatures. Move container plants indoors.

Sunburn can damage many plants, especially newly planted trees. Young citrus, fruit trees, ash, magnolia and many others can suffer severe bark burn unless the trunk is protected. Use a latex paint or commercial wrap to prevent damage.

Plants often need protection from weather extremes. Here, the trunks of young citrus trees are wrapped to prevent sunburn. And, when tender plants such as citrus are planted in containers, they can be moved to a protected location when frost threatens.

Iron chlorosis is the inability of plants to absorb iron in the soil. It is most often seen on new growth, such as this mulberry tree.

Plant Problems

Developing good gardening habits as described in the preceding section will help prevent plant problems before they start. Following soil preparation guidelines and maintaining a regular program of watering, fertilizing, weeding and pruning will keep plants growing vigorously. Healthy plants are resistant to attacks from pests and diseases. And, by caring for your plants regularly, you become more familar with them. If a problem occurs, in most cases you will be able to spot it and treat it promptly.

WHY PLANTS FAIL

A plant's decline and eventual death is usually caused by one of three things: improper growing conditions, insect pests or diseases. Pests and diseases are sometimes unavoidable, but landscape plants are dependent on you to supply a proper environment for growth. Some of the more basic causes of plant decline are: too high or too low temperatures, too much or too little water and soil that is inadequate for proper root growth. See chart, opposite.

Proper plant selection is a major factor in avoiding these problems. Because of weather extremes of hot, arid climates, site adaptation is especially important. Many plants common to temperate climates will die here in short order without the aid of an unusually persistent gardener.

COMMON SENSE PEST CONTROL

Every insect you see is not out to destroy your landscape plants; many are beneficial. In addition to the well-known ladybug and praying mantis, there are many parasitic wasps, spiders and helpful insects, including the common black ground beetle, that prey on plant-damaging pests. Using these "good guys" is an important part of the developing science known as *Integrated Pest Management* (IPM). The basic principle of IPM is to use the simplest, least disruptive means of control first, reserving the most poisonous remedy as "the last resort."

Many pests can be controlled without sprays if they are discovered early. There are times when sprays are necessary to save valuable plants. Some chemical controls are suggested in the following, but because treatments change so rapidly, it is best to consult a reputable nurseryman or your county agent.

If you do use chemicals to control pests, do it thoughtfully. Read the label and follow directions.

COMMON PLANT PESTS

Knowing a little about the habits of insects and other pests will help you control them. Weather has a great influence on their behavior. Winds aid their spread. Erratic climatic variations often cause unpredictable insect infestations.

Another significant factor in arid areas is the lack of vegetation during most of the year. Winter and spring rains cause nonresidential plant growth, resulting in a corresponding increase in insect populations. With the onset of hot, dry weather, natural vegetation quickly dies and dries up. The insects then migrate to your irrigated landscape to survive.

Rabbits and other rodents can be serious plant pests. They especially love tender young plants. Repellent sprays are not that effective and have to be applied after each rain or irrigation. The best protection is a barrier such as chicken wire held in place by stakes. When the plant has reached a size when it can withstand attacks, the barrier can be removed.

The following descriptions cover many common pests found in arid climates. We have followed the recommendations of the United States Department of Agriculture regarding use of chemical insecticides.

Aphids. These small, innocent-looking insects can reproduce very fast to build up huge populations. Their thin bodies leave them vulnerable to dehydration, but they can be very damaging during mild weather. See photo page 38.
- "Honeydew," a substance secreted by aphids, protects them somewhat from drying. A mild detergent solution—one tablespoon per gallon of water—will dissolve the honeydew so the aphids are more likely to dry out.
- Strong blasts of water from the hose, done frequently, will sometimes control.
- Aphids overwinter in the egg stage. Eggs are laid on twigs and bark of trees and shrubs, and are nearly invisible to the eye. These eggs are highly susceptible to a dormant oil spray, usually applied before a plant buds out in spring. Summer oils may be applied to some evergreen and growing plants. Read the product label.
- Insecticides such as malathion and diazinon will stop aphids from feeding and protect the plant from reinfestation for a few days. Depending on your plants and the degree of aphid damage you will tolerate, weekly applications may be necessary. This is especially true if the weather is suitable to their reproduction. Spray intervals will be noted on the product label.

Spider mites. Unlike aphids, these tiny, spiderlike pests thrive in hot, dry, dusty weather. Expect them anytime of year, but they are more active during warm periods. They are particularly damaging to Italian cypress, junipers, loquat, arborvitae and pyracantha, but attack many other landscape plants as well. Early detection of spider mites requires an experienced, observant eye. Leaves will begin to appear a little dusty and there may be a speckling of tiny yellow spots. Turn the leaves over and look for a fine, silvery webbing. Rub your fingers over the leaf to feel for a gritty residue. Finally, look for the tiny mites with a small magnifying glass.
- Just as aphids are destroyed by drying, spider mites are discouraged by moisture. Washing plants weekly during summer with a forceful blast from the hose will help prevent population build-ups. But this is only a preventive measure. By the time much webbing is evident, some chemical control will be necessary. Products containing Kelthane (dicofol) are usually recommended. If the webbing is very thick, it may actually repel water. The addition of a commercial wetting agent, called a *surfactant,* such as Exhalt 800 or Spray Mate, will enable the spray to penetrate the webbing and work more effectively.

Climbing cutworm, ashworm or fall cankerworm. This gray, 1/2 to 1-inch long caterpillar frequently attacks ash trees. It feeds on leaves during the night and hides in bark crevices during the day.
- If your trees are regularly attacked by this pest, make a barrier on the lower tree trunk so the worms cannot pass. Do this in mid-March or early April. The product Tanglefoot, a sticky substance, works well for this purpose.
- A bacterial control for caterpillars,

Problems Caused by Poor Growing Conditions

Symptom	Probable Cause	Remedy
Poor growth, perhaps stunted. Sparse foliage.	Inadequate planting hole, lack of nutrients, compacted soil lacking humus, poor drainage, salt buildup. These generalized symptoms may indicate Texas root rot (if decline is sudden), nematode damage, crown gall or other problems.	Revitalize plant by cultivating around root area, adding appropriate amounts of balanced fertilizer. Make 4 to 5-inch earth basin for water at drip line. Add 1 to 2-inch layer of organic material as mulch. Water slowly and deeply.
Poor growth. Leaves generally yellowish and oldest leaves mostly yellow. Lowest leaves may fall.	Likely a deficiency of nitrogen. Possibly a result of reduced light levels, especially if growth is leggy and spindly.	Use balanced fertilizer or one that primarily supplies nitrogen. Apply according to label directions. If plant is in shade, reduce shade, if possible, or move to a sunnier location.
Poor growth, as above, but primarily new leaves are yellow. Topmost leaves become pale yellow while veins remain green.	Iron deficiency in soil, or more likely, inability of the plant roots to absorb what iron is present (chlorosis). In desert gardens, typical causes are high soil pH or excess lime.	Spray iron chelate on leaves to the point of runoff—chelated iron can be absorbed by plants regardless of pH. Or, add chelates to soil and cultivate root zone. Or, incorporate ferrous sulfate. See page 35. In any case, add an organic mulch and water deeply.
Same symptoms of iron chlorosis as above. In addition, soil wet, even waterlogged. Soil around roots has a sulfurous odor. Roots rotting.	Poor drainage or overwatering. Sour odor caused by lack of oxygen in root zone.	If drainage is blocked by a layer of hard, impermeable soil, dig or drill through layer to allow water to pass through. Or, sacrifice plant and install artificial drainage. Or, build raised bed to provide good drainage.
Wilting, loss of luster, dry or rolled leaf edges. Poor growth over a long period. Some branch and twig dieback.	Too little water.	Prevent by never allowing the soil to completely dry out, even in winter. Check for soil moisture even in cool, rainy periods.
Leaf edges look dry or burned. Some wilting. Poor growth.	Damage from drought, usually aggravated by excess soil salts. Or, potassium deficiency, or excessive boron. Too high or too low temperatures.	Prevent by watering less frequently, but more deeply. Provide at least one slow, deep soak a month during growing season to flush salts from root zone. Especially important for container plants.
Wilting and collapse. Burned spots on leaves, scalded bark—especially those plant parts facing south or west. Leaves have yellow or orange dry spots and look scorched.	Sunburn and heat injury. Caused by sudden high temperatures and low humidity, or long hours of afternoon sun. Reflective surfaces also contribute. New plants, container plants and others not thoroughly acclimated to a new site are most susceptible.	Use plants adapted to intense heat. Allow new plants to adjust gradually by providing shade protection the first season. Keep water supply steady. Protect sensitive bark with wrap or paint with white water-base paint. Move container plants to shade during periods of high temperatures. See photo, page 35.
New plants or transplants wilt soon after planting.	Transplant shock. Occurs if too much of a plant's root system is removed, bruised or dried out. Young plants shock easily in temperature extremes.	Some larger plants benefit by removing some top growth to compensate for the decreased root system. Trim or thin parts not essential to plant structure. Provide some shade. Spray with a substance called *antitranspirant* to reduce moisture loss. Application of vitamin B, available at nurseries, may help.
Bud, flower or fruit drop.	Some is normal, especially on citrus in June. Excessive drop is a sign of plant stress. There are many possible causes. Sudden changes in temperature or soil moisture are most common. Heavy winds can cause major damage.	Maintain steady moisture supply. Provide some wind protection. Feed 4 to 6 weeks *before* flowering. Do not overfertilize.

Bacillus thurenginesis, is widely available under a variety of trade names such as Dipel, Thuricide and Bactur. The worms ingest the bacterial spores by eating leaf tissue that has been sprayed. Apply spray during evenings early in the season before the worms are full size. It works slowly but is effective.

● Chemical insecticide controls for ashworms are carbaryl (Sevin) and methoxychlor. If the trees are tall, these or any other sprays expected to reach tall trees are best applied by a professional with a high-pressure sprayer.

Tent caterpillars. They're easy to spot: They make tentlike structures of webbing usually at branch tips high in trees. Cottonwood and willow trees are favorite hosts. The caterpillars are protected within the tent and feed there until time to pupate.

● The tents of these pests can be removed from branches by using a long pole that has either a cone-shaped brush or several nails on one end. The webs can be wound on the pole and then removed from the tree and burned. If you have a pole pruner it may be simpler to prune away branch tips supporting the tents.

● A strong spray of detergent water may help break up the tents, making the caterpillars more susceptible to other predatory insects, weather and insecticide sprays. *Bacillus thurenginesis* is effective if the spray can penetrate the webbing. Malathion or methoxychlor will control.

Borers. Of the various kinds, the flat-headed apple borer is frequently a problem. They attack roses, pyracantha, fruit trees and others, but are usually considered a secondary pest. They follow other problems such as sunburn of the trunk, pruning wounds and the like. They seldom invade undamaged bark.

Palo verde borer. (California prionus). The adult is 3 to 4 inches long and 3/4 inch wide. It is reddish brown to black with long antennae. Larvae are 4 to 5 inches long and an inch or more in diameter. They feed on roots of the palo verde and deciduous fruit and shade trees, causing extensive damage. They may feed for 4 or 5 years before reaching maturity. Symptoms are a decline in the plant's vigor and eventual die-back of limbs. You may also notice 1/2 to 3/4-inch holes in the soil near trees after rains. These are holes through which adults have emerged. Adults may re-enter these same holes to lay more eggs.

• Grubs of the palo verde borer are well insulated by several inches of soil, so control is difficult. Insecticides may be pressure injected at about the time adult beetles emerge—late spring to early summer and again in early fall. The adult beetles are night fliers and lay eggs on bark at the base of trees or in debris around the chosen host. An insecticide treatment around the base of trees just after egg-laying time, June 20 to 30, may poison the newly hatched larvae before they enter the ground.

• If you suspect this pest is damaging your trees, check with your local county extension service for the latest control methods.

Agave weevils. Large century plants and other agaves are susceptible to damage caused by this stout, 3/4-inch long weevil—a snout-nosed beetle—and its grubs. They bore into the soft tissues of the crown, often just underground, and damage or kill the plant.

• Drench the soil around the plant's base and through the center of the plant with the currently recommended insecticide. Treatment in the early stages may save the plant. An annual preventive treatment, applied in early summer, is recommended.

Scale. These usually small, hard-shelled, immobile pests cluster on plants. Some appear as tiny dots; others look like oysters. Cacti are favorites of scale.

• Forceful sprays of insecticide plus detergent are helpful. Some kinds are susceptible to summer and winter oil sprays, but oil sprays will damage certain cactus plants.

The recommended insecticide is Sevin. Orthene, a systemic, is also effective. If the scale is particularly thick, it may be best to scrape off as many as possible with a stiff brush before applying any controls. Take care not to cut into soft inner growth.

Grape leaf skeletonizers. These tiny, yellow, black-striped worms literally "skeletonize" the leaves of grape vines and other plants. Several generations may appear in one summer.

• Small, initial infestations may be picked off by hand.

• *Bacillis thurenginesis* (Dipel, Thuricide) is a safe biological control but works slowly. Sevin, malathion or diazinon will also control.

Whiteflies. These annoying, pure white, flying insects suck vital juices from plants. They reproduce fast, so the key to their management is early control while numbers are small. Once a large population is established, it is virtually impossible to eliminate them until inclement weather reduces their numbers.

• Sprays of detergent water help control. See Aphids. Be careful that the detergent doesn't injure plants.

• A commercial spreader-sticker such as Exhalt 800 is another control method. Materials of this type essentially glue the tiny whiteflies to leaves.

• Whiteflies are attracted to the color yellow. Sticky yellow boards can be used as flypaper near infested plants.

• The botanical insecticide pyrethrin is relatively effective if used frequently. The synthetic imitation of pyrethrin, Resmenthrin, provides more effective control.

• Dormant oil sprays (sprays applied when plants are dormant and not growing) in the winter reduce the number of whitefly eggs.

• Insecticides to use include the systemics Meta-Systox R and aceph-ate. Use them only on nonedible plants. Use malathion if you spray your vegetable garden. Study product labels first.

• Whatever method you use, be persistent. Try to spray every 4 to 7 days, or as the product label directs. It is all-important to cover the plant thoroughly with the spray, especially the under-sides of leaves.

DISEASES

Arid climates are unfavorable for many common plant diseases. Most require considerable humidity to develop. The following diseases are those most likely to be a problem in desert home landscapes.

Texas root rot or cotton root rot. This fungus disease requires special attention. Left untreated, it can gradually spread through the soil, killing most of the plants in its path. The only plants

Aphids attached to underside of apple leaf. See page 36.

Agave americana destroyed by agave weevil. See text above.

Grape leaf skeletonizers on grape leaf. See text above.

that are immune are the grasses or grasslike plants—plants with fibrous roots such as palms, bananas and bamboos. Resistant plants are the pines, eucalyptus, citrus with sour orange rootstock, cypress, oleander, pomegranate, jasmine, myrtle, pyracantha, elderberry, rosemary, olive and juniper.

Plants afflicted with Texas root rot often wilt and die rapidly. Or, sometimes plants decline the first summer and die the second. Action must be taken immediately to save the plant. To treat:

1. Loosen top soil to the drip line.
2. Cover the ground with 2 inches of composted steer manure.
3. Scatter ammonium sulfate on top of manure, 1 pound to every 10 square feet of surface area.
4. Add an equal amount of soil sulfur on top of manure and ammonium sulfate, digging into the top 6 to 8 inches of soil.
5. Soak the soil to 3 feet deep.
6. If the plant is affected and leaves are wilted and dry, reduce foliage by about one-half to compensate for the root loss. Prune only secondary branches, preserving the basic scaffold.

Armillaria root rot, oak root fungus. This disease is encouraged by excessive soil moisture at the plant's crown. Tree declines in vigor, foliage yellows and leaves drop. Fan-shaped plaques (see photo) appear between bark and wood at trunk.
● Prevent by not overwatering. If

your plant contracts this disease, contact your county extension agent for treatment.

Root-knot nematodes. These microscopic, wormlike animals parasitize plant roots, reducing plant growth and health.
● Use nematode-resistant plants when possible. If you suspect nematodes are damaging your plants, consult an expert. Diagnosis and control require professional help.

Crown gall. A very common plant disease caused by a bacteria. It can destroy large fruit trees such as peach and almond, and also plagues roses. Round, woody, lumpy growth can be seen near the soil line, although in some cases the galls form below ground or well above the soil line. Growth is poor, foliage is thin and plants eventually succumb.
● Check plants for any suspicious bumps before taking them home from the nursery. Remove growth and add organic matter and perhaps sulfur to the soil to lower the pH—crown gall does not thrive in very acidic soil.

Aleppo pine blight. Sections of foliage suddenly dehydrate on sun-exposed sides of actively growing trees. Affected needles first appear whitish green then turn brown. In more severe cases, sap may exude from affected twigs. Some dieback usually occurs, but many blighted branches survive. The "blight" seems to be a weather-induced shock to tender, actively growing foliage. It occurs during early winter, and is apparently

caused by shifts between day and night temperature extremes. It seems to be worsened by drying winds, because it is typically found on the sun-exposed, windward sides of trees. Also, blighting seems to be more common to trees receiving frequent, shallow irrigation, especially in lawn areas.
● To prevent the blight from occurring, avoid light, frequent watering. Water deeply and infrequently.

Powdery mildew. This well-known disease is all too common in arid regions. It is more prevalent in spring and fall than summer. Plants commonly attacked are roses, crepe myrtles, euonymous, grapes, zinnias and dahlias.
● A form of sulfur, lime-sulfur, wettable sulfur or sulfur dust, is a good preventive, but it will not eradicate the disease once it is established. Useful fungicides include Acti-dione PM and Triforine.

Verticillum wilt. This disease is most active in cool, wet soils, especially in spring, but sometimes shows up in fall. It is introduced to the soil by infected plants. Symptoms are branches that die back progressively on one side of the tree from base to tip. When cut, the wood beneath the bark shows a dark, discolored layer.
● Purchase healthy plants. Treat by promoting plant vigor—work 3 inches of organic matter into top few inches of soil around the drip line. Feed with light applications of ammonium sulfate every 6 weeks during growing season.

Whiteflies. See page 38.
Photo courtesy University of Arizona.

Cottony-cushion scale. See page 38.
Photo courtesy University of Arizona.

Armillaria mellia, oak root fungus, on oleander. See text above.

Here is the heart of this book. The plant photographs and descriptions are designed to make it easy for you to create your landscape under the sun.

Plants for Arid Lands

This chapter describes over 300 plant species adapted to growing in warm, arid environments. Plants are listed in alphabetical order, according to botanical name. You may find these names cumbersome at first, but there are sound reasons for their use: A given plant may have many common names, or different plants may be known by the same name. Common names also change from region to region within the country, or the world.

From a practical viewpoint, the most important aspect of using scientific names is that you know exactly what you are getting. This is not always the case with common names. If you select a juniper for a low ground cover, it would make a lot of difference if you planted *Juniperus horizontalis* with a maximum height of 18 inches, or *Juniperus chinensis* 'Kaizuka', which will reach up to 15 feet in height. If you only know plants by their common names, see the index for reference.

A blooming palo verde tree adds color to this desert landscape. Above: Texas ranger works well as an attractive, easy-care plant for this entryway.

A Close Look at a Botanical Name

Genus: Plants consisting of one or more species sharing many characteristics, usually flowers and fruit. Species within a genus have more in common with each other compared to others. The plural is *genera*.

Species: Plants with certain discernible differences from other plants within the same genus. Their characteristics normally continue from generation to generation.

Variations in species that occur in nature are *varieties*. A variety is usually quite different from its species. For example, the honey locust *Gleditsia triacanthos* has thorny branches. The variety *Gleditsia triacanthos inermis* has no thorns.

ASPARAGUS densiflorus 'Sprengeri'
Family: *Liliaceae*
(A. sprengeri).

Former botanical name

Family: The broadest classification used in this book. Plants within a family share some general characteristics, but differ enough that they can be further categorized into *genera*.

Cultivar: The word comes from the combination of the words *culti-*vated *var*·iety, meaning, they are cultivated (bred) for their desirable characteristics.

41

Cultural Requirements

Growing plants is not an exact science, but rather a learned art. It depends on a myriad of changing conditions, but mostly, it depends upon you, the grower and caretaker.

The boxed cultural guide with each plant description gives you the plant's basic requirements as to soil, sun, water, temperature and maintenance. It also indicates the zone or zones where that plant is best adapted. We must stress that the adaptations are only guides, and not hard and fast rules. Temper the information with your own experiences, and with the guidance of your nurseryman and local cooperative extension agent.

The following are explanations of the cultural terms used in the plant descriptions. The glossary at right gives definitions of additional terms.

SUN

Full Shade—No direct sun. Areas under overhangs, roofs or interiors.
Part or Filtered Shade—Dappled sun. Under trees or lath cover.
Open Shade—In shade with open sky.
Afternoon Shadow—East sides of buildings or near trees where shade is cast during afternoon hours.
Full Sun—Full exposure to the sun at all times, away from reflected heat.
Reflected Sun—Near south or west walls or pavement which collects and reflects the sun.

IRRIGATION

Ample—Soil is damp to the touch.
Moderate—Irrigate when the top two inches of soil is dry.
Occasional—Deep, widely spaced waterings, sometimes referred to in the text as *little*.
None—No water in addition to rainfall. In very dry climates or during drought, water may be required.
Note: Even drought-tolerant plants require water when first planted. Be sure plants are established before stopping irrigation.

SOIL

Tolerant—Plant will grow in a range of soil conditions.
Well Prepared or Improved Garden Soil—Ample planting pit with 30 to 50 percent humus (organic amendment) added. Water drains freely.
Average Garden Soil—Soil of established gardens.
Light Soil—Sandy or gravelly soil with fast drainage, with little organic matter.

Heavy Soil—Soil with high amount of clay or silt and has poor drainage.

TEMPERATURE

Tolerant—Accepts a wide range. Plants generally tolerate heat and average cold of their adapted zones.
Heat Tolerant—Plants will grow in hot locations, even in the low zone.
Tender—Sensitive to cold and may be damaged or killed.
Hardy—Adapted to cold of the climate zones covered in this book. A plant may be hardy, but not heat-tolerant.
Temperature Ranges—whenever possible, a low temperature range has been given. There are so many variables—whether a plant has hardened off, unseasonable frosts, microclimates—that these cannot be considered absolute.

MAINTENANCE

Constant—Continual care of some sort. Plants to be fussed over.
Periodic or Occasional—Requires some regular care, but tolerant.
Seasonal—Needs attention at certain times of the year: feeding, pruning and cleaning.
None—Can largely be ignored once established. Occasional pruning, grooming and irrigation will usually improve any plant's health and appearance.

Glossary of Terms

Alkaline: Soils with high pH, 7.0 or above.
Bare root: Dormant plants are sold and planted this way at certain times of the year, usually late winter or early spring.
Bordeaux: An effective plant disease fungicide available at nurseries.
Bracts: Colorful, specialized leaves surrounding the true flowers of some plants.
Buttress roots: Roots extending out on the surface of the ground at the base of trees.
Caliche (hardpan): Impervious, rocklike layer of lime in the soil.
Catkins: Long, slender, fuzzy, usually greenish flowers.
Chlorosis: Abnormal yellowing of plants, often caused by iron or nitrogen deficiency.
Crown: Top of a tree where it branches out, or where roots join trunk.
Deciduous: Plants that shed all of their leaves each year, usually all at once.
Espalier: Plants pruned and trained to grow flat against a wall or fence.
Evergreen: Plant foliage remains green through the year for more than one season.
Foliage: The mass of leaves on a plant.
Foundation plant: Plants used at the bases of buildings to integrate plants and structures with the site.
Herbaceous: A plant with no woody parts.
Humus: Organic matter that is in, or added to the soil.
Leach: To wash salts out of the soil with a long, slow irrigation.
Leader: Main, dominant shoot of a plant.
Leggy: Abnormally tall and spindly plant, usually caused by a lack of light or overcrowding.
Microclimate: Small area within a climate zone where temperatures are modified due to terrain, sun exposure, wind flow and other factors.
Mulch: Layer of material laid on the soil. Used to retain moisture, modify soil temperature and prevent weeds.
Naturalize: Plants become so well adapted to a given area that they reseed naturally and become part of the natural landscape.
Overstory: Tall tree or plant above smaller plants.
Refurbish: To replant an area, especially one that has been neglected or abused.
Rejuvenate: To renew and create a more youthful growing pattern, usually done by pruning.
Rhizomes: Thick horizontal stems that spread laterally through or on the soil.
Specimen: Single plant used alone in the landscape as an eyecatcher.
Standard: Tree or shrub trained into a small, treelike form with an erect, single trunk, often with a clipped, rounded crown.
Systemic: Pesticide that is absorbed by the plant, supplying internal protection against pest attacks.
Sucker: Vigorous shoots that sprout from the base, trunks or major branches of plants.
Topiary: Method of pruning which shapes plants into artificial forms.
Transitional: Area between a heavily irrigated landscape and a drier landscape.
Underplant: Smaller plant beneath a taller plant.

ABELIA grandiflora

Family: *Caprifoliaceae*

This graceful, mounding shrub from China is very much at home in the garden or along a boundary. Glossy, fine-textured, pointed leaves are dark green with a bronze cast. They densely cover branches that arch upward and outward from the ground, then downward. Numerous pinkish white, bell-shaped flowers appear from early summer through fall. A moderate grower to 6 feet, sometimes higher, spreading 5 feet or more unless trimmed.

Cultivars: 'Prostrata' is a low prostrate form featuring white flowers. It grows 18 to 24 inches high and spreads to 4 feet. Use as a ground or bank cover, or as a filler plant in the filtered shade of trees. 'Edward Goucher' is pink-flowering and has bronze-green foliage growing to 4 or 5 feet. Use in groups or in foreground.

Special design features: Woodsy effect.

Uses: General-purpose shrub for background, boundary, visual screen, space definition, or as a large filler or foundation plant for large structures. A neat clean plant for poolsides or close-up viewing.

Disadvantages: Iron chlorosis in alkaline soils.

Planting and care: Best planted fall to spring from containers. For a solid cover, space larger plants 4 to 6 feet apart, prostrate 3 to 4 feet apart. Prune branches selectively, and clean out twiggy growth in late winter to keep its open, graceful, arching form. Avoid shearing—it alters the natural shape and lessens bloom. If the plant has been damaged by frost or is in need of rejuvenation it may be cut back to the ground. Established plants heavily pruned will regrow and bloom the same year.

All zones
Evergreen to deciduous in colder zones
Soil: Average garden soil.
Sun: Filtered or open shade, morning sun in low deserts. Accepts full sun in high zone, except when placed against reflective surfaces.
Water: Moderate.
Temperature: May lose leaves at 15°F (−10°C). Fast recovery.
Maintenance: Occasional to low.

Abelia grandiflora

Abelia grandiflora

ACACIA abyssinica

Family: *Leguminosae*

This tropical-looking tree from Ethiopia resembles the jacaranda, but is smaller and hardier. It can make a patio look like a vacation spot. Erect and open-branched with feathery lush green foliage, the Abyssinian acacia grows at a moderate rate to 20 to 25 feet high, sometimes higher. Spreads to 30 feet if given room. To train as a single trunk the first branches should be 8 feet or more above ground. Creamy yellow puffball blooms that are mildly fragrant appear midspring. Brown pods follow. Heavily furled bark on old trees is an attraction.

Special design features: Tropical or subtropical effect.

Uses: Patio, lawn, garden or large planter subject. Specimen, grouping or row.

Disadvantages: Leaf midribs are difficult to sweep up after they fall.

Planting and care: Plant any time from containers. Space 15 feet or more apart for a row. Prune dead twigs from interior. Older trees may require a heavy pruning to remove sucker growth from the interior of tree.

Acacia abyssinica

Middle and low zones
Evergreen in mild climates, deciduous with hard frosts
Soil: Average garden soil with good drainage. Accepts shallow soil or limited soil space.
Sun: Part, full or reflected sun.
Water: Moderate for fast growth. Meager irrigation slows development. Accepts ample water in flower beds or lawns. Tolerates periods of drought but will show stress and become unattractive if kept too dry.
Temperature: Loses leaves when temperature suddenly drops to mid or low 20's °F (−7°C). Usually recovers quickly in spring.
Maintenance: Occasional.

Acacia abyssinica

ACACIA constricta
Family: *Leguminosae*

Mescat acacia • Whitethorn acacia

The tough and durable whitethorn acacia is native to the deserts of the Southwest. It might not be a tree you would choose, but it is worth preserving if already in place. It tolerates a wide range of unfavorable conditions, including heat and cold, drought, poor, stony or alkaline soils and hot winds. It naturally forms a thicket of dark thorny branches 6 to 8 feet high, spreading 6 to 8 feet wide. Pruned and trained, it becomes an attractive single or multiple-trunked tree that grows to 18 feet and spreads as wide. Twigs and branches of first new growth have white thorns about 1/2 inch long, but slower growth on more mature plants will not. Occasionally, plants will have no thorns at all. Feathery, finely cut foliage appears in midspring and provides filtered shade. Small, fragrant, puffball blooms appear mid to late spring for several weeks and sometimes again after summer rains.

Special design features: Subtropical or desert effect. Flowers attract birds.

Uses: Transitional or natural gardens, barrier plant on banks for cover or erosion control. Replanting disturbed desert areas or for reinforcing sparse desert areas.

Disadvantages: Profuse seeding. Difficult to eradicate once established. Not particularly attractive in winter. Caterpillars may defoliate leaves in spring, but leaves quickly come out again. Occasional infestations of desert mistletoe. Because of thorns, they should not be planted near walkways.

Planting and care: Plant any time from containers or grow from seed. Space 5 to 6 feet apart for thicket or barrier, 12 feet for desert grove. To form a tree, stake one or more main leaders and gradually remove side branches up to head height.

All zones
Deciduous
Soil: Tolerant.
Sun: Part, full or reflected sun.
Water: None to moderate once established. Prefers occasional deep irrigation, especially in areas with less than 10 inches annual rainfall.
Maintenance: None for trees in a natural landscape.

Acacia constricta

Acacia constricta

ACACIA farnesiana
Family: *Leguminosae*

Popinac • Sweet acacia

Sweet acacia is native to the tropics and the subtropics of the Americas, and is grown in warm climates throughout the world. It is a multiple-branched, vase-shaped shrub with ferny foliage growing 8 to 10 feet tall. Trained as a single or multiple-trunked tree, it grows 15 to 20 feet high with a canopy-shaped crown spreading 15 to 20 feet. A profusion of very fragrant, yellow puffball flowers appears in late fall and lasts over a long period. Small beans follow the flowers.

Special design features: Tropical or subtropical effect.

Uses: Patio or street tree, transitional, tropical or desert groupings.

Disadvantages: Beans may litter pavement. Sweet acacias recover slowly from cold damage, and require complete retraining. Caterpillars may defoliate and leave webs in spring, but tree leafs out again.

Planting and care: Plant fall to spring from containers. Prune side branches and stake selected canes to develop a tree shape. Water generously for fast growth, sparsely for small tree.

Acacia smallii is nearly the same plant as the above, but with one important distinction—it is hardier to cold. Because the two species look alike, they are often confused with each other, even at the nursery. If you live in the high or middle zone, be certain to purchase *A. smallii.*

Low zone
Evergreen to partly deciduous
Soil: Tolerates nearly any soil. Accepts alkaline soil.
Sun: Part, full or reflected sun. Enjoys even the hottest locations.
Water: Accepts occasional deep irrigation, or total neglect if rainfall is 10 to 12 inches annually. Supplemental irrigation is needed in drier areas. Grows fastest with moderate or ample water.
Temperature: Sweet acacias lose leaves and new growth is damaged at 20°F (−7°C). Severe damage occurs at 15°F (−10°C). Small's acacia survives to below 15°F (−10°C).
Maintenance: Low, except pruning and training needed to develop a tree form.

Acacia farnesiana

ACACIA greggii

Family: *Leguminosae*

This native of the southwestern United States and northern Mexico naturally forms a rugged tangle of thorny branches 6 to 8 feet high. With training, it can become a 10-foot tree with a spread of 12 feet or more. Feathery, gray-green foliage appears in midspring. Branches and twigs are covered with small thorns resembling the sharp, curved claws of a cat, which take hold with tenacity if you brush them. Profusions of fuzzy yellow catkins appear in late spring or early summer, followed by flat rust-colored beans up to 3 inches long. A second but less spectacular bloom may appear after summer rains. Rough, dark bark and picturesque form of plants trained into multiple-trunked trees make this an attractive landscape subject, with or without foliage. This tough plant withstands unfavorable conditions of drought, heat, cold and poor soil.

Special design features: Desert to subtropical effect when in leaf.

Uses: Desert, natural or transitional gardens. Trained plant makes a handsome silhouette. Refurbish the desert in disturbed areas, for harboring wildlife, or for banks, property edges or as a shrub forming an impenetrable barrier. Provides filtered shade for a patio or courtyard when trained as a tree.

Disadvantages: Needs training to be attractive at close range. Thorns snag easily, making it difficult to work with or be near if branches are low. Mistletoe infestations. Attacked by caterpillars in spring.

Planting and care: Plant from containers any time or start from seed. Scratch seeds on one side to allow water penetration.

All zones

Deciduous

Soil: Tolerant. Accepts alkaline conditions.

Sun: Full to reflected sun.

Water: Once established, requires no water in areas receiving 10 to 12 inches of annual rainfall. Grows best and fastest with occasional deep soakings.

Temperature: Hardy. Accepts heat.

Maintenance: Low to none unless you are training as a tree.

Acacia greggii

ACACIA stenophylla

Family: *Leguminosae*

Erect to bending in form with an open crown, this unusual Australian tree has a sparse showing of dark, gray-green threadlike leaves 1 foot in length that produce a weeping effect. Leaf threads are not true leaves, but function as such. The tree grows at a moderate rate to 20 feet tall and spreads to 15 feet. Sparse, creamy white, puffball blooms appear from fall to spring. This tree is remarkably tough and tolerant of drought, heat and poor soil conditions.

Special design features: Interesting silhouette. Weeping effect. Character plant.

Uses: Small grove creates an interesting grouping. Silhouette plant against structures or sky. Transitional plant. Visual screen or break—distinct and effective but not dense. Dusty situations. Light shade beneath allows other plants to grow, but looks best with bare earth or rock mulch. Specimen plant.

Disadvantages: None known.

Planting and care: Plant from containers any time, or start from seed. Prune side branches to obtain desired shape. If frozen, wait until new spring growth reveals the extent of damage. Then remove damaged growth.

Middle and low zones, marginal in high zone

Evergreen

Soil: Tolerant of poor soils, but shows better growth and development in well-prepared soil.

Sun: Full to reflected sun.

Water: Moderate until established to desired size, then occasional deep irrigation.

Temperature: Hardy to 20°F (−7°C). New growth damaged below 20°F (−7°C).

Maintenance: Low to none.

Acacia stenophylla

Acacia stenophylla

ACANTHUS mollis

Family: *Acanthaceae*

This native of the Mediterranean region is the classical plant of Greek culture. It is prized for its lustrous, deeply lobed leaves reaching 2 feet in length, supported by purplish green stems. Plants form large mounds of bold, deep green foliage 6 to 8 feet wide in a few years. They spread by underground rhizomes. Flower spikes rise 2 to 3 feet above the plant in early summer. Whitish, lilac or pale pink tubular blossoms are interesting in form but drab in color when compared to the dramatic foliage. Plants become dormant in summer after bloom and foliage dies to the ground, but leaf out again when the weather cools. Prefers some humidity and shelter from the elements.

Cultivars: 'Latifolia' is hardier than the species and has larger leaves.
Special design features: Lush bold foliage. Tropical effect in winter. Combines with bamboos, ferns and other verdant plants.
Uses: Filler plant or ground cover 2 to 3 feet high in protected areas such as atriums, entryways, shady planters or underneath trees protected from wind and hot sun. Roots are shallow and survive with limited soil depth.
Disadvantages: Alkaline soils may burn. Spreading fleshy roots need to be contained to keep under control. Subject to invasions of slugs and snails.
Planting and care: Plant rhizomes in late summer before foliage appears, or from containers any time. Divide old plantings and reset while dormant. Space 3 to 6 feet apart. To delay dormant period, remove flower stalks and water abundantly. When fading leaves become unattractive, cut them off near the ground. Use bait for slugs and snails.

All zones

Herbaceous perennial

Soil: Tolerant of soil quality, but needs good surface drainage. Prefers moist garden soil. Alkaline or dry conditions make plants unattractive.
Sun: Full, open or filtered shade, or morning sun. Stays dormant longer in hotter locations.
Water: Needs ample water during growing season, requires little in summer. Do not allow it to dry out.
Temperature: Foliage damaged at 25°F (−4°C) or below, but plant produces more leaves immediately. Usually not bothered by cold when grown in semi-sheltered areas.
Maintenance: Occasional grooming.

Acanthus mollis

AGAVE species

Family: *Agavaceae*

Because of their striking form and general tolerance to cold, heat, sun, drought and poor or alkaline soils, agaves are some of the most useful plants in the hot desert. Size varies from 1 to 6 feet tall with many interesting leaf and color variations. The fleshy pointed leaves are usually armed with thorns at the tips and sometimes grow in rosettes along edges around the center. This form allows them to capture minute amounts of rain, guiding it to the roots. New growth and eventually the flower stalk rises from the center of the plant. Desert agaves such as *A. americana* can survive in areas receiving a mere 3 inches of rain per year. While they are useful under such conditions, they are not necessarily attractive.

Special design features: Bold form and foliage contrast.
Uses: Outstanding plant for harsh environments or difficult situations where other plants fail, such as in shallow soils. Specimen plant, or mass for a bold effect against structures. Rows for a barrier, sculptural space definer or windscreen. Desert or natural gardens with other desert or wild plants. Banks for erosion control. Many are excellent container plants.
Disadvantages: Lower leaves may become dry or untidy and need to be removed. Grubs can kill unless found ear-

---

All zones with exceptions

Evergreen

Soil: Tolerates most types except water-logged.
Sun: Accepts part shade. Some species prefer it in middle or low zones. Usually planted in full or reflected sun.
Water: Little to none in areas with more than 3 inches annual rainfall or where plants receive extra runoff. Where summers are dry supply some water to keep leaves plump.
Temperature: Damaged when temperature remains below 20°F (−7°C) for a long period. Slow to recover.
Maintenance: Little to none unless you wish to groom.
Note: Agaves described here are only a few of the many interesting and useful landscape plants in this family.

Agave americana

ly or prevented by chemical treatment.(See page 38.) Some types produce numerous offsets at base, which can become a nuisance. Once an agave blooms, it dies and should be removed. This can be difficult and costly if the plant is large. Plants can also become too large for allotted space. Be sure plants in the ground are where you want them and have enough space to grow to maturity. Sharp tips of agaves near walkways or other use areas need to be cut off to avoid accidents. Better yet, plant away from frequented areas.

Planting and care: Plant any time from containers or offsets. In dry climates supply young plants with water occasionally until established or when plants show stress by wilting.

Agave americana
Century plant, magay _____

Most important of the desert agaves, this bold, sculptural plant has grayish blue-green leaf blades that reach 3 to 5 feet in length. It grows up to 6 feet in height and as wide or wider before the 20-foot bloom stalk makes its appearance. Stalks grow fast—6 to 14 inches a day—and look like giant asparagus until the horizontal arms open outward and the greenish yellow blooms appear. The woody stalk has an interesting silhouette and is often left standing even after it dies. This agave is a prolific producer of offsets. Leaf blades are well armed with saw-toothed edges and sharp points at the tips and should be placed at least 6 feet away from frequented areas. Leaves and entire plant may be injured in cold winter areas.

Agave americana 'Marginata'
Variegated century plant _____

'Marginata' is similar in form to *A. americana*, but slightly more refined and

Agave weberi

smaller in scale. Its narrower leaf blades are gray-green with a yellow stripe down each margin. It seems more susceptible to grubs and should be treated annually to prevent infestations. This agave looks nice in the garden as a contrast in color and form to other plants. It blends well in desert or tropical groupings and is an attractive container plant. Moderate water is acceptable, with occasional deep soakings. It survives without any irrigation in areas with 10 inches or more annual rainfall. Like *A. americana*, it is well armed and should not be placed near paths or walkways. Space plants 3 to 6 feet apart for a mass effect.

Agave vilmoriniana
Octopus agave, midas agave _____

Blue-gray with a sprawling, twisted form and leaf blades more slender than *A. americana*, the octopus agave is a striking specimen plant in the ground or in containers. It is one of the few agaves without points on leaf ends. It seldom reaches over 3 or 4 feet high but spreads to 6 feet wide.

Blends with tropical or desert settings or with rocks and other structural materials. Give moderate water in summer and protect from cold and reflected sun. Treat to prevent grubs. Accepts shade and does nicely in garden situations. Space 3 to 6 feet apart for massing. Use only in the middle and low zones because it is more sensitive to cold.

Agave weberi
Smooth-edged agave _____

Similar to *A. americana* but slightly smaller and more refined, the smooth-edged agave is a deeper blue-green with soft, almost luminous leaves. Leaves to 4 feet long by 8 inches wide are armed only at the tip and lack the saw-toothed edges of most desert agaves. Plant is noticeably sensitive to light and closes up when conditions are too hot or too bright, but more tolerant of cold than even *A. americana*. It spawns offsets, but not as vigorously as *A. americana*. Needs some supplemental irrigation, especially in hot summer areas. Space 3 to 6 feet apart for massing.

Agave americana 'Marginata'

Agave vilmoriniana

AILANTHUS altissima

Family: *Simaroubaceae*

Ailanthus • Tree-of-heaven • Copal tree • Varnish tree

This is the tree from *A Tree Grows in Brooklyn*, one that grows and thrives in pollution and other poor conditions. Versatile with a handsome form and medium green foliage, it grows rapidly to 20 feet, spreading 15 feet. Later it grows more slowly to an eventual 50 feet by 40 feet, if the soil is deep. These trees will accommodate the space they are given, and are often seen in a row against buildings or as a grove of saplings. Often the grove multiplies due to its tendency to produce suckers. Decorative seeds appear in summer on females after a spring bloom. While not as refined as some trees, this Chinese native should not be overlooked as a fast-growing shade tree.

Special design features: Vertical trunk with interesting twig structure. Heavy shade. Deciduous—gives winter sun and summer shade. Golden fall color. Tropical effect, especially in rows.

Uses: Ailanthus is often condemned as a "weed tree," and, although it does have its faults, there are few trees that thrive under such harsh conditions. It grows almost anywhere—city tree, sanitary land fills, street or yard. Plant several together to make a sapling grove for a narrow space. Try it where conditions are poor but shade is needed. Naturalizes in alleys or along walls where it gets extra water.

Disadvantages: Suckers. Large trees can develop buttress roots and heaves structures if set too close. Fertilized females reseed profusely.

Planting and care: Plant any time of year from containers or bare root in winter. It can also be grown from viable seed. Trees have stout trunks and usually do not need pruning and training. If pruning is necessary, trim in late winter during dormant season. Remove suckers near base unless a sapling grove is desired. Suckers in lawns can be mowed.

All zones

Deciduous

Soil: Widely adaptable.
Sun: Part shade to full or reflected sun.
Water: Moderate to ample. Drought-tolerant for periods when established. Grows with runoff in areas receiving 10 inches or more of annual rainfall.
Temperature: Hardy. Tolerates heat.
Maintenance: Low to seasonal. None in areas where spread of plants isn't a problem.

Ailanthus altissima

ALBIZIA julibrissin

Family: *Leguminosae*

Silk tree • Mimosa

This small tree grows somewhat rapidly to 10 to 15 feet tall, with a canopy to 20 feet wide. Flat spreading crown of bright green, ferny leaves and a constant summer bloom of fuzzy, pink flowers are attractions. It is decorative even when branches are bare, with 3-inch papery seed pods hanging on through winter. Native to Asia.

Cultivars: 'Rubra' has flowers that are a deeper pink than the species.

Special design features: Airy tropical effect. Summer color. Filtered shade.

Uses: Specimen or grouping for screening from above, or canopy tree over flower beds or plants which need some shade in summer. Lawn, patio or street tree. Silhouette plant against structures.

Disadvantages: Flowers and pods litter. Plants are sometimes hard to get started and do poorly the first few years after planting. Leaves drop in midfall and do not reappear until mid or late spring, making a long dormant period. Trees are sometimes short-lived.

Planting and care: Plant from containers in spring. Stake carefully and train by pinching out low buds and side branchlets to make a high crown. Trim off seed pods if they are objectionable. To create a canopy, prune tree so scaffold branches begin about 8 feet from the ground.

All zones

Deciduous

Soil: Accepts poor or gravelly soil, but prefers improved garden soil, with good drainage and depth. Avoid heavy clay.
Sun: Full to reflected sun.
Water: Moderate.
Temperature: Hardy to cold. Accepts heat.
Maintenance: Seasonal grooming as desired.

Albizia julibrissin

ALOE barbadensis

Family: *Liliaceae*

(A. perfoliata vera, A. vera). Aloe vera, originally from North Africa, was brought by Spanish padres to the New World and planted in mission herb gardens. It has been used for centuries as a treatment for burns and skin afflictions. Leaves of spiky succulent rosettes grow upward to 12 or 18 inches. They have sharp tips and whitish to reddish teeth along the margins. In spring, bloom stalks rise 3 feet above the plant, bearing dense, arrow-shaped clusters of yellowish or orange flowers. One plant quickly spawns many offsets, forming a dense clump in a few years. Clumps of old plantings may reach widths up to 10 feet if not divided. Plants are very drought and heat-resistant.

Special design features: Spiky, greenish, succulent leaves. May have two or more bloom periods. Flowers are loved by hummingbirds.

Uses: Desert or rock gardens and transitional areas. Attractive as scattered clumps with bare earth and rocks. Attractive in containers.

Disadvantages: No pests or diseases except an occasional infestation of a thrip which deforms plants. Old clumps may look unkempt after blooming and need to be groomed.

Planting and care: Plant or transplant from containers or divisions in spring. Set single plants about 18 to 24 inches apart to establish a dense clump in 2 or 3 years, after which time the clumps may be separated. Discard plants damaged by thrips.

> **Middle and low zones, sunny protected microclimates in high zone**
>
> **Evergreen**
>
> **Soil:** Tolerant.
> **Sun:** Open or part shade. Full or reflected sun.
> **Water:** Occasional sprinkling keeps the leaves succulent.
> **Temperature:** Nipped by cold below 25°F (−4°C). Thrives in heat.
> **Maintenance:** None to occasional.

Aloe barbadensis

Aloe barbadensis

ALOE saponaria

Family: *Liliaceae*

(A. latifolia). This aloe from South Africa is a low, clumping plant with pointed succulent leaves growing in a rosette form. Leaves are pale to medium green tinged red with dull white spots. Horny brown teeth edge the leaf blades, which are sometimes dry and twisted at the tips, especially in dry or cold weather. Plants show more red in cold weather, or if exposed to strong sunlight, or if irrigation is meager. Plants grown in the shade are more green and succulent. Scarlet to yellowish flowers appear two or three times a year. Purplish stalks add a vertical dimension rising 18 to 30 inches above the 9 to 12-inch plant. One plant produces a dense clump in 3 or 4 years by sending out numerous offsets.

Special design features: Tropical or desert effect. Spiky foliage. Periodic bloom lends color and a vertical effect.

Uses: Desert, wild or rock gardens. Transitional areas. Ground or bank cover. Underplant for palms or yuccas.

Disadvantages: Occasionally attacked by an unidentified thrip that deforms the plant or bloom. Plants located in direct or reflected or summer sun will have a dry and unkempt look.

Planting and care: Divide clumps or plant from containers in spring or summer. Space new plants about 18 inches apart to allow for offsets. Divide crowded clumps. Remove old flower stalks. Immediately discard plants deformed by thrips.

Aloe arborescens and *Aloe ferox* are two of the tree-type aloes that are sometimes seen in the desert. Of the two, *A. ferox* seems to be the hardiest and has the most spectacular bloom. *A. arborescens* produces a bright red flower spike that often blooms in midwinter. This and a number of the more tender aloes are often grown as container plants in colder areas so they can be brought indoors when frost threatens. Nearly all the aloes make very good container plants and can live for years in a small amount of soil.

> **Middle and low zones**
>
> **Evergreen**
>
> **Soil:** Tolerant. Prefers gravelly or loose soils with good drainage.
> **Sun:** Part shade to full or reflected sun. Plants in shade appear more tropical and succulent.
> **Water:** A little supplemental water keeps the leaves plump. Accepts moderate water if drainage is good.
> **Temperature:** Subject to damage in long hard freeze, but recovers fast. Unestablished new plantings may die out in a hard freeze.
> **Maintenance:** Occasional removal of old bloom and dividing clumps.

Aloe saponaria

ANTIGONON leptopus

Family: *Polygonaceae*

Queen's wreath • Coral vine • Confederate vine

A festive, summer vine with a tropical feeling, the queen's wreath festoons over fences, trellises and walls in even the hottest locations. It grows bigger each year, dazzling observers with its shocking pink flowers. This tendril-climbing vine from Mexico sends out shoots in late spring, blooms by midspring or midsummer if vine does not freeze to the ground each year. Leaves are 4 inches long, shaped like hearts or arrowheads. Sprays of blossoms concentrate where they get the most sun and last until cool weather. With frost or winter drought the vine becomes dormant and the top dies back to the ground, but is easily pulled down and discarded.

Cultivars: 'Album' has white flowers.

Special design features: Tropical effect. Summer color.

Uses: Trellises or porch posts. A fine plant for hot south and west sides, or as a shade cover over sun-sensitive plants.

Disadvantages: Flowers attract bees in late summer and fall.

Planting and care: Plant from containers in spring in even the hottest locations. Provide supports for it to climb. When damaged by frost, pull down dry growth.

Antigonon leptopus

Antigonon leptopus

> **All zones**
>
> **Evergreen to dormant with frost or in dry winter areas**
>
> **Soil:** Tolerant.
> **Sun:** Full to reflected sun.
> **Water:** Moderate.
> **Temperature:** Mulch roots where temperatures go below 20°F (−7°C).
> **Maintenance:** Very little.

ARAUCARIA bidwillii

Family: *Auricariaceae*

Bunya-bunya

This genus once grew widely throughout the world, but is now found in nature only in the Southern Hemisphere. It is the genus of the ancient trees of the petrified forest. Bunya-bunya, an Australian native, is an erect conifer with horizontal branches drooping downward at the tips. Juvenile leaves are a shiny dark green with sharply pointed tips prickly to the touch. Mature leaves are small woody ovals up to 1/2 inch long that grow in overlapping spirals. These leaves do not appear for a number of years, until the tree has attained height and maturity. It is not known how tall a bunya-bunya will get in the desert, but specimens in California reach 80 feet with broadly spreading crowns.

Special design features: Interesting form and structure. Bold, deep green vertical.

Uses: Excellent container plant for several years before it outgrows the space. Specimen or emphasis plant for large public spaces or for impressive entrances to residential or institutional projects. Silhouette plant against tall structures. Mature trees are thought to be too large for the average residence, but not for many, many years.

Disadvantages: Slow grower. Prickly juvenile foliage can be annoying near frequented areas. If walks are established near trees later, remove lower branches. Foliage may brown out in extreme or prolonged cold or if soil becomes too dry. Iron chlorosis in alkaline soils. Heavy cones on old trees fall and could injure people under tree.

Planting and care: Plant from containers in spring when danger of frost has past. Be sure tree has an ample planting hole with good drainage. This ensures room for maximum root growth. Stake trunks on young trees until they are self-supporting.

> **Middle zone, protected locations in high zone**
>
> **Evergreen**
>
> **Soil:** Tolerant, but needs good drainage and adequate soil moisture. Prefers improved garden soil.
> **Sun:** Part shade to full sun.
> **Water:** Moderate to ample, especially when young.
> **Temperature:** Foliage on young trees or container plants browns out in intense or prolonged cold below 25°F (−4°C). Older trees in the ground tolerate much colder temperatures without damage.
> **Maintenance:** Low.

Araucaria bidwillii

ARECASTRUM romanzoffianum

Queen palm

Family: *Palmae*

(Cocos romanzoffiana, C. plumosa). A graceful and refined palm of residential scale, the queen palm gives a festive feeling to any garden setting. This South American native is very erect in form, with a smooth, gray, ringed trunk to 12 inches in diameter. Feathery fronds are shiny medium green with slender papery fillaments 2 feet or more in length. The luxuriant crown spreads 10 or 15 feet wide, more reminiscent of a tropical seashore than the desert. In the desert, these palms grow at a moderate to fast rate 20 to 25 feet high, occasionally reaching 30 to 40 feet with crowns 20 to 30 feet wide. Palms may produce small plumes with blossoms followed by miniature coconutlike fruit.

Special design features: Airy vertical form. Tropical or subtropical effect. Light shade. Little or no litter.

Uses: Specimen, grove or row. Silhouette plant. Tropical groupings around swimming pools. Small microclimates, south sides or protected intimate gardens for close-up viewing. Large planters. Attractive in courtyards of tall buildings or in the home garden.

Disadvantages: Leaves develop dry straw-colored filaments from hot winds, cold or age. Avoid planting in high wind areas. Fronds may become pale with iron chlorosis. Fruiting streamers look messy.

Planting and care: Plant from containers in spring or summer, especially in middle zone. Space 8 to 15 feet apart for a grove, 15 to 20 feet or more for a row. Feed regularly and provide extra iron to promote fastest growth and more attractive appearance. Protect from hot dry winds in low elevation deserts.

Arecastrum romanzoffianum

Low zone, protected microclimates in middle zone

Evergreen

Soil: Prefers well-prepared garden soil with depth, but tolerates a range of soils, including alkaline situations, if supplied with good drainage.
Sun: Part shade to full sun. Accepts reflected sun in middle zone.
Water: Moderate to ample.
Temperature: Foliage damage at about 25°F (−4°C). Recovers from freezes of 20°F (−7°C) or below, but takes all summer to recover.
Maintenance: Some grooming of leaves and fruiting parts.

ARUNDO donax

Giant reed • Carrizo • Cana brava

Family: *Gramineae*

Giant reed is a tall, coarse grass which grows rapidly 10 to 15 feet high, spreading by underground rhizomes to form large stands. Flat, light green leaf blades nearly 3 inches wide hug the bamboolike vertical canes. Whitish plumes 2 feet or more in length bloom at stalk tops in summer. It naturalizes in ditches and other low places where water collects. This reed is native to the Mediterranean region and was brought by the Spaniards to the New World as a building material. It is believed to have been used for ceilings until they discovered saguaro ribs did a better job. It is still used for mud and wattle construction and for roofing adobe houses in Mexico. It is similar to and sometimes confused with *Phragmites communis*, commonly known as carizzo, a smaller native reed used by Southwest Indians for a multitude of purposes.

Special design features: Tropical, rustic or waterside effect. Strong vertical. A large scale plant.

Uses: Quick screen or windbreak. Use at the outer edges of property or near duck ponds or other informal water situations. Mass for erosion control.

Disadvantages: Invasive and prone to litter. Difficult to dig up and eradicate once established. Dry appearance in cold winters when it goes dormant.

Planting and care: Best planted or transplanted in spring after frost. Space small clumps 3 feet apart allowing for spreading to make a solid screen. Cut out messy or dry shafts and leaves to groom. May be cut to the ground to rejuvenate. Plant where the roots can be contained to avoid invasiveness. In the coldest areas mulch roots in winter.

All zones

Evergreen to dormant

Soil: Tolerant. Prefers rich moist soil, but grows wherever it can find enough moisture.
Sun: Part shade to full or reflected sun.
Water: Accepts any amount. Needs occasional irrigation in the driest areas and to look its best. Tolerates periods of drought, but prefers ample water.
Temperature: Damaged below 28°F (−2°C), drying out to straw color, but recovers quickly in spring. With adequate moisture accepts heat, growing vigorously.
Maintenance: None to occasional depending on personal taste and location.

Arundo donax

ASPARAGUS densiflorus 'Sprengeri'

Family: *Liliaceae* _____ Sprenger asparagus

(A. sprengeri). This festive plant of the Victorian Age is back in style. It is a multiple-branched trailing plant with small, bright, yellow-green, needlelike leaves. It grows fast in spring and moderately the rest of the year. On the ground it mounds 12 to 24 inches high, or cascades over walls, banks or containers with branches trailing to 4 feet or more. A sparse spring bloom of tiny white flowers may produce a few bright red fall berries. Roots develop fleshy nodules, allowing the plant to withstand periods of drought or neglect.

Cultivars: *A. densiflorus* 'Sprengeri Compacta' is like the regular 'Sprengeri' with a more compact mounding form and tighter growth habit.

Special design features: Ferny effect. Color and texture contrast to dark evergreens.

Uses: Containers, especially hanging pots. Attractive trailing and spilling over planter walls or banks. For low borders or edgings, especially framing colorful flower beds. A billowing underplant, ground cover or foundation plant.

Disadvantages: Iron chlorosis. Sometimes develops dry stems and foliage if neglected or if growing conditions are too hot or too dry.

Planting and care: Plant from containers in spring after danger of frost has passed, or in protected areas any time. Space 30 inches apart for a ground or bank cover. Groom by cutting back the longest trailers or removing dried ones. Plants may be cut back severely to rejuvenate. Give extra iron for deeper green foliage.

Middle and low zones

Evergreen

Soil: Tolerant. Best in enriched soil with good drainage.

Sun: Any exposure, but becomes stringy in deep shade.

Water: Moderate to ample. Established plants tolerate some drought.

Temperature: Freezes back at about 24°F (−3°C). Recovers quickly in warm weather. Since plants are usually grown in more protected places, or they are brought indoors in winter, freezing is seldom a problem.

Maintenance: Minimum.

Asparagus densiflorus 'Sprengeri'

ASPIDISTRA elatior

Family: *Liliaceae* _____ Cast-iron plant • Barroom plant

(A. lurida). The cast-iron plant was often grown in Victorian barrooms, where it survived like "cast-iron" despite poor light and neglect. This Chinese native is again becoming popular because of its remarkable ability to survive deep shade, poor soil, cold and periods of neglect. Distinctive, wide, pointed, leathery leaves arch upward and outward 12 to 30 inches on grooved stalks. They are dark shiny green with interesting parallel veins.

Plants grow slowly and spread by underground rhizomes. Inconspicuous brown flowers close to the ground appear in spring.

Cultivars: 'Variegata' has green and white striped leaves.

Special design features: Tolerates less light than almost any other plant so it can be grown in dark corners. Tropical effect. Bold foliage. Attractive close-up.

Uses: A touch of green for dark places such as under outside stairwells, in recessed passageways, or under extended overhangs on north sides. Does well in planters or containers with limited root room. Neat plant for shady poolsides.

Disadvantages: Variegated form loses variegation in very rich soil. Sunburns.

Planting and care: Plant from containers any time. Dust or wipe leaves occasionally to keep attractive. Feed in spring and summer for most attractive appearance.

All zones

Evergreen

Soil: Tolerant. Prefers enriched soil with good drainage. Variegated plants should be grown in poor soil to keep color.

Sun: Deep to open shade. No direct sun.

Water: Moderate, but withstands periods of neglect.

Temperature: Hardy to cold. Accepts heat with shade and water.

Maintenance: None to occasional grooming.

Aspidistra elatior

Aspidistra elatior

ATRIPLEX species

Family: *Cenopodiaceae*

The saltbushes are an important group of somewhat weedy plants which grow naturally in the deserts of the world. They are especially well adapted to desert conditions, withstanding saline-alkaline soils, and surviving on as little as 3 inches of annual rainfall. Their often spongy leaves are used for water storage. As the water is used over an extended period the leaves contract, turning from gray to white or silver. This color change occurs when the plant absorbs and deposits salts from the soil through the leaf pores. The lighter color helps reflect the sun's intense rays. These adaptations make the saltbush plant group extremely important for vegetating the driest land, and also as a browse for animals. With a little extra water from runoff or irrigation, saltbushes do well.

Special design features: Informal, rustic.

Uses: Valuable for replanting open areas, or in desert or wild gardens where little or no care is given. Ground or bank cover.

Planting and care: Plant from containers or cuttings or sow from seed in spring. Space randomly for a natural look. Protect young plants from cold and irrigate until they are well established. Prune or trim any time.

Disadvantages: Some species are difficult to obtain. Sometimes rangy or unkempt.

All zones except as noted

Evergreen to deciduous

Soil: Tolerant of most soils. Most species require good drainage.
Sun: Full to reflected sun.
Water: Moderate to dry. Looks best with deep periodic irrigation.
Temperature: Variable according to plant origin—hardy to somewhat tender. Tolerant of heat.
Maintenance: None, except an occasional pruning or trimming.

Atriplex canescens
Four-wing saltbush, chamiso, cenzo, saltsage

This multiple-branched, gray-green shrub is found throughout the western part of the United States and Mexico at elevations from 150 to 7,000 feet. Plants are fairly slow growers, reaching 3 to 6 feet high spreading 4 to 8 feet wide. They may be sparse or densely foliated with narrow leaves 1/2 to 2 inches long. Fruits cluster at the ends of young branches, forming golden cascades.

Set out seedlings in spring; young plants are tender to cold. Or start from cuttings in spring. Space 1-1/2 to 4 feet apart for a hedge or screen which may be clipped or shaped any time. Water until established. Grows fastest and looks best with regular irrigation.

Atriplex lentiformis
Quail bush, lens-scale, white thistle, big saltbush

This saltbush is a showy shrub native to the low and middle deserts of the Southwest. Given ample water it grows at a moderate to fast rate to 12 feet high, spreading 8 to 15 feet wide. It grows slowly to about 5 feet in dry areas without supplemental irrigation. Bare in winter but handsome in summer, hollylike blue-gray leaves are 1-1/2 inches long on slender, sometimes spiny branches pale gray to whitish in color. Clusters of decorative fruit show at the branch tips late spring to early summer. Tolerant of high heat and cold. Can survive on 3-1/2 inches of rainfall a year in deep soils where its deep root system can penetrate to draw on subsurface moisture. Responds best with deep but infrequent irrigation.

Atriplex lentiformis 'Breweri'
Brewer saltbush

This is the best of the saltbushes for garden use. There is almost a lush quality to this gray-foliaged shrub with its soft, triangular, 2-inch leaves which densely cover the plant. A slow to moderate grower, this mounding plant reaches 3 to 8 feet high and 6 to 8 feet wide, depending on growing conditions. Heavy decorative clusters of golden fruit appear at the top of the foliage in late spring. Neglected plants growing under difficult circumstances have a rangy appearance. Attractive with some shaping; it makes an excellent clipped hedge. Tolerant of high heat and cold.

Atriplex semibacata
Australian saltbush

A drought-tolerant and widely adaptable plant, the Australian saltbush was introduced into California in the 1880's where it has naturalized. It was later brought to Arizona and New Mexico, probably carried in the wool of sheep from California or Australia. Of the varieties found in the United States, the Soil Conservation Service has found 'Corto' to possess the most desirable combination of characteristics: uniform growth habit, drought tolerance and cold hardiness. This semiherbaceous, semiprostrate plant grows rapidly from seed to 8 to 10 inches high with a spread of 6 feet. Leaves are gray-green and numerous. Inconspicuous flowers produce fleshy fruit which turn red at maturity and are attractive in the fall. Plants reseed themselves and spread under favorable conditions. An evergreen perennial in warm areas. Older plants are killed back to the crown at 20°F (−7°C) or below. Young plants should recover in spring from winter temperatures of 17°F (−9°C); probably killed at 10°F (−12°C).

Atriplex canescens

Atriplex lentiformis

Atriplex lentiformis 'Breweri'

Atriplex semibacata

AUCUBA japonica
Family: *Gornacea*

Japanese aucuba • Japanese laurel

This glossy-foliaged plant from Japan and the Himalayas is erect in form and slow growing to 6, 8, sometimes 10 feet high and nearly as wide. The long, slender, dense, toothed leaves are attractive viewed up-close in the shaded, sheltered gardens where it is best adapted. Female plants bear tiny maroon flowers in spring, followed by clusters of red 3/4-inch berries fall through winter if male plants are present. The species is dark green, but for a bright spot of interest try one of the following cultivars.

Cultivars: 'Variegata', the gold dust plant, is perhaps the most widely planted and best known of the aucubas. 'Sulphur', green leaves with wide yellow edges. 'Pictura' (sometimes called 'Aureo-maculata'), a female with gold at the leaf center and dark green leaf edges with yellow flecks. Cultivars with dark green leaves are 'Nana' ('Dwarf Female'), to 3 feet high, and 'Longifolia' ('Salicifolia'), a female with long tooth-edged leaves.

Special design features: Lush tropical feeling. Attractive at close range. Variegated forms brighten up shady spots. Can be grown indoors.

Uses: Sheltered and shaded gardens where an arrangement of especially attractive plants is desired for year-round viewing. Atriums, entryways, dark corners, planters and containers. Variegated forms brighten dark spaces and add contrast.

Disadvantages: Sunburn when hit by the sun's direct rays in middle and low deserts. Subject to infestations of mealybug and scale.

Planting and care: Plant any time from containers. If shaping is desired, cut at the leaf joints. Feed monthly for fastest growth. Wash leaves occasionally to keep the natural, polished look.

Aucuba japonica

> **All zones when protected from sun and wind**
>
> **Evergreen**
>
> **Soil:** Tolerant except of alkaline soils. Prefers a rich planting mix.
> **Sun:** Full or open shade. Likes north sides or overhead structures. Accepts filtered shade in high deserts.
> **Water:** Ample.
> **Temperature:** Accepts heat with water. Hardy to 5°F (−15°C).
> **Maintenance:** Little.

BACCHARIS pilularis
Family: *Compositae*

Dwarf coyote brush • Chaparral broom

A dense planting of this low, mounding California native looks like a patch of giant moss from a distance. Small, closely set, toothed leaves of bright green densely cover this trailing, ground-hugging plant. Plants grow at a moderate rate 6 to 24 inches high, trailing 3 to 6 feet wide or wider. Fastest growth is in the cool of the year. Plants look best in spring and may merely hold their own in summer. To avoid the cottony blossom fluffs the females produce, seek plants grown from male cuttings. The species is somewhat rangy and is not planted as often as the cultivar below.

Cultivars: 'Twin Peaks' is a male cutting-grown selection, very dense and low.

Special design features: Dense mounding to billowing plant. Oriental effect. Attractive poolside appearance with water and rocks and adapted waterside plants.

Uses: Ground cover and rock gardens. Low edging or filler plant. Erosion control and fire retardant.

Disadvantages: Subject to infestations of red spider mite in June or July. Branches become woody and need to be cut back. Plants seem to be weak in hot summer areas and may die out or brown suddenly.

Planting and care: Plant from flats or containers any time except in the heat of summer. Space 30 inches apart for cover. In fall remove old woody branches and thin to rejuvenate. Fill in bare spots. Spray for red spider mite.

Baccharis pilularis

> **All zones**
>
> **Evergreen**
>
> **Soil:** Tolerant. Accepts damp situation in cool weather, but needs good drainage in summer.
> **Sun:** Does best with full sun. Accepts part shade or reflected sun.
> **Water:** Little to moderate. Looks best with winter irrigation and infrequent waterings in summer; allow to dry out in between.
> **Temperature:** Nearly hardy to cold. Some frost damage in the coldest areas. Tolerant of heat.
> **Maintenance:** Little to occasional.

BACCHARIS sarothroides

Desert broom • Rosin bush • Broom baccharis

Family: *Compositae*

This bright green rounded shrub is native to the Southwest, Mexico and Baja. The desert broom grows easily in broken or disturbed soil, often as a volunteer, naturalizing in areas receiving a little extra runoff. It is fast growing, reaching 3 to 9 feet in height and as wide, with a dense to open habit depending on conditions. The resinous foliage has a crisp, broomlike appearance. Plants bloom in the fall; the males bear flat inconspicuous flowers. The females develop attractive plumelike buds at branch tips that open to release a myriad of white silky seeds that float away with the breeze. These seeds produce a profuse amount of plants. Plants grown from male cuttings avoid this "seedy" problem. This is a plant of wide tolerances. It adapts to a design and takes on the character of its surroundings.

Special design features: One of the brightest green plants in the desert.

Uses: For revegetating disturbed areas, erosion control on banks. Massed or in rows as visual or windscreens. Desert, wild or transitional gardens. May be clipped as a hedge.

Disadvantages: Difficult to eradicate once established. The profusion of floating seeds is a mess in pools. Reseeding is sometimes a problem. Shortlived.

Planting and care: Plant any time of year from containers. Space 2 to 3 feet apart for a clipped hedge, 3 to 5 feet for a row or mass planting. Plants in ground do not transplant well unless very small. A light pruning will produce a thicker, more attractive plant and will remove seed plumes. Shear as desired.

All zones

Evergreen

Soil: Adaptable. Accepts most conditions including alkaline soils.
Sun: Part shade to full or reflected sun.
Water: Accepts any amount of water from none to ample. Looks best with some. Needs occasional irrigation until established and if annual rainfall is below 10 inches a year.
Temperature: Hardy. Accepts heat.
Maintenance: Remove unwanted seedlings. Do this when plants are small and the ground is moist.

Baccharis sarothroides

Baccharis sarothroides

BAMBUSA species

Bamboo

Family: *Graminae*

Bamboos are perennial grasses of great ornamental value. They range in size from low, ground-hugging plants to the giant timber bamboo, with woody canes 35 feet or more in height with a diameter up to 6 inches. Bamboo canes may be hollow or solid, and are divided into sections called *internodes* by joints called *nodes*. Nodes on the upper part of the plant produce buds that develop into leaf-bearing branches. In larger bamboos these divide into secondary branches which bear the flat leaf blades.

Bamboos are of two kinds—running and clumping. Running bamboos, represented in this book under *Phyllostachys*, have underground rhizomes that rapidly grow horizontally to varying distances from the original plant before rising from the ground to start new stands. Large groves can shoot up in a number of years unless the rhizomes are controlled. Running bamboos are generally hardier to cold than clumping bamboos. Clumping bamboos, more tropical in nature and more tender to cold, expand outwardly cane by cane as new shoots spring up at the clump edges. They do not send out horizontal stems or rhizomes. In a new planting canes are small at first, no wider in diameter than the original shoot. Old plantings develop into tight clumps with the canes growing to maximum size. Clumping bamboos are represented under *Bambusa*.

Bamboos usually flower only once after many years of growth. They usually die after flowering, but have in rare instances been known to survive and to recover after several years of intensive care. Whole stands and even divisions taken from them over the years will flower all at once, no matter where they are located.

While it is often said that bamboo is difficult to eradicate, this is not necessarily so. To eliminate bamboo, cut out all canes and trim off all shoots as they start to grow. The plants will exhaust their vigor and die. In the desert the easiest way is to eliminate watering.

Special design features: As a landscape plant bamboos give a very special mood and character to a place, evoking mystery, intimacy, the tropics or the Orient, depending on the setting. The rustling sound of bamboo leaves in the wind is an added dimension.

Note: Most bamboos are from the tropics and the subtropics. Some are temperate zone plants hardy enough to grow in even the northern states. Because conditions of generally lower humidity and greater seasonal temperature fluctuations are hard on bamboo in arid lands, only a few of the more successful are listed here.

Bambusa glaucescens 'Alphonse Karr'

(B. verticillaga). This lush, medium-sized bamboo is an outstanding clumping variety, exceptionally graceful and attractive. Shafts usually reach 10 to 15 feet in height, to 30 feet with optimum conditions. Cane diameters are usually between 1/2 and 3/4 inch. 'Alphonse Karr' is recognizable by the pink and green stripes on the bases of the young stems.

Bambusa glaucescens 'Alphonse Karr'

Special design features: Oriental, tropical or jungle effect. Adapted to waterside situations. Handsome silhouette. Rustle of leaves is pleasant to the ears. Verdant plant for the summer garden.

Uses: As a feature plant in protected intimate gardens sheltered from cold. Planters and containers.

Disadvantages: Looks scraggly in cold winters. Greedy roots. Leaves litter swimming pools, but are attractive as a ground mulch.

Planting and care: Plant from containers in spring when danger of frost has passed. Or, divide old clumps in spring and replant. When dividing cut away a segment with at least 3 shoots and cut those shoots back to 2 or 3 feet. Be sure segment contains numerous roots. Replant where desired, setting 2 to 3 feet apart if you want the planting to fill in. Keep constantly damp until new growth begins. For fastest growth, feed and water regularly as you would a lawn. Do not cultivate or walk near clumps, as new shoots may be damaged.

Low zone

Evergreen

Soil: Tolerant. Prefers improved garden soil. Dislikes alkaline situations.
Sun: Part shade to full or reflected sun.
Water: Moderate to ample.
Temperature: 15°F (−10°C) is the lowest temperature limit.
Maintenance: Regular care to look best.

Bambusa glaucescens riviereorum
Chinese goddess bamboo _____

(B. argentea nana, B. multiplex). A dwarf bamboo with solid stems and fernlike foliage, the Chinese goddess bamboo grows only 4 to 6 feet high, with occasional plants reaching 10 feet in height under favorable conditions. This plant is unusually refined and graceful in the summer garden, especially for close-up viewing in a small space. Use as a background for garden sculpture or lanterns, or as a specimen. Combines well with masonry or natural materials. Leaves often have dry edges, and may be completely straw-colored in winter. Plants sometimes look unkempt, especially in cold areas. The cultural requirements are the same as for 'Alphonse Karr'; low temperature range is also 15°F (−10°C). It requires part shade in the desert and shelter from hot dry winds.

Bambusa oldhamii
Oldham bamboo, clump giant timber bamboo _____

(Sinocalamus oldhamii, Dendrocalamus latiflorus). This bamboo is native to China and Taiwan. It is a tall, rustling, grasslike plant that gives a cool feeling of retreat and contemplation. Once established, it is a fast grower with canes reaching 15 to 25 feet, occasionally to 40 feet with a diameter of 3 inches. New canes suddenly

Bambusa oldhamii

pop up and grow at an amazing rate, but are never any larger in diameter than the original sprout. Canes are deep green and sometimes turn a golden hue when older. The medium green or dry leaves give a variegated effect. A clumping bamboo, it produces no runners. The plant spreads laterally and sends up even taller canes as the clump matures. Cultivation should be kept at a minimum around bamboo clumps during the growth period to avoid injuring the new sprouts.

Special design features: Oriental, tropical, or jungle effect. Dappled light creates aura of mystery. Strong vertical. Handsome silhouette. Rustles in wind. Strong accent. Dramatic near water.

Uses: A fast hedge or screen. For high protected courtyards away from winter chill, or against tall buildings on southeast or south sides. Anywhere in mild winter areas.

Disadvantages: Looks poorly during cold winters. Greedy roots. Produces too much litter for use near swimming pools, but makes a durable ground mulch otherwise. In middle zone a cold spell during fall growing period may damage young canes.

Planting and care: Plant from containers in spring when danger from frost has passed. For fastest growth, feed monthly and soak with water weekly. Trim leaves to show off stems. Divide clumps in spring or thin canes if clumps become too large or dense. When dividing clumps, cut away a segment with at least three shoots, trim back to 3 feet tall. Be sure clump contains numerous roots. Replant where desired and keep constantly damp until new growth begins. Give extra iron if plant becomes chlorotic or to improve color.

Low zone

Evergreen clumping

Soil: Tolerant. Avoid alkaline conditions.
Sun: Partial shade to full or reflected sun.
Water: Moderate to ample.
Temperature: Foliage damaged at 20°F (−7°C). New or developing canes may be injured around 28°F (−2°C).
Maintenance: Regular care to look best.

Bambusa species

BAUHINIA species

Family: *Leguminosae*

Flowering trees and shrubs from the edge of the tropics, orchid trees present a spectacular display of delicate, 2-inch, orchid-like flowers for a long period. Large, bright green leaves shaped like round-winged butterflies cover plants most of the year, dropping for a short period after the bloom. An exception is summer-blooming *B. forficata*, which may be evergreen in warm areas. Because of their spectacular flowers and handsome foliage, orchid trees are very desirable in the right location. They are not attractive when frozen back regularly because they bloom on last year's wood and must go through a recovery period. This is especially true of *B. variegata*. Plants will survive, but may not have many seasons of bloom. If you have a warm sunny spot you may want to grow this tree. Give it added protection during the first few years when it is most sensitive to cold.

Special design features: Tropical effect with a spectacular bloom.

Uses: Patio or courtyard. Street trees in mild winter areas.

Disadvantages: Frosted plants go through a long, unattractive, recovery period. Sometimes older foliage looks mangy in between blossom periods. Susceptible to iron chlorosis.

Planting and care: Plant from containers in spring when weather warms up. Stake and prune young trees carefully. With cold-damaged trees, wait until foliage reappears before pruning, so that you don't trim living wood. Pale trees may need extra feeding in alkaline soils. When leaves are burned from soil salts, give long slow soakings to leach soil.

Low zone and very protected locations in middle zone

Briefly deciduous

Soil: Tolerant. Leaves may burn in alkaline conditions.
Sun: Part shade to full or reflected sun.
Water: Moderate to ample.
Temperature: *B. blakeana* hardy to 25°F (−4°C). *B. forficata* hardy to below 20°F (−7°C). *B. variegata* hardy to 22°F (−4°C).
Maintenance: Seasonal.

Bauhinia forficata
White butterfly orchid tree

(*B. corniculata, B. candicans*). A small tree with medium green foliage, the white butterfly orchid tree is a moderate grower to 20 feet with a wide flat crown. Plant silhouettes are often picturesque, with a leaning trunk and angled branches, thorny at the joints, becoming twisted with age. Delicate white flowers resembling butterfly orchids may appear in late spring, continuing as a pleasant sprinkling through summer. Flowers in clusters open at night, lasting into the following day. Hardy to below 20°F (−7°C), but becomes deciduous.

This South American native is the best orchid tree for the middle zone since blooms appear in spring after frosts. It is also a hardier plant.

Bauhinia punctata

(*B. galpinii*). This is a typical bauhinia with red orchidlike flowers. It can be a sprawling shrub or trained as a small tree. Its

Bauhinia forficata

almost vining character lends itself well to espalier; the abundance and vibrant color of the blossoms allows it to be used like a bougainvillea. Thrives in hot locations, and is as hardy as *B. variegata*.

Bauhinia variegata
Purple orchid tree

This native to India and China grows at a moderate rate to 30 feet with a broad bushy crown of medium green foliage. Most attractive when grown with multiple trunks, but will grow as a standard as well. Bloom begins in midwinter and lasts into spring. Bloom is massive, covering tree with 3-inch or wider orchidlike flowers. Each broad, overlapping petal ranges in color from lavender to magenta and purple, sometimes white. Central petal is marked dark purple. Bloom often continues while leaves are falling, which is sometimes unattractive. The most spectacular blooms occur when leaves have fallen early from cold or prolonged drought and masses of flowers grace the crown. New leaves appear as bloom ends. Trees may have heavy crop of beans. Hardy to 22°F (−4°C).

Best in low zone, but performs satisfactorily in warm protected locations in middle zone.

Bauhinia variegata 'Candida', white orchid tree, is nearly the same tree as *B. variegata*, but worthy of separate mention. Its white flowers sometimes have a purple streak in center petal.

Bauhinia variegata 'Candida'

Bauhinia variegata

BOUGAINVILLEA species
Family: *Nyctaginaceae*

One of the most delightful and rewarding of the color plants is bougainvillea, a native of Brazil. In bloom it creates beauty and flamboyant color in any environment and is now grown and appreciated in warm climates throughout the world. Bougainvillea color is not blossoms, but *bracts*, specialized leaves, which enclose the two or three small yellow blossoms. They appear in masses of purple, magenta, red, orange, yellow or white. Most bougainvilleas are woody vines that naturally climb by hooking their sharp thorns into supports. Some are newly developed shrubs. In cultivation vining types can be trained as large, sprawling, shrubby masses over banks or trellises or formed into standard trees.

Bougainvillea 'Rainbow Gold'

Cultivars: (vines). *B. spectabilis*, (sometimes *B. brasiliensis*), purple. A vigorous grower in cool summer areas, generally considered to be the hardiest.

'Barbara Karst', red in sun to magenta in shade. Hardiest of the reds, it recovers quickly from frost and grows vigorously and blooms early in the season. Loves the desert summer.

'San Diego Red', ('Scarlet O'Hara', 'American Red') similar in color to 'Barbara Karst', has larger leaves and fewer but larger blooms in clusters. One of the hardier strains, recovers quickly from cold, but blooms later than 'Barbara Karst'.

'Texas Dawn' is one of the closest to pink bougainvilleas yet developed. Vigorous grower, it blooms profusely over a long period in summer after 'Barbara Karst' has finished. Appreciates high heat and some humidity.

'Rainbow Gold', burnt orange, blooms vigorously.

'Orange King', tender to frost, but loves long hot summers. More open in growth, with bracts that turn from orange to copper to pink.

'California Gold' bears golden bracts profusely for a long period. More sensitive to cold than red bougainvilleas.

'Jamaica White', white bracts, tinged pink in cooler weather. A fairly new cultivar, it is not known how it performs in winters of the middle zone.

Shrubs: 'La Jolla', red bracts. Performs well and recovers from freezes in the middle zone even in containers.

'Crimson Jewel', red bracts cover densely over a long period.

'Convent' (also called 'Panama Queen'), shrubby with large clusters of magenta-purple bracts over a long season. There are many more cultivars available for landscape use, some with double bracts.

Special design features: Riot of tropical color. Promotes a festive mood.

Uses: Color or emphasis plants for warm locations, such as south and west walls. Striking on trellises. Hot banks where they will not be frozen back. Train as trees in warm areas.

Disadvantages: Freezing. Root balls tend to fall apart when plants are taken from container or transplanted. Aggressive grower—it can take over in frost-free areas.

Planting and care: Set out new plants in spring after frosts are over. Remove from containers carefully to keep root ball intact. To avoid problem of a crumbling root ball, containers can be sliced or perforated and bottoms removed before planting and then set in ground. This will allow roots to spread. Metal containers will rust away with time. Supply spring and summer feedings but overfertilized or overwatered plants bloom little, if at all. Train vines by tying to supports. Cut off unwanted branches. To train vines as bank covers, tie to hairpin-type stakes in the ground. To train as a vine or as a tree in a frost-free climate, tie to a pole and prune back occasionally to a basic framework to form a strong structure.

Low zone, protected locations in middle zone

Evergreen in mild areas

Soil: Tolerant. Needs good drainage.
Sun: *B. spectabilis* prefers cool summers. All others appreciate heat of full or reflected sun. Plants tolerate partial shade but do not bloom well.
Water: Moderate to ample. Requires little water once established.
Temperature: Damaged between 30° and 25°F (−4°C) depending on situation and hardiness of variety. Most plants stop blooming as nights cool, but reflected heat from a wall can extend bloom period. Established plants damaged by cold usually recover quickly in warm weather.
Maintenance: Very little except where training is required. Some clean-up of fallen bracts.

Bougainvillea species

Bougainvillea species

BRACHYCHITON populneus

Bottle tree • Kurrajong tree

Family: *Sterculiaceae*

(B. diversifolia, Sterculia diversifolia). The bottle tree from Australia has a vertical form with a pyramidal crown. A moderate to fast grower, it can reach a height of 20 feet in 5 or 6 years, eventually reaching 30 or even 50 feet in warm areas. Crown of young trees is narrow and dense and becomes wider on older trees. Foliage is shiny and lobed, rustling like a poplar in the wind. The trunk, covered with interestingly patterned greenish bark, is thick at the base but suddenly tapers inward several feet above the ground, resembling a bottle. Clusters of small bell-shaped flowers are dusted pink, appear in May or June followed by decorative woody seed pods shaped like boats. Tree revels in heat.

Special design features: Refined and pleasing at close range. Vertical form of bright green.

Uses: Specimen tree or sapling grove. Row plantings make an attractive and effective screen or background. Does well where space is limited, but can heave sidewalks or paving. Roadside tree or windbreak.

Disadvantages: Susceptible to Texas root rot, iron chlorosis and cold damage. Fuzz inside seed pods is prickly and can be irritating to skin.

Planting and care: Plant from containers in spring. Root systems are fragile, so handle with care. Young trees may need staking until trunks thicken. Give extra iron to avoid chlorosis.

Brachychiton populneus

Brachychiton populneus

Low zone and most of middle zone
Evergreen
Soil: Tolerant. Needs good drainage. May yellow in alkaline soils.
Sun: Full to reflected sun.
Water: Occasional deep waterings preferred, although it seems to do well in lawns with moderate or ample water.
Temperature: Young trees killed and foliage damaged on older trees at about 18°F (−8°C). Fast recovery in spring.
Maintenance: Little to none.

BRAHEA armata

Mexican blue palm • Big blue hesper palm • Rock palm

Family: *Palmae*

(Erythea armata). This fan palm from Baja, California, is very much at home in the desert garden. It tolerates greater extremes of heat, cold, wind and other adverse conditions than most palms. Grows slowly to 25 or 30 feet, sometimes higher, with a crown spread of 8 to 10 feet. Crown is wider than high, formed of stiff waxy fronds which retain color until they bend down to the trunk. Fragrant, whitish flower garlands to 18 feet long in summer mature to reddish brown and produce hard berrylike fruit. Not as graceful as other palms as it matures because of heavy trunk, small head and stiff leaves. However, younger plants are lush and attractive and remain so for many years.

Special design features: Interesting blue-gray foliage. Informal. Vertical with age.

Uses: Specimen, grove or row. Residential, street or public area. Informal areas, desert, or wild gardens. Complements other desert plants.

Disadvantages: Does not seem to have any special problems.

Planting and care: Plant any time from containers, but best in spring. Transplant spring or early summer. Larger plants are difficult to transplant successfully.

Brahea armata

Brahea edulis is another fan palm that is often used in these areas. In appearance, it is much like *B. armata,* with lighter, silver-blue leaves. It is a moderate grower that stays in scale with the average home landscape, with a maximum height of approximately 30 feet. It is an excellent plant for a tropical effect.

Middle and low zone, protected microclimates in high zone
Evergreen
Soil: Tolerant. Grows naturally in limestone soils. Prefers improved garden soil.
Sun: Tolerates part shade to full or reflected sun.
Water: Drought tolerant. For best results give deep, widely spaced irrigation.
Temperature: Hardy to 15°F (−10°C).
Maintenance: Some clean-up of old fronds and fruit.

BUTIA capitata
Family: *Palmae*

Pindo palm • Jelly palm • Brazil butia palm

(Cocos australis, C. campestris). This is a graceful, small-scale feather palm, originally from Brazil and Argentina. One of the three hardiest exotic palms. It can be used in a small area for many years without outgrowing the space, growing slowly to 10, sometimes 20 feet with a spread of 10 to 12 feet. Vertical trunk and gray-green fronds deeply *recurve* as they bend downward. Palms are most attractive in youth before they develop the rather heavy trunk with vertical leaf bases which sometimes become marred by fungus. Try to keep leaf cuts the same length to create an even pattern, or skim them off to create a smooth, fibrous trunk. Female trees produce large, edible, pineapple-flavored fruit, used to make tasty jelly or preserves.

Special design features: Light festive appearance. Refined eyecatcher in youth. Tropical effect with tropical plants.

Uses: Emphasis or accent plant for containers, tubs, planters or gardens. Because of its gray color, combines with desert plants or contrasts with tropical plants. Resistant to Texas root rot.

Disadvantages: Plants may be hard to locate. Trunks on older specimens sometimes awkward and unattractive in appearance. Subject to infestations of root knot nematodes and bud rot. Leaf stumps and trunks sometimes get a fungus which eats and deforms them. Poor drainage and alkaline soils sometimes cause iron chlorosis.

Planting and care: Plant from containers any time of year, but transplant in warm weather. Remove old, drooping or dry fronds and spent blossoms. Do not wet crown to avoid infecting palm with bud or trunk rot. Plants need careful grooming as they age to keep a tidy appearance. Give extra iron if foliage pales.

Middle and low zones, warmer microclimates of high zone

Evergreen

Soil: Tolerant, but needs good drainage. Best in improved garden soil.
Sun: Full to reflected sun.
Water: Moderate.
Temperature: Hardy to 15°F (−10°C).
Maintenance: Occasional trimming.

Butia capitata

BUXUS microphylla japonica
Family: *Buxaceae*

Japanese boxwood

This shrub from Japan has been used for centuries as an outstanding hedge plant for formal or contained gardens—it accepts clipping very well. Slow growing to an eventual 4 to 6 feet high and as wide unless contained, it makes a dense, rounded form if allowed to grow naturally. When trimmed, Japanese boxwood will remain a low, compact hedge for a long time. Shiny, bright green leaves are small and rounded and cover plant to ground.

Special design features: Dependable and durable formal plant.

Uses: Specimen or hedge plant for low or medium hedges or edgings. Excellent in formal or contained gardens as a box hedge, space divider or topiary plant.

Disadvantages: May sunburn if clipped too closely late in the season. Subject to browning out in spots from reflected heat or yellowing from iron chlorosis.

Planting and care: Plant any time from flats or containers. For continuous planting, space plants from flats 6 to 9 inches apart, gallon plants 9 to 12 inches apart. Begin clipping and shaping plants when they are young. Give regular feedings and extra iron to encourage green foliage during growing season, especially in alkaline situations. Any heavy clippings should be done in fall or late winter before heat arrives. Shear as needed to keep neat.

All zones

Evergreen

Soil: Tolerant. Prefers improved garden soil. Accepts some degree of alkalinity.
Sun: Open shade, part shade, full sun. Burns in reflected sun.
Water: Moderate.
Temperature: Hardy. Accepts heat with adequate water.
Maintenance: Moderate amount of clipping and feeding.

Buxus microphylla japonica

Buxus microphylla japonica

CAESALPINIA gilliesii

Yellow bird of paradise • Mexican bird of paradise

Family: Leguminosae

(Poinciana gilliesii). Angular and picturesque, this exotic plant bears numerous pyramidal clusters of yellow flowers with long red stamens that look like tropical birds. Generally open in form with a slender trunk, branches bear finely cut medium green leaflets. Branches, leaves and fruit may be flecked with brown specks. Flat pods 4 to 5 inches long follow spring to summer bloom. Bird of paradise is generally deciduous except in the warmest places, but is never its best in winter. With water it grows rapidly to 6 feet or more high. Usually a lower and slower-growing plant, especially where it has to struggle. A native of South America and Mexico, it is widely adapted to the Southwest, where it has naturalized.

Special design features: Warm weather color. Desert, tropical or subtropical effects.

Uses: Specimen or mass. Accent or silhouette plant, alone or in combination with tropical or desert plants.

Disadvantages: Brown fuzz on stems. Twigs and seed pods produce unpleasant odor, and seed pods are poisonous. Plants may become pests in some places because they reseed profusely.

Planting and care: May be grown from seed. Plant from containers any time.

Irrigate until established, especially during first summer. Prune to keep groomed and to remove pods. May be cut back severely in winter, its most unattractive season. It will grow and bloom as a thick, low shrub in spring, or allow it to become a tall and open plant. Survives drought and neglect, but may go dormant. Needs supplemental water during bloom season to look best and to have a good show of flowers. Plants grown in an arid environment stay small and seldom bloom.

All zones
Deciduous
Soil: Tolerant of a wide variety of soils.
Sun: Full and reflected sun, also part shade.
Water: Accepts any amount. Needs some supplemental water in the desert to look well.
Temperature: Hardy to about 10°F (−12°C) where it may die back to the ground. Recovers quickly in warm weather.
Maintenance: Little to none.

Caesalpinia gilliesii

Caesalpinia gilliesii

CAESALPINIA pulcherrima

Red bird of paradise • Dwarf poinciana • Flower fence

Family: Leguminosae

(Poinciana pulcherrima). The brilliant orange-red and yellow flower clusters and lush, medium green, ferny foliage of this summer bloomer add a vibrant, tropical effect to the landscape. A native from Mexico and the West Indies, it thrives in the hottest places where little else will grow, blooming continuously during the warm season. A vase-shaped shrub with large, finely cut, feathery leaves and

Caesalpinia pulcherrima

sometimes prickly branches, dwarf poinciana is slow to get started, but becomes a vigorous grower and profuse bloomer after a year or two. Once roots are established, a plant cut back in winter can attain a height of 3 feet or more early in the season and bloom by early summer. Established untrimmed plants in areas seldom nipped by frost may reach over 6 feet in height and become almost treelike. Spectacular flowers in flat pyramidal clusters at branch tips produce flat 3 to 6-inch seed pods that when ripe, pop open and scatter the seeds.

Special design features: Warm seasonal color. Tropical effect.

Uses: Specimen as accent, mass for effect. Rows for screening. Natural, desert, tropical and rock gardens. Spruces up bare areas.

Disadvantages: Fruit is poisonous. Plants slow to start and may not grow much or bloom the first year or two. This is especially true in dry summer areas lacking sufficient water. Unattractive in winter. Sometimes gets Texas root rot.

Planting and care: Start from seed or plant from containers in spring when frost is past. Space 3 to 5 feet apart for rows or massing. If plant is twiggy in appearance in winter, cut back 4 to 6 inches above the ground. This also encourages branching.

Middle and low zones, high zone in protected locations
Evergreen in warm climates
Soil: Tolerant of wide range of soils.
Sun: Full to reflected sun.
Water: Moderate to ample during flowering. Occasional during dormant season. Tolerates small amounts during warm season, but little bloom.
Temperature: Becomes purplish in cool of winter. Freezes to ground around 30°F (−1°C). Mulching roots and lower stems in cooler areas will help plants recover from much lower temperatures.
Maintenance: Little.

Caesalpinia pulcherrima

CALLISTEMON citrinus

Bottlebrush • Scarlet bottlebrush • Lemon bottlebrush

Family: *Myrtaceae*

(C. lanceolatus). The bottlebrush is a narrow-leafed shrub of fast to moderate growth 8 to 12 feet high. It spreads to 6 feet, with stiff upright or arching branches. Erect, loose, open and rugged could be used to describe this Australian native. Perhaps most important are the words *spectacular bloom*. A heavy bloom of red "bottlebrush" flowers to 6 inches long appear in midspring; some blooms appear periodically over the warm season. Small woody seed capsules remain on the twigs, adding interest.

All zones, protect from cold in high zone

Evergreen

Soil: Widely tolerant. Accepts somewhat alkaline soils with good drainage and extra iron.
Sun: Part shade. Prefers full to reflected sun.
Water: Little to moderate. Once established seldom needs supplemental irrigation in areas with 10 inches or more annual rainfall with extra runoff. Accepts ample water in garden situations, but sometimes develops chlorosis.
Temperature: Hardy to about 20°F (−7°C), badly damaged below.
Maintenance: Little to none.

Cultivars: 'Jeffers' is smaller, to about 6 feet high and more erect in form and refined in appearance. Its magenta flowers fade to lavender.
Special design features: Informal if untrimmed. Periodic color.
Uses: Specimen, standard, screen or espalier. Handsome silhouette plant. Transitional gardens. Combines with desert or tropical plants.
Disadvantages: Subject to iron chlorosis

Callistemon citrinus

in alkaline soils. Slow to recover if badly frozen and often seems to show iron deficiency with new growth. Woody and coarse with age.
Planting and care: Plant from containers any time; best in spring in high zone. Space 4 feet apart for hedge or screen. Space 'Jeffers' 3 feet apart. Prune lightly to shape and train. Feed extra iron to keep foliage healthy green. Otherwise, plants tolerate neglect.

Callistemon citrinus

CALLISTEMON viminalis

Weeping bottlebrush

Family: *Myrtaceae*

This Australian native has an erect trunk and an irregular crown with branches that bend deeply downward in a weeping effect. It grows at a moderate or fast rate 15 to 20 feet high and may eventually reach 30 feet or more under optimum conditions. Crown is usually narrow, no more than 15 feet wide, with slender 4-inch leaves tinged bronze when young. Heavy 6-inch, deep red bottlebrush flowers appear in spring and cycle into bloom one or more times during the warm season.
Cultivars: 'Captain Cook' is a dwarf variety said to grow to only 4 feet high, however, in conditions to its liking, it has been seen up to 8 feet. Attractive for borders or low screens.
Special design features: Interesting picturesque form. Contained size. Sporadic color.
Uses: Patio and street trees. Silhouette plant against masonry. Best used where protected from high winds and cold, such as in courtyards or beside tall buildings.
Disadvantages: Subject to iron chlorosis, which seems to be worse on overwatered plants or plants recovering from cold damage. Damaged by wind. Leaves brown in cold winters. Stringy look of bare branches as they recurve at the top of the tree is sometimes unattractive.
Planting and care: Plant from containers

in spring. Stake young trees for support until a strong trunk develops. Prune to shape and thin heavy branches occasionally to prevent top-heaviness and wind damage. Feed with iron to keep green and prevent chlorosis. Added nitrogen promotes attractive growth.

Middle and low zones

Evergreen

Soil: Tolerant. Needs good drainage. Prefers prepared garden soil with low alkalinity.
Sun: Part shade to full or reflected sun.
Water: Prefers moderate.
Temperature: May be damaged at 20°F (−7°C), but may tolerate 15°F (−10°C) with only foliage damage. Recovery from damage is moderately fast.
Maintenance: Occasional.

A few other bottlebrushes deserve mention. A number of varieties are offered as *C. phoeniceus*, the fiery bottlebrush. It is a shrub 6 to 8 feet tall and quite similar to *C. citrinus*. The stiff bottlebrush, *C. rigidus*, is a unique plant quite stiff in appearance, making a tree to 20 feet with a spread of 10 feet. It has sharp leaves and gray-green to purplish flowers. All species of bottlebrush seem to do well in desert gardens.

Callistemon viminalis

CAMPSIS radicans
Trumpet creeper ● Trumpet vine ● Cow itch ● Trumpet honeysuckle

Family: *Bignoniaceae*

(Bignonia radicans, Tecoma radicans). Trumpet creeper is a rampant-growing, self-attaching or sprawling vine native to the eastern United States. Stunning clusters of 3-inch, orange to red, trumpet-shaped flowers appear from midsummer until frost. Considered a weed in the temperate zone, it is appreciated in arid zones for summer greenery, color and tolerance of a wide range of growing conditions. A fast grower to 20, sometimes 40 feet high and as wide. Rich green leaves are divided into large numerous leaflets. Old branches become woody, covered with cinnamon-brown bark. Plants climb by means of aerial rootlets and spread by suckering and rerooting where branches touch the ground.

Cultivars: *C. tagliabuana* 'Mme. Galen', a hybrid, is the best known, bearing salmon flowers.

Special design features: Rampant summer color and cover.

Uses: Bold vine for rustic or large scale situations. Excellent with evergreens, which mask its winter dormancy. Wall, trellis, arbor or fence climber. Attractive color at property edges. With careful and frequent pruning and pinching it can be formed into a hedge or featured shrub.

Disadvantages: Vigorous plants may become rank and invasive. Difficult to eradicate once established, or to contain in small areas.

Planting and care: Grow from seed or plant from containers any time. Transplant or set out bare root in winter. Train and cut as desired. No other special care required.

Campsis radicans

All zones

Deciduous

Soil: Tolerant.
Sun: Part shade, full or reflected sun.
Water: Needs some to moderate irrigation during summer growth. Tolerates ample water and grows vigorously.
Temperature: Tolerant of heat and cold.
Maintenance: Widely variable depending on situation and amount of containment desired.

CARISSA grandiflora
Natal plum ● Amatungula

Family: *Apocynaceae*

(C. macrocarpa). Natal plum is a pleasing plant with dense, dark green, almost succulent leathery leaves, fragrant white flowers and edible fruit. This South African native revels in heat. A fast grower to 5 to 7 feet and as wide, with short, green, thorny stems supporting closely set, shiny, rounded 3-inch leaves. Waxy, starlike flowers are scattered over plant through the warm season. Flowers are often followed by 1-inch, red, beet-shaped fruit, delicious in salads.

Cultivars: 'Fancy', to 6 feet high, is an ample flower and fruit producer. It makes an excellent espalier. 'Ruby Point' also reaches up to 6 feet. New leaves have red color through the growing season. Low forms include: 'Boxwood Beauty', to 2 feet high and as wide, very compact like a large-leafed boxwood, but seldom needs pruning or shaping. Useful as low hedge or divider and for formal training. 'Prostrata', to 2 feet high, trails to about 3 feet wide. Use as a ground cover if vertical branches are trimmed off. 'Green Carpet' reaches to 18 inches high, spreading 4 feet. It has smaller leaves than most, and makes an excellent ground cover. 'Tuttle' ('Nana Compacta Tuttlei') is compact and spreading, to 2 to 3 feet high and 3 feet wide. Good producer of flowers and fruit.

Special design features: Dense dark green. Refined and handsome when viewed close-up. Fits with formal or informal designs. Oriental effect.

Uses: All are handsome, refined barrier plants. Taller forms make clipped or unclipped hedges and screens in warm areas, foundation or wall plants against south or west walls under overhangs in middle zone, or against any wall in lower zone. Low plants can be clipped and shaped into low hedges or forms, trained on walls or used as ground covers in warm areas (or where protected from cold).

Disadvantages: Winter browning in areas of frost.

Planting and care: Plant in spring when danger from frost is past. Space larger plants 4 feet apart for continuous planting; low plants 18 to 36 inches apart depending on spread of cultivar. Cut back vertical shoots on ground covers to encourage spreading. Do any heavy pruning in early spring.

Carissa grandiflora

Low zone, protected places in middle zone

Evergreen

Soil: Average garden soil with good drainage.
Sun: Widely tolerant. Prefers full or reflected sun. Growth is more open in shade, with fewer flowers and fruit.
Water: Moderate to ample. Tolerates some drought when established, but looks poor.
Temperature: Damaged at 28°F (−2°C). Recovers quickly in spring.
Maintenance: Looks great with almost no care, except an occasional trimming.

CARNEGIEA gigantea
Family: Cactaceae

Saguaro • Sahuaro • Arizona giant • Giant cactus

(Cereus giganteus). This is the spectacular, well-known columnar cactus native to southern Arizona and northern Mexico. Saguaros grow very slowly to a height of 60 feet or more, and are long-lived, up to 250 years. It takes many years for plants to reach just inches in height. When 60 to 75 years old and about 20 feet tall, they may sprout budding branches, or arms. A saguaro may develop a single arm or several. As many as 15 have been counted on a single specimen. In late spring or early summer, wreaths of white flowers appear at the top and branch tips, opening at night, and lasting into the next afternoon. Summer fruits are red. They split open and the black seeds are eaten by birds. Plants are heavily protected by law, and require tags from the United States Commission of Agriculture for legal purchase or to be transplanted from one place to another. Seeds are nearly 100 percent viable, but to raise your own saguaro from seed you will need great patience and a long life!

Special design features: Strong dramatic vertical. Spectacular skyline silhouette.

Uses: Desert or natural gardens as an emphasis plant.

Disadvantages: Damage by vandalism and lightning. Bacterial necrosis mainly a problem affecting older specimens. It is fatal if not treated in the early stages.

Planting and care: Large specimens are very heavy and should be transplanted by professionals with special equipment. If you want to move smaller plants (up to 6 feet), begin by digging around and bracing with padded boards to form a sling along

Carnegiea gigantea

its entire length. Retain as many roots as possible. Set into a wide and usually shallow hole so cactus is no deeper than it was before. Keep propped until plant develops new roots. To encourage growth, occasionally irrigate a wide area around base in summer. If discolorations develop seek information and diagnosis; it may be the beginning of bacterial necrosis which will cause death in a few years. Advanced symptom is a black, foul-smelling liquid oozing from irregular brown spots on the trunk. To treat, dig out the affected flesh with a spoon. Shape bottom of wound at base so it will drain. Wash entire area with a solution of one part household bleach to nine parts water.

> **Middle and low zones**
>
> **Evergreen**
>
> **Soil:** Prefers rocky or loose soil with good drainage.
> **Sun:** Part shade to full or reflected sun.
> **Water:** None to occasional. Supply some supplemental water during extended droughts, especially in summer. Plants receiving a little extra water grow faster.
> **Temperature:** Hardy in the zones where it grows naturally.
> **Maintenance:** None.

CARPOBROTUS edulis
Family: Aizoceae

Hottentot fig • Fig marigold • Common ice plant

(Mesembryanthemum edule). The hottentot fig is a coarse-textured ground and bank cover native to South Africa. Its light green, succulent leaves are bold, angular, three-sided and curving to 4 inches long and 1/2 inch thick. Woody branches trail to 3 feet or more, rooting as they go. Older plantings mound 12 to 18 inches high due to accumulation of old runners under-

Carpobrotus edulis

neath. Creamy to yellow, yellowish pink or lavender flowers bloom spring to summer and produce small fruits, edible, but not very tasty.

Special design features: Bold-textured, succulent, mounding ground cover.

Uses: Casual transitional plant. Ground or shallow bank cover. Containers, rock gardens and near pools.

Carpobrotus chilensis

Disadvantages: Bare spots. Woody fire hazard. Rodents eat it.

Planting and care: Set from cuttings, flats, or containers in spring after frosts are over. Space 18 to 24 inches apart. Fill in bare areas by pinching off cuttings from established plants and setting them in the ground where needed. This is usually necessary every year.

C. chilensis is a native coastal iceplant found in California, Baja and Chile. It is much like the hottentot fig only more refined, and can be used the same way. It is not as vigorous or tolerant of desert conditions, however. A unique feature is purple blossoms which emit a fruity fragrance that smells like something good to eat.

> **Middle and low zones**
>
> **Evergreen**
>
> **Soil:** Prefers sandy loose soil with good drainage. Tolerates alkaline conditions.
> **Sun:** Part shade, full or reflected sun.
> **Water:** Needs supplemental water in arid areas.
> **Temperature:** Accepts heat, but may be damaged by cold in the 20's°F (−7°C).
> **Maintenance:** Occasional.

CARYA illinoensis

Family: *Juglandaceae*

(C. oliviformis, Hicoria pecan). This native of the south central United States to northeastern Mexico grows surprisingly well in the deserts if supplied with deep soil and irrigation. A handsome, well-structured tree, it grows at a moderate rate up to 70 feet, with an equal spread in favorable situations. Most commonly seen as a garden tree 25 to 40 feet high but can be kept smaller by pruning. Compound medium to deep green leaves are long and have 11 to 17 pointed leaflets to 7 inches long. Trees are usually very erect with single trunks and billowing crowns. Inconspicuous flowers in spring produce a nut borne in a husk in late summer or fall. Pecans have recently become a cash crop in the Southwest. Plant at least two trees (for pollination) for a better crop of nuts.

Cultivars: Numerous are available— 'Western Schley', commonly called 'Western', is said to be best for desert climates. Plants from New Mexico or California are best for desert areas.

Special design features: Dominant green statement in the landscape. Woodsy look reminiscent of "back home." Shade tree which produces edible nuts. Handsome silhouette in any season. Yellow fall color.

Uses: Quality shade tree which does well in lawns. Slower growing than some, but worth waiting for. Groves or specimens in lawns or gardens. Especially attractive when given room to spread. South sides of buildings where it gives shade in summer, allows sun in winter.

Disadvantages: Brittle branches. Subject to attack by aphids and a number of diseases, including Texas root rot. Nuts sometimes attacked by a weevil.

Planting and care: Plant from containers any time or bare root in winter. Space trees 25 to 30 feet apart for grove. Prune to remove dead branches on older trees or to form strong structure on younger trees. Commercial trees are tip-pruned to keep them dense and easier to harvest. Good subject for drip irrigation. Seek information from your county agent if you are interested in raising nut crops.

All zones

Deciduous

Soil: Don't plant in rocky or caliche soils. Deep soil is preferred so tree may send down taproot for fullest development.
Sun: Full to reflected sun.
Water: Moderate to ample.
Temperature: Hardy. Takes desert heat with ample water.
Maintenance: Prune dead branches and control weeds underneath tree.

Carya illinoensis

CASSIA artemisioides

Family: *Leguminosae*

Feathery cassia ● Silvery cassia ● Wormwood senna ● Old man senna

This Australian native produces a breathtaking display of yellow flowers in late winter or early spring. A rounded shrub, it grows fast up to 6 feet high with gray to gray-green needlelike foliage. While unobtrusive most of the year, in early spring the plant covers itself with a profusion of fragrant yellow flowers about 1/2 inch across, followed by flat pods to 3 inches long.

Special design features: Early season color.

Uses: Specimen or mass planting. Informal unclipped screen or space divider. Foundation plant. Desert or wild gardens where it requires no maintenance and sometimes naturalizes.

Disadvantages: Can be damaged by cold just as it begins to bloom. Beans and resulting litter are a problem only where a neat, manicured effect is desired.

Planting and care: Plant from containers in spring. Space 3 feet or more apart in a sunny protected area. Prune lightly after bloom to remove pods if they are objectionable. Prune also to promote new growth and bushy form.

Cassia artemisioides

Low zone, warmer microclimates of middle zone

Evergreen

Soil: Tolerant, but needs good drainage.
Sun: Part shade to full or reflected sun.
Water: Prefers moderate at least until established. Tolerates periods of drought and occasional irrigation, but assumes the rugged, woody appearance of unirrigated native plants.
Temperature: Damaged below 20°F (−7°C), blossoms damaged at about 28°F(−2°C).
Maintenance: Occasional pruning.

Cassia artemisioides

CASSIA wislizeni
Family: *Leguminosae*

Shrubby senna

This shrub is native to the dry slopes and mesas of the Southwest and northern Mexico. Nondescript most of the year until it bursts into bloom. Numerous, clear, rich, yellow flowers appear continuously over the warm season, usually from June to September, but sometimes from February to October in warm areas. A slow to moderate grower from 3 to 6 feet high, its performance depends on available moisture. Dark rather rigid branches present an interesting growth pattern, especially when in leaf and flower. Leaves are gray-green, finely divided, and make a lacy background for small flowers followed by slender pods.

Special design features: Summer color and specimen interest. Desert feeling.

Uses: Desert or wild gardens as specimen or part of grouping.

Disadvantages: Bleak during winter period.

Planting and care: Plant from seed or containers any time. Irrigate until established and occasionally afterwards, especially if annual rainfall is less than 10 inches a year and to encourage the bloom. Plants stop blooming if they get too dry.

Cassia wislizeni

All zones

Deciduous

Soil: Prefers rocky or loose sandy soils with good drainage.
Sun: Part shade to full or reflected sun.
Water: Occasional to none depending on situation.
Temperature: Hardy to cold.
Maintenance: Little to none.

Cassia wislizeni

CASUARINA species
Family: *Casuarinaceae*

Beefwood ● Australian pine ● She oak ● Bull oak

The fringy *Casuarina* species from Australia are some of the toughest trees for the desert, withstanding heat, drought, wind and poor soil. They are erect and slender in form with dark bark and dark gray-green "needles"—actually tiny jointed branchlets. Young trees are pyramidal and rather dense, with foliage reaching nearly to the ground. Trees become more open and airy with age, shedding lower branches for several feet above the ground. Trees look something like a pine, thus its common name Australian pine. They are also sometimes confused with tamarisk due to the similiarity of their foliage. All species produce little woody cones about 1/2 to 1 inch long on female plants.

Special design features: Erect, slender, fringy.

Uses: Street trees and windscreens. Tall-growing visual or shade screens when massed or planted in rows. Specimen for small yards or patios where a narrow, upright tree is needed. A good choice for drier property edges or for rehabilitation of landfills or other problem areas.

Disadvantages: Sometimes woody and sparse appearance with age and with dead branches where neglected.

Planting and care: Plant from containers in any season, but best in spring in middle zone to allow for establishment before winter. Prune any time to remove dead branches on older trees. An occasional pruning helps keep trees neat and appealing in appearance.

Casuarina cunninghamiana
Australian pine

This is the hardiest and probably the most handsome of the group. A fast grower to 70 feet with a strong, rather dense growth of fine branches covered with dark green foliage. Trees look similar to pines, and produce fruit that look like cones. See photo, top of page 67. It does better in cooler areas.

Casuarina stricta
Beefwood, she oak

(*C. quadrivalvis*). This tree is very erect, open and slender with shorter branches and a more open crown than the others. A very fast grower to 30 feet, but 20 feet is a more common height. Because of its pliant branches this is not a tree for windy places or it will grow lopsided. Quite hardy and does well in middle zone winters, where it serves nicely as a street tree. Because of its rough appearance this is not a tree for close-up viewing. Hardy to the middle and low zones.

Casuarina stricta

66

Casuarina equisetifolia
Horsetail tree, south sea ironwood, mile tree

The horsetail tree grows to about 80 feet with favorable conditions. It grows in the wild from southeast Asia to the Pacific Islands to various parts of Australia as a seashore tree. It withstands any amount of brackish soil and water, and any other rugged situation except cold. For low zones only. Excellent for the seashore or for reclaiming sand dunes.

Middle and low zones

Evergreen

Soil: Widely tolerant. Accepts sand, clay or even wet, brackish or alkaline soil. Grows best and fastest in well-prepared garden soil.
Sun: Part shade to full or reflected sun.
Water: Best with deep spaced irrigation. Once established, trees survive in areas receiving 10 or more inches of annual rainfall without supplemental water, but grow faster and look better with some irrigation.
Temperature: *C. cunninghamiana* and *C. stricta* are hardy to about 15°F (−10°C) or below. *C. equisetifolia* is more tender, damaged at about 25°F (−4°C).
Maintenance: Occasional to none.

Casuarina cunninghamiana

CATALPA speciosa
Family: *Bignoniaceae*

Western catalpa • Cigar tree • Indian bean

A tree familiar to many from the temperate zone, the western catalpa is native to the central United States. It is fast growing to 40 feet or more, pyramidal in shape, and has very large heart-shaped leaves reaching up to a foot in length. This bold, large-scale tree is sometimes weedy, and seldom reaches its full height in the desert, but is valued for its tropical appearance and striking flowers. Well adapted to extremes of heat and cold. Early careful training and staking can make a tall-trunked, umbrella-crowned form. Trumpet-shaped flowers of pure white with yellow and brown markings appear in clusters above foliage in late spring and early summer, followed in fall by foot-long beans which look like long cigars.

All zones

Deciduous

Soil: Tolerant. Chlorosis can be a problem in some areas.
Sun: Part shade to full or reflected sun.
Water: Best with moderate. Accepts ample.
Temperature: Tolerant.
Maintenance: Occasional pruning and some clean-up in spring, summer and fall.

Special design features: Tropical effect with bold foliage. Spring-summer bloom.
Uses: Park, lawn, street or garden tree for shade. South sides of buildings where it provides abundant shade in summer, sun in winter.
Disadvantages: Flowers litter spring through summer. Bean and leaf litter in autumn. High winds tatter leaves and may break the brittle branches.

Catalpa speciosa

Planting and care: Plant from containers any time or bare root in winter. Important to thin and prune as necessary to shape and promote growth when tree is young as well as to allow wind to pass through branches.

Catalpa speciosa

CEDRUS atlantica

Family: *Pinaceae*

Atlas cedar

(C. libani atlantica). One of the three or four true cedars, the atlas cedar comes from North Africa. The blue atlas cedar, *C. atlantica* 'Glauca', is probably best known. A slow to moderate grower to 100 feet in its native Atlas Mountains, it is not yet known whether this tree will attain such heights under cultivation in lower desert climates and soil. Tree is irregular in youth, but grows into a flat-topped, pyramidal giant with stiff, upwardly horizontal branches. Branches are clothed in tuftlike clusters of tiny blue-green needles. Trees can be obtained trained as a symmetrical form or untrained with a rustic and irregular shape.

Cultivars: 'Glauca' is a grafted selection with silvery blue foliage and pyramidal form. A rustic selection, more often used as an accent or specimen in large containers. 'Glauca Pendula' is a weeping kind with sprawling cascading branches.

Special design features: Rustic, Oriental or woodsy effect. Blue color of some varieties.

Uses: Species is a strong vertical and skyline tree for large spaces such as parks or large lawns. Mass as a tall screen near bodies of water to filter the air. This tree is seldom seen, but deserves wider use. The rustic blue Atlas cedar is more commonly used as an accent or specimen in a planter or container. It will look nice for a few years, until it becomes root bound and begins to decline. In the ground it develops into a gnarled, rustic small tree.

Planting and care: Plant from containers any time. Set species 16 to 30 feet apart for a grove. No special care, pests or problems have been noted so far with this uncommon tree. Pinch tips to shorten long heavy branches on young trees.

Middle and high zones

Evergreen

Soil: Prefers well prepared garden soil. Avoid light gravelly soils.
Water: Prefers moderate. Established trees accept short periods of drought.
Sun: Part shade to full or reflected sun.
Temperature: Tolerant of heat and cold.
Maintenance: Occasional grooming and trimming.

Cedrus atlantica 'Glauca'

CEDRUS deodara

Family: *Pinaceae*

Deodar cedar

The mostly widely planted of the true cedars, the deodar makes a dramatic statement in the landscape. Graceful and pyramidal with a youthful form reminiscent of a Christmas tree, this Himalayan native grows at a moderate to fast rate to 80 feet with a 40-foot spread at the base. Silhouette is partly open so you can see the sky through the horizontal branches, which bend downward near the ends. Older trees have a bobbing leader or main branch which leans to one side. Silvery gray to gray-green foliage darkens with age. Needles 2 inches long grow in tufts along branches. Cones are up to 5 inches long. Individual cultivars vary in form and color.

Cultivars: 'Prostrata' will grow flat on the ground or hang over planters. 'Compacta' is slow growing, rounded and dense. 'Pendula' has a weeping form.

Special design features: Tall accent or silhouette. Skyline tree. Foliage contrast with broadleaf evergreens or lighter green or gray plants.

Uses: Accent or specimen tree, or mass for effect. Park or large open area for screen, accent, or bold form. Plants set close together can be trimmed as a hedge.

Planting and care: Plant from containers any time. Prune in late spring to shape or to encourage dense bushy growth. Plant area under tree with a needle-absorbing ground cover such as a low juniper or leave unraked to create a mulch of needles. Give extra iron if tree seems pale or yellows from chlorosis.

Middle and high zones, marginal in low zone

Evergreen

Soil: Best in heavy soils, especially well prepared garden soil. Avoid light or gravelly soils.
Sun: Full to reflected sun.
Water: Prefers moderate. Accepts deep, widely spaced irrigation.
Temperature: Hardy to 5°F (−15°C).
Maintenance: Little to none.

Another true cedar that should be used more is *C. libani*, cedar of Lebanon. It is a fine tree for the desert zones. Erect and conical in youth, but develops into a broad-headed form with several main trunks at maturity. A mature tree will reach 80 feet with a 40-foot spread at the base. Foliage is bright green in youth, becoming dark green with age. Requires ordinary garden care, with the same cultural requirements as *C. deodora*.

Cedrus deodara

CELTIS pallida

Family: *Ulmaceae*

(C. tala pallida). This densely branched, thorny, desert shrub from the Southwest and northern Mexico forms dense thickets, creating a visual screen as well as an impenetrable barrier. Form is mounding and irregular. Each jutting branch has many sharp spines and dull, medium green leaves 1 to 3 inches long. Rough-textured foliage somewhat resembles that of Chinese elm. Becomes deciduous from drought or cold. Desert hackberry grows slowly 3 to 10 feet high, depending on conditions. Inconspicuous spring blooms produce small orange berries loved by birds.

All zones

Evergreen to deciduous

Soil: Adaptable. Prefers loose gravelly soils.
Sun: Part shade to full or reflected sun.
Water: None needed once established, but grows faster with occasional irrigation. In deserts receiving less than 10 inches of annual rainfall supplemental irrigation is necessary. In extreme drought loses leaves.
Temperature: Loses leaves in severe cold below 20°F (−7°C). Recovers quickly in warm weather.
Maintenance: Little to none.

Special design features: Rugged informal plant.
Uses: Barrier or visual screen. Rehabilitation of the desert. Desert gardens where birds seek it for shelter and for food in season. Combines with other desert plants.

Planting and care: Grows easily from seed, but slowly. May be purchased in 1 or 5-gallon containers and planted any time of year. Needs no pruning unless a specific shaping is desired. Tolerant of heat, drought, wind and poor soil, and requires no special care.

Celtis pallida

CELTIS reticulata

Family: *Ulmaceae*

(C. douglasii). Not a true desert plant in the drought-resistant sense, netleaf hackberry grows along streams and washes at elevations from 2,500 to 6,000 feet throughout much of the Southwest and northern Mexico. A craggy tree with pendulous branches and interesting, nubby, light gray bark, it grows at a moderate rate to 25 or 30 feet and spreads as wide. Leaf form may vary in size and shape and in density depending on individual tree and conditions. Insignificant spring flowers produce orange to purple 1/4-inch berries in clusters late summer into winter, loved by birds.

Special design features: Picturesque character plant, almost Oriental in form, especially when bare. Informal.

Uses: Specimen tree in desert or wild gardens where it receives some irrigation. Bird gardens, parks, playgrounds or lawns.

Disadvantages: Reseeds profusely where there is some concentration of moisture. In periods of drought tree dies back, becomes sparse and unkempt.

Planting and care: Plant from containers any time or bare root in winter. Space 20 to 30 feet apart for a row. Prune only to remove dead branches or to shape.

Middle and high zones, low zone with extra water

Deciduous

Soil: Tolerant, but prefers loose sandy soil with some depth.
Sun: Full to reflected sun.
Water: Needs moderate irrigation or deep evenly spaced soakings in spring and summer, especially in lower zone or areas with less than 10 inches of annual rainfall.
Temperature: Tolerant of heat and cold.
Maintenance: Occasional in garden situations.

Celtis reticulata (spring)

Celtis reticulata (winter)

CERATONIA siliqua
Family: *Leguminosae*

Carob • St. John's bread • Algarroba • Locust bean

The carob is native to the eastern part of the Mediterranean. In the United States carob powder is sold as a substitute for chocolate. Usually seen as a very dark green, dense, wide-spreading tree, sometimes pyramidal in form. The carob has crinkled leathery leaflets and a heavy trunk. Individuals grow at a slow to moderate rate 30 to 40 feet tall and spread their crowns 25 to 30 feet. Trees can grow to 20 feet in 10 years. Males have attractive pinkish blossoms in spring with a pungent odor some find objectionable. Females have small red blossoms that produce dark flat pods about 12 inches long. Carobs grow in the same temperature ranges as citrus; they revel in heat and accept the hottest situations.

Special design features: Dense shade with interesting foliage. Desert oasis.

Uses: Street or park tree or large patios or courtyards. In large areas it can be grown in closely set rows as a hedge or screen. Not for narrow planting areas.

Disadvantages: Buttress roots will heave paving if planted too close. Subject to Texas root rot and wind breakage. Beans and flowers litter.

Planting and care: Grow from seed or plant young trees from containers in spring after frost danger is past. Space 35 to 40 feet apart for row planting, 10 to 15 feet apart for a hedge. Trees naturally branch low, so they accept severe pruning and staking when young to make a high crown. Protect cold-sensitive young trees for the first few years in winter. Trees frosted when young often develop into multiple-trunked trees.

Ceratonia siliqua

Best in low zone, marginal in middle zone

Evergreen

Soil: Prefers sandy soil with good drainage. Accepts some alkalinity.
Sun: Part shade to full or reflected sun.
Water: Drought tolerant, but grows fastest and looks best with deep widely spaced irrigation.
Temperature: Young trees are tender to cold below 22°F (−4°C). Mature trees may suffer only superficial foliage damage at that temperature and are known to survive freezes to 18°F (−8°C) and below.

CERCIDIUM floridum
Family: *Leguminosae*

Blue palo verde

(C. torreyanum, Parkinsonia torreyana). This tree is the first palo verde to bloom in spring, becoming a striking mass of golden flowers. Native to the southwestern United States and northern Mexico, it grows along washes and on sandy alluvial fans. A wide-spreading tree with a rounded crown and a naturally low branching habit, the blue palo verde grows fast up to 25 feet high and as wide. Bark and foliage are blue-green, but lower trunks of older trees become rough and gray. Tiny compound leaves with rounded leaflets shed in cold or drought. Midspring bloom is followed by numerous 2 to 3-inch seed pods.

Special design features: Spring color, desert or subtropical effect.

Uses: Street, patio or garden tree. Specimen, grouping, row or silhouette plant. Desert gardens or lawns.

Disadvantages: Litter of flowers, beans and seeds on pavements. Roots are sometimes attacked by larvae of the palo verde beetle. (See page 37.) Mistletoe and distorted clumps of growth called witches broom (caused by a mite) occasionally infest the crowns. Older trees are not as attractive as younger ones. Trees tend to be short-lived, from 20 to 40 years.

Planting and care: Plant seed that has been *scarified*, scratched, so water can enter the seed. Sow in deep containers, because plants are almost impossible to transplant from the open ground. Plant from containers any time. Space 20 feet or more apart for street tree or row planting, 15 feet for a grove. Young trees need careful staking and tying. Prune branches on young trees high if they are near walks, and to reveal their interesting structure, especially on multiple-trunked specimens. Cut out infestations of mistletoe immediately, preferably by cutting off the branch at some distance below infestation. Remove dead branches and witches broom to groom. Beans may be cut off before they fall.

Cercidium floridum

All zones, hardy to 4,000 feet

Deciduous

Soil: Tolerant of most soils, but needs good drainage. Prefers sandy soil. Accepts some alkalinity.
Sun: Full to reflected sun.
Water: Best with some irrigation for fastest growth and to look best, especially in dry summer areas. Accepts spaced irrigation or lawn watering. Tolerates periods of drought but may lose leaves.
Temperature: Hardy to about 12°F (−11°C). Accepts heat when adequate water is supplied.
Maintenance: Occasional, once trained.

CERCIDIUM microphyllum
Little-leaf palo verde • Foothill palo verde • Mesa palo verde
Family: *Leguminosae*

(Parkinsonia microphyllum). Usually smaller and tougher than the blue palo verde, the little-leaf palo verde has yellow-green bark and leaves, and blooms a little later in the season. Native to the rocky foothills and mesas to the 4,000 foot level in the Southwest and northwestern Mexico. It overlaps the range of the blue palo verde, and the two are often found growing in close proximity. Little-leaf palo verdes are irregular and craggy in form, with rounded crowns and fleshy-appearing trunks that divide near the ground. Numerous twigs support tiny rounded leaflets which drop in periods of cold or drought. If there is a long drought the tree

will self-prune and small branches die and drop off. It is slow growing to 10 or 12 feet, much faster and larger with irrigation. Yellow flowers cover the tree in midspring for a long period, followed by 3-inch long beans. Because of its slower growth, the little-leaf is not widely available in the nursery trade. If you have one it can be trained into a charming patio tree.

Special design features: Spring color. Desert effect.

Uses: Desert gardens and patios.

Disadvantages: Heavy litter of beans, leaves and flowers. Larvae of palo verde beetle (prionid beetle) may eat the roots and cause branches to die, sometimes killing the whole tree. Mistletoe infestations.

Planting and care: Plant seeds in containers any time. Scratch or sandpaper seed so that it will absorb water. Space 10 feet or more apart for massing. Prune as desired to shape. Remove mistletoe by cutting off the branch well below the point of infestation if possible, at a crotch.

All zones

Deciduous

Soil: Tolerant. Prefers loose gravelly soil with good drainage.

Sun: Full to reflected sun.

Water: Little to none in areas of 12 inches or more annual rainfall. New plantings require regular irrigation. Grows faster and looks better with some supplemental irrigation; requires irrigation in low rainfall areas.

Temperature: Hardy to cold, tolerant of heat.

Maintenance: Occasional grooming.

Cercidium microphyllum

CERCIS canadensis
Eastern redbud • Redbud
Family: *Leguminosae*

One of the most spectacular of the spring-flowering trees, eastern redbud grows well, has few maintenance problems and supplies summer shade. A moderate grower 20 to 35 feet with horizontal branches and a wide, spreading crown. In early spring before the heart-shaped, medium green leaves appear profusions of pealike, bluish pink blossoms cover branches, twigs, even the trunk. Flat reddish brown bean pods to 3-1/2 inches long may follow. Foliage turns golden in fall before dropping. Native to the central and eastern parts of the United States.

Cultivars: 'Rosea' has brilliant pink flowers. 'Forest Pansy' has purple foliage. 'Alba' has white flowers. A similar plant usually grown as a shrub to 10 feet is 'Mexicana', Mexican redbud. It produces a spectacular spring bloom.

Special design features: One of the best trees for spring and fall color, summer shade and winter sun. Woodsy effect.

Uses: Patio, street or lawn tree. Place trees on north or east sides in low zone where there is some protection from sun.

Disadvantages: Suffers sunburn in low zone or from reflected heat.

Planting and care: Plant bare root in winter or from containers any time. May be started from seed, but slow. Give regular garden care and corrective pruning.

Cercis canadensis

Middle and high zones, cooler locations in low zone

Deciduous

Soil: Improved garden soil with good drainage.

Sun: Part shade to full sun.

Water: Moderate, but will accept ample water, such as in lawns.

Temperature: Hardy to cold. Needs some protection from drying winds and beating sun of the low desert.

Maintenance: Little.

Cercis canadensis

C. chinensis, Chinese redbud, shares many features with the above. Flower color is the same and leaf shape is very similar. In form it is different, staying low and shrubby. Sometimes planted for a woodsy effect, providing the same early spring color as *C. canadensis.*

CHAMAEROPS humilis

Mediterranean fan palm • European fan palm

Family: *Palmae*

The Mediterranean fan palm is a well-behaved palm originating from southern Europe. It is always attractive and lends itself to a wide range of uses. A small-scale palm, it has a rounded head of fanlike leaves 4 to 5 feet or more across. It suckers and forms a low clump of several heads at the tips of several trunks which lean outward from the center. If numerous heads are undesirable, they may be removed to develop a single-trunked palm. This is excellent for a narrow planting space. Development is usually slow, about 6 inches a year. Fronds vary in color from deep green to grayish or yellowish green. Leaf stems are thorny and have lots of fiber at the base. Trunks 6 to 8 inches in diameter emerge as old fans are removed. In time, trunks grow to a height of 10 or 20 feet and lean outward so that a single plant may sprawl to 20 feet or so. Bloom is unobtrusive. Fruit below leaves around trunk looks like small, shiny black beads.

Special design features: Bold accent, tropical effect.

Uses: Containers and small planting areas. Linear barrier or screen. Transitional plants. Specimen or mass as an underplanting for taller palms.

Disadvantages: Young plants somewhat tender to cold. Occasionally attacked by heart rot fungus during humid periods in late summer.

Planting and care: May be planted from containers any time, or transplanted during warm season. Accepts neglect without complaint, but grows faster and larger with ample feedings and water.

All zones

Evergreen

Soil: Tolerant. Accepts alkaline conditions. Prefers rich moist soils.

Sun: Full, open or filtered shade. Part or full sun.

Water: Accepts periods of drought when established, but stays stunted. Prefers moderate water. Does nicely with deep, widely spaced irrigation. Tolerates ample water such as in lawn area if the drainage is good.

Temperature: Tolerant of heat and temperatures as low as 12° to 15°F (−10°C). Has survived brief spells at 6°F (−15°C).

Maintenance: Rarely needs attention unless you are trimming it into one or a limited number of trunks.

Chamaerops humilis

CHILOPSIS linearis

Desert willow • Flowering willow • Willowleaf catalpa

Family: *Bignoniaceae*

(Bignonia linearis). This shrub or tree is native to the southwestern quarter of the United States and south into Mexico, growing along dry washes to an elevation of 4,000 feet. Not a true willow but a plant with long weeping leaves, it can be found as a sprawling shrub, or as a tree reaching 30 feet. Form is usually open, with a twiggy structure, smooth gray bark and slender leaves, either straight or curved. Constant production of fragrant, trumpet-shaped flowers begins in April and continues until late summer. Flowers are shades of white, lavender and pink, followed by slender pods 4 to 12 inches long, which may hang on branches all winter. Plants grow rapidly with adequate water and soil, and are tolerant of heat, cold, drought and wind. Plants bloom the first year.

Cultivars: 'Barranco' is a seedling from Texas tested and released by the Plant Materials Center, Los Lunas, New Mexico. It is more upright in form with curved leaves, and ample flowers in rich shades of pink to lavender.

Special design features: Weeping form, summer bloom. Evokes waterside effect of desert oasis.

Uses: Specimen or grove in patio, lawn or garden. Desert or wild gardens, or transitional areas. Windbreaks, visual screens. In dense stands for erosion control. Streets, roadsides, median strips. Plant with cold-hardy evergreens to mask its winter dormancy.

Disadvantages: Suckers from roots. Some litter from seed pods. Not very attractive in winter.

Planting and care: Seeds planted shallowly in moist soil in May will emerge in a week if soil is kept moist. Plant from containers any time. May also be planted bare root in winter. Trees with ample water will grow 3 feet a year for the first few years before slowing. To form a tree, remove basal suckers and train a strong central leader. Slender leaders may need staking until they become strong. Prune to shape as desired in dormant season.

Chilopsis linearis

Chilopsis linearis

All zones

Deciduous

Soil: Adaptable. Prefers deep loose soils with good drainage.

Sun: Part shade to full or reflected sun.

Water: Moderate to ample in summer. Periodic soakings are satisfactory.

Temperature: Hardy to cold and tolerant of heat.

Maintenance: None in natural areas. Occasional in patios and gardens.

CHORISIA speciosa
Family: *Bombaceae*

This is the tree from which kapok (used to stuff pillows) is harvested. Erect in form, the floss-silk tree grows rapidly at first, then slowly to 30 feet, sometimes to 60 feet with a crown spreading 30 feet or more. The bright green, slender trunk sometimes thickens, turns gray with age, and may develop a pattern of gray thorny projections which disappear on some older individuals. Large bright green leaflets radiate from the center of the palmate leaves like a fan. Abundant lilylike flowers in fall may be rose to orchid to wine in color with brown-flecked white centers. Seed capsules to 8 inches hold the kapok. Plants grown from seed will vary, but excellent grafted selections are available in containers which will produce plants of predictable size, form and flower color. A striking tree originating from Brazil and Argentina now widely cultivated in the warmer regions of the world.

Special design features: Tropical feeling. Fall flowers.

Uses: Featured tree for courtyards or south sides of buildings where protected from cold. Not considered a lawn tree, but seems to accept lawn conditions in the desert if it has good drainage.

Disadvantages: Susceptible to cold—it may recover rapidly, but could be perma-nently distorted. Spines on trunk.

Planting and care: Plant from containers after danger of frost has passed and weather has warmed up. Give ample room and fast drainage for roots. The tree naturally prunes itself by dropping branches as it grows, so prune lower branches beforehand to prevent them from falling on someone. Protect trunks on young trees from sunburn. Set away from walks because of spines on trunk. A much hardier tree is *C. insignis.* It is white flowering instead of pink, common-ly called white floss silk tree. Flowers and foliage are similar to the above tree, but color is perhaps not as bright green. It has the same heavy thorny trunk. This is a tree that does well in both the middle and low zones, with cultural requirements the same as *C. speciosa.*

Chorisia speciosa

> **Low zone, middle zone where protected**
>
> **Evergreen to briefly deciduous**
>
> **Soil:** Prefers prepared garden soil. Fast drainage is important.
> **Sun:** Part shade to full sun.
> **Water:** Ample in summer. Reduce water in September to encourage bloom and harden tree for winter. Supply occasional deep irrigation during the cool of the year. With too little water it becomes drought deciduous.
> **Temperature:** Foliage may drop at 27°F (−2°C). Wood badly damaged at about 20°F (−7°C).
> **Maintenance:** General garden care.

CITRUS
Family: *Rutaceae*

One of the delights of living in a warm and sunny climate is growing citrus. Plants of the citrus family are mostly evergreen spiny shrubs, or small to medium-sized trees, useful for fruit and flower production and for landscape use. Most originated in south or southeast Asia and the Malay Peninsula, and have been planted and widely developed for centuries in relatively frost-free areas of the world. Few planted today are of original stock, being hybrids, seedlings or bud sports. Most are grafted on sturdy roots adapted to different soil conditions and have resistance to disease.

Special design features: Deep, glossy green foliage. Colorful fruit contrasting with the foliage. Spring fragrance. Lush tropical or subtropical effect.

Uses: Fruit-producing plants. Specimen, patio or street tree. Grove, row or screen. Clipped as hedges, standards or topiary plants. Large containers.

LIMITATIONS OF COLD, HEAT AND SOIL

Citrus may be planted in almost any climate as outdoor-indoor container plants as long as they are taken inside during severe winters, in the manner of the European orangeries. Cold is the prime controlling factor when planting citrus out-doors for landscape use or for fruit production. Heat also causes problems. When summer heat begins, plants are under great stress and "June drop" may occur, a situation which startles many gardeners when they find dozens of small, green, immature fruits on the ground. Leaf burn occurs in the desert during hot summers. Other limiting factors are excessively alkaline or rocky soil, poor drainage and salty water.

Hardiest citrus are the ornamentals: calamondin, sour orange and bouquet orange, in that order. They may be planted where the temperature does not fall below 20°F (−5°C). They have, however, survived several days of below freezing temperatures which at times dipped to 18°F (−8°C) and possibly lower. While mature trees of fruiting varieties have survived temperatures 20°F (−5°C) and lower, fruit may be injured below 29°F (−1°C). The hardiest fruit-bearing citrus is kumquat, then grapefruit, then tangerines (mandarins and tangeloes) followed by sweet oranges. Lemons are sensitive to 29° to 30°F (−1°C), and limes are most sensitive, being primarily tropical plants that enjoy not only higher temperatures, but higher humidity.

It is important to realize that the low temperature tolerance is not the absolute factor in determining hardiness to cold, but how *long* the temperature stays down and *how often* low temperatures occur.

COMMON CITRUS PROBLEMS

Cold damage: Leaves may freeze and fall off, but tree will recover. If the sap freezes branches may rupture, which usually kills them. If the trunk ruptures, the tree may die. Do not trim cold-damaged trees until new growth pinpoints dead wood.

Fruit drop: Trees have many more blossoms and set much more fruit than they can actually produce, so a certain amount of fruit must drop off. Unless excessive, fruit drop should not be alarming, as only 5 percent of the initial flowering on a healthy tree can develop into a normal crop. In late spring or early summer fruits are about the size of a pea and are lightly attached to branches so that they readily fall off in times of stress. When a mild spring turns to summer heat, the tree *is* put under stress. If hot dry winds coincide with insufficient watering, the tree is placed under a great strain. Stress at this time can be caused not only by heat and insufficient water, but lack of fertilizer at proper times in spring. Application of fertilizer after flowering and before the fruit is golf-ball size sometimes causes *more* stress and more fruit to drop. It is during critical fruit development that you should be sure watering is uniform and sufficient.

Salt burn: Appearance of brown spots 1/8 to 1/4 inch across on the leaves indicates need for soil leaching.

Sunburn: Trees with insufficient moisture may develop dry leaves as well as drop fruit in times of high summer heat. Also, leaves and fruit on south side of trees are most likely to get burned.

Gummosis: A fungus disease caused by exposing trunks of citrus trees to too much moisture. It is also called brown rot gummosis and foot rot. It is a brown, gummy discoloration of the inner layer of the tree, the *cambium*. If it goes completely around the trunk it may kill the plant. Prevent by never letting trunk stand in water. Build a dike around the trunk to keep irrigation water from wetting it. If the tree is infected, remove discolored bark and wood to a point just beyond the infection, and treat with a Bordeaux paste mixture.

Pests: Aphids may appear at growing tips in early spring months. Tiny citrus thrips cause crinkled leathery leaves and scarred fruit. They can be damaging to young trees. Orange dog is a curious but not dangerous pest which looks like a bird dropping on a leaf. It is the worm stage of the swallowtail butterfly. Red or brown scale should be watched for and treated. Citrus red mite is recognizable by the red eggs it lays along side the top midrib of a leaf. Eggs look as if they are strapped to the leaf by tiny cables. Or, shake branches over a white sheet of paper. Mites are nearly microscopic, but will appear as tiny specks.

If you recognize any of these problems or experience any others, immediately

Fruit damage from sunburn.

consult a reputable nurseryman, your county agent or the nearest extension service for the latest method of treatment. See also the section on Plant Problems, page 36.

PLANTING CITRUS
Time to plant: March or April when frost danger is over.

Place: If you are in a borderline area, locate plants in a warm microclimate. See page 9.

How to plant: Prepare the hole *before* you buy the tree. Holes should be at least twice the size of the root ball and should have good drainage. Check drainage by filling hole with 4 inches of water. If water is gone in 4 hours the drainage is acceptable. If drainage is poor, dig a "chimney" through impervious soil layers to gravel. Fill hole part way with sandy loam and organic material combined with soil from hole. Leave enough depth so you can put the tree in the center and fill around it.

Selecting a tree: Trees 1 or 2 years old are best for easy handling. They should be about 3 to 4 feet tall with a trunk 5/8 inch in diameter. Be sure the crown does not appear too large for the root ball, which should be no less than 18 inches across and about 24 inches deep. The root ball may be covered with burlap, tar paper or planted in a plastic or metal container. Trees in plastic containers are reported to recover from transplant shock more quickly than others. Larger trees may be planted, but require special handling not covered here.

Putting the plant in the hole: The method depends on how roots are contained. Be sure hole has been soaked so the soil has settled and will not sink once tree is planted. Plant should be set in the

Gummosis damage on citrus trunk.

center of the hole, making sure the bud union is buried no deeper than when it was in the container. Plants in plastic containers should be removed from the container before being set in the hole. Wet roots thoroughly and remove carefully so as not to shatter root ball. Plants with roots in burlap may be set in the hole and burlap loosened from the trunk and spread out. It will rot in the hole and does not have to be removed. Plants with roots wrapped in tar paper may also be set in the hole and the tar paper can then be cut off. Fill soil in around roots, pressing it down to be sure there are no air pockets. Form a basin with a 3 or 4-inch high mound or berm at the edge to hold irrigation water. The basin should be as wide as the hole and should be enlarged as tree grows. Water newly planted trees immediately.

CARE OF YOUNG CITRUS TREES
Irrigation: Frequency depends on soil type, temperature and humidity. For a sandy loam irrigate about once every week or ten days in summer and every two to three weeks in winter. In hot desert areas, twice a week or more may be necessary in summer. The important thing to remember is, *never* let new trees dry out. Fill the basin slowly. Do not spray leaves because they will absorb salts in the water and may burn. Slightly wilted leaves indicate immediate need for irrigation. Heavy clay soils need less frequent irrigation; sandy soils need more. In very hot weather (110°F, 42°C) newly planted trees may need water every day. Taper off frequency of irrigation in fall to harden tree for winter.

Sun protection: Protect bare trunks from the sun by white-washing with a water-based paint or by wrapping with burlap for the first year or more until branches shade the trunk. See photo, page 35.

Pruning: Citrus require little pruning, but you should remove suckers growing below the bud union. Pinch the rapidly growing leader to encourage bushiness and uniform growth. Wide arching branches sometimes need cutting if they grow too long.

Frost protection: In colder areas, wrap trunk with burlap to keep from freezing. Put temporary framework over tree for the first three or four winters while tree is at its most tender. Cover with a sheet or light blanket during the night, remove covering during the day. Avoid using plastic, and don't allow covering to touch tree. Trees not uncovered during the day may prematurely start new growth that will be extra tender. For mildly cold areas, hang flood lights, trouble lights or Christmas tree lights in the tree for added warmth. Petroleum coke blocks are also available.

Feeding: Except in very poor soils, do not feed the first year. To be safe, it's much better to underfeed than overfeed, which can burn or kill the plant. The second year sprinkle a good citrus fertilizer on the soil to the drip line and irrigate immediately. Follow the package directions for amounts and application methods. Do this four times during the growing season starting in late winter or early spring. Thereafter feed the tree three times a year with a balanced citrus fertilizer available at nurseries. Feed the end of February or the first of March, in May, and around the first of September.

CARE OF MATURE CITRUS TREES
A citrus tree is considered mature at 12 years of age. If you move into a house with trees in the landscape and they are not small or dwarf varieties, consider them mature and treat them as such.

Irrigation: Soak heavily the first part of February, then as needed. It is best to supply a constant supply of moisture, especially during fruit production. Establish a schedule short of allowing leaf wilt, as this creates stress. For sandy loam soils, every 3 to 4 weeks may be enough in winter, water about every 10 days in summer. High temperatures and low humidity may call for more. The basin should be filled slowly, but water should not be allowed to stand nor keep the trunk wet. Continue to protect trunk from moisture with an inner soil dike near its base.

Feeding: Use a balanced, commercial fertilizer made especially for citrus and follow the directions on the package label. Feed three times a year: at the end of February or first of March; May when the fruit is golf-ball size, and around the first of September.

Pruning: Prune only as needed to remove dead wood, inside shoots or suckers below graft or bud union. Skirt of lower branches may be removed if you prefer, but they shade trunk and produce much of the fruit. If trunk is exposed to the sun, cover with water-based paint or wrap with burlap to keep it from being burned.

GENERAL CARE
Tree basins should be enlarged as the tree grows. The berm of earth should be just beyond the drip line—the outer reach of branches—and should be 2 to 4 inches high. Maintain an inner dike a short distance from the trunk to prevent wood from becoming damp, which can cause gummosis fungus. Keep grass and weeds pulled and prevent competition for water by not planting in the basin around tree. Mulch the basin in summer with ground bark, straw or other material to preserve moisture, cool soil and discourage weeds. Remove mulch in winter to allow sun to warm soil.

In areas with highly alkaline-saline soils or water, it is necessary to leach soil at least once a year to remove damaging salts. In January give a long, slow soak of 8 to 12 inches of water. This will wet soil to 6 feet or more and carry salt accumulations away from roots. Some growers recommend leaching in February. Feed after

soaking so the soil is thoroughly damp and the roots are completely hydrated. Then water again to wash some of the dissolved fertilizer down. Some growers recommend a second deep watering in March. Afterwards you should begin a summer watering schedule as the weather warms up.

CITRUS IN CONTAINERS
Growing citrus in containers greatly extends the range in which they can be grown. And, where soil problems are severe, containers provide a near-perfect growing medium. Citrus in containers are also attractive additions to patios, decks or entryways—anywhere you would normally locate a container plant. Due to the restricted area for root growth plants are smaller than those grown in the ground, but they still produce tasty fruit.

By far the most valuable aspect of containers is their mobility. In the upper zone where winter cold prevents citrus from being grown outdoors, plants in containers can be moved indoors when temperatures drop. Many gardeners in all parts of the United States are discovering that citrus can be a very attractive indoor plant. Choose a bright location for your indoor citrus, and keep an eye out for pests which may like the tender foliage. If you follow this indoors-outdoors routine, make the transition from one place to another gradually. For example, don't move a tree that has been indoors all winter directly into bright spring sun. Sudden changes in temperatures and sunlight can defoliate a tree.

Container culture: Growing citrus in containers calls for special cultural requirements. Plants generally require more frequent watering and fertilization. Although most native soils contain ample amounts of the necessary micronutrients, container or potting soil mixes sometimes do not. If regular applications of nitrogen fertilizers do not improve a chlorotic condition (yellowing of plant leaves), it is probably due to a lack of iron, maganese or zinc. Each of these elements can be purchased in chelated or sulfated forms at nurseries. Some citrus fertilizers already contain them.

SELECTING A TREE
Grafted trees are named varieties of predictable fruit type and quality. While disagreement exists on the best root stock for home garden plants, some experts feel plants for home use should be on sour orange root stock for grapefruit, orange, tangerine (mandarin) or tangelo. This root stock is best for heavier soils, most resistant to cold, and the sturdiest. It also keeps the tree at a manageable size. Commercial orchards often use macrophylla stock. Trees on this stock grow rapidly and become very large. They are also not as cold resistant. Lemons should be grafted to rough lemon stock.

To avoid disappointment showing up years later, it is important to take care in the selection of a citrus tree, especially one for fruit production. There are certain precautions you can take to be sure your tree is free of disease and is on a suitable root stock. And, you also want to be sure you are choosing the correct kind or variety of citrus. Make your purchase from a reliable nursery. Also, ask your nurseryman about tagged trees. Tagging is a new program initiated by the Crop Improvement Association in some states. Trees certified by the association have a blue tag, stating certain information and includes a plant registration number. It also warrants the tree to be free from disease. The tag is put on at the time of the grafting and is non-removable. Sometimes, a tree's identity is unknown if its paper label falls off. Be certain the tree you select has the original label, or your orange may taste suspiciously like a lemon.

A SELECTION OF CITRUS
Because of the botanical confusion resulting from citrus hybridizing, we have listed common names first, botanical names second. Plants mentioned are by no means the entire citrus selection, but are reliable performers in the climates covered in this book. They are generally listed in order of hardiness from the toughest to the most tender.

Dwarf 'Washington' navel in container.

Calamondin, calamondin lime, sour acid mandarin
Citrofortunella mitis _____

Usually seen as a dwarf with a columnar or rounded form. Growing at a moderate rate 8 to 10 feet tall and 6 to 8 feet wide, this plant produces decorative 1-inch orange fruit by the hundreds. Fruit is usually sour, sweeter ones are merely tart. Fruit sets over a period from March through December, and remains for a long period on the tree.

Uses: Outstanding ornamental value as a standard tree, street tree or near patios. Excellent as large container plants or as clean plants around swimming pools. Clip or use unclipped as a hedge or screen spaced 4 or 5 feet apart. Good in limited spaces and for close-up viewing.

Kumquat
Fortunella margarita _____

A close relative of citrus with oval or round orange-colored fruit to 1-1/2 inches long ripening November through March. Fruits have sweet skin and tart flesh, which makes a flavorful marmalade. Individuals grow at a moderate to slow rate from 6 to 25 feet. 'Nagami' is probably the most common; 'Meiwa' is best for eating.

Uses: Grafted dwarfs make excellent, colorful container plants. Seedling-grown types may be used as clipped standards, espaliers, shrubs or small trees where there is limited space. They are excellent around pools. Attractive for close-up viewing. Accepts pruning and shaping.

Sour orange, seville orange, bitter orange, bigarade
Citrus aurantium _____

Grown in Spain for the rough and bitter fruit shipped to England for marmalade, sour orange was brought by the Spaniards to the New World. A seedling tree in the original botanical form, it is a durable, decorative, vigorous grower to 15 to 20 feet, rarely 30 feet and spreading 10 to 15 feet. White waxy, spring flowers produce an unsurpassed fragrance for several weeks in spring. Spiny crowns are densely foliated with dark green leaves 4 inches in length, which are fragrant when crushed. Colorful orange fruits ripen in fall, remaining on the plant as a cheery accent in winter.

Uses: Outstanding ornamental tree for the garden, poolside, patio or street. Clips well into formal shapes. Well mannered for smaller spaces. The fruit is very acid, but can be used for pies, drinks, garnishes and marmalade.

'Bouquet' orange, 'Bouquet des Fleurs', Bergamot orange
Citrus aurantium 'Bergamia' _____

This is a moderate grower reaching 8 to 10 feet tall and spreading about 8 feet wide. Distinctive curling leaves cover the tree to the ground. Showy fragrant flowers in clusters are followed by yellowish fruit suitable for marmalade, but are not outstandingly attractive. This plant is very hardy and tolerates much heat.

Uses: Distinctive ornamental shrub. Space 5 to 6 feet apart as a large hedge or windbreak. Good plant for containers.

FRUIT-PRODUCING TREE FORMS
Citrus as fruit trees is considered standard for the low zone. In the high zone it should be grown only when it can be brought indoors or given ample shelter in winter. If fruit is your goal in the middle zone, place it in a warm spot away from the winter frosts, such as on south or southwest sides of buildings where it will receive winter sun and shelter from winds.

The following plants are listed in a general order of hardiness, from most to least. Mature sizes given for grapefruit,

Nagami kumquat

tangerines, tangeloes and oranges are for grafts grown on sour orange root stock.

Grapefruit
Citrus paradisi _____

Grapefruit has become very popular in the home garden; it is one of the easiest citrus trees to grow. And grapefruit grown in the desert probably produces the finest fruit found anywhere in the world. Fruit is basically of two kinds—pink or white (yellow), and ripens November to June. Pink selections are 'Redblush' and 'Ruby Red'. 'Marsh', the white selection, is highly acid, and nearly seedless. Best in May or June when it contains more sugar. Both kinds reach full flavor with a long period of high summer heat and keep well on the tree. Trees reach about 12 to 20 feet with a spread of 20 to 24 feet. All are vigorous growers that bear consistently.

Tangerine, mandarin orange
Citrus reticulata _____

There are many varieties of this plant. Most popular and widely planted is 'Clementine', the Algerian tangerine. Other fine varieties include 'Fairchild', 'Kara' and 'Kinnow'. Trees reach 12 to 15 feet

Calamondin

Sour orange

'Bouquet des Fleurs'

spreading 16 to 20 feet. Fruit ripens November through January, is deep orange, seedy and easy to peel. Yields are irregular, usually with a heavy set about every other year. Tangerines benefit from cross-pollination with other tangerines or tangeloes. They are generally best eaten as soon as they are ripe.

Tangelo

The tangelo is a cross between a tangerine and a grapefruit. Most popular is 'Minneola' which produces large, flavorful, red-orange fruit that is easy to peel. Fruit ripens in February and March, and stores well on the tree for two months. Fruit is more sensitive to cold than the tree and may be lost in cold winter areas. 'Orlando' is less vigorous, but is slightly more cold tolerant. Yellow-orange fruit ripen from November to January. Trees often grow rapidly with distinctive cupped leaves and are regular heavy producers of fruit. If you have room, plant one of each of the above for cross pollination. Trees are about the same size and have the same growth rate as the tangerine.

Sweet oranges
Citrus sinensis _____

Oranges found in grocery stores and grown in groves for juice and eating originated in China and South Vietnam. They are classified in four groups based on fruit characteristics (naval and blood orange) and on geographical ancestry (Spanish oranges or Mediterranean oranges). They are widely cultivated in subtropical and tropical areas of the world and are considered to be among the most prized of the world's fruits.

"Arizona Sweets" is the name given to a group of Valencia-type oranges that does well in the desert. They are probably easiest to grow and most dependable for the home gardener. 'Diller' is said to be

Satsuma mandarin

'Minneola' tangelo

hardiest of all, producing small seedy fruit excellent for juice. Fruit ripens in November and December before heavy frost can damage it. Tree size is 15 to 20 feet by 20 to 24 feet. 'Trovita', a California variety, is vigorous, tolerant of cold and heat and a dependable producer of thin-skinned fruit which ripens in early spring. In warm areas, grow 'Valencia'. It ripens March through May and produces medium-sized fruit good for juice and eating. Trees grow to 20 or 25 feet with a 20 to 24-foot spread.

'Washington' navel does very well as a crop tree in the warm interior valleys of California, but is often a disappointing producer in the desert home garden. It may flower abundantly, but sheds most of the flowers and produces a few small fruit which split readily. It does best in medium to heavy soils, very poorly in sandy soil. Fruit ripens November through February. Allow 16 to 20 feet for spread.

'Robertson' navel is similar to 'Washington' but smaller, thus it takes up less space. It has similar needs as 'Washington' and tends to fruit in clusters. Fruit ripens 2 to 3 weeks earlier than 'Washington' with a fairly ample production.

One dwarf citrus variety that deserves mention is the 'Shamouti', grown on dwarf root stock for home garden use. It was developed in Palestine, and is reported to produce large crops of big seedless oranges. It grows wider than high with large leaves.

The blood orange is also worth planting. The best varieties for the desert are 'Sanguinella' and 'Moro'. These are vigorous, open trees with fruit which ripens in late spring. The pulp of ripe blood oranges is suffused with red or pink. To have fruit of tasty quality, pick early while still slightly tart.

Grapefruit

'Washington' navel orange

Lemon
Citrus lemon

Lemon is fastest growing of the citrus and quite tender to frost. Grafted trees are best on rough lemon rootstock. 'Meyer' is one of the hardier lemons with round, thick-skinned, orange-tinted fruit, but is currently banned in some states—Arizona for one—because it hosts citrus quick decline, a serious virus disease. Disease-free plants are being developed and may again be available. 'Lisbon' is a dense, thorny tree of vigorous growth that likes the high heat of the desert summer. It grows 20 to 25 feet high and 22 to 26 feet wide. Fruit ripens in fall but may be picked early when it becomes juicy so as to avoid frost damage. 'Eureka' is the standard fruit of the market. It is similar to 'Lisbon' but smaller, more open, and less vigorous. Fruit is borne throughout the year. 'Ponderosa' is actually a lemon-citron hybrid, and produces giant bumpy fruits with thick skin and a mild flavor. It grows fast, bears early, and is angular and open with large leaves. Trees grow rapidly to 10 feet.

'Lisbon' lemon

Lime
Citrus aurantifolia

Lime is most tender of the citrus and is usually seen as a thorny bush 12 to 15 feet in height. There are basically two kinds of limes. The 'Key' lime, also known as 'Mexican' and 'West Indian', and the Persian lime. 'Bearss', a seedling of the Persian lime, is said to be hardier. It grows well in the low zone and in protected areas in the middle zone. The yellow fruits are seedless, very acid and aromatic. It can grow in the same climate as lemons. Form is open in youth, dense at maturity.

'Bearss' lime

CLYTOSTOMA callistegioides
Family: Bignoniaceae

Violet trumpet vine ● Love-charm

(*Bignonia callistegioides, B. speciosa, B. violacea*). The violet trumpet vine is a woody, tendril-climbing vine from southern Brazil and Argentina. It looks good close-up and is especially pleasing in spring when it produces profusions of 3-inch lavender blossoms. Lustrous green leaves densely cover the plant. It needs training and support to climb, and it grows fast to cover an area approximately 12 by 20 feet.

Special design features: Exuberant woodland or tropical effect, with a lush spring bloom.

Uses: Northern and eastern exposures, or southern exposures where it has air circulation, such as on a porch post under an overhang. Train to drape over cool walls, trellises, fences and posts.

Disadvantages: Iron chlorosis and winter chlorosis—a condition in early spring when the plant foliage is ready to grow but the ground has not warmed enough for the roots to function fully. Unattractive in winter if the temperature goes below freezing; leaves may fall or become pale with dry spots.

Planting and care: Plant in spring after frosts. Feed established plants in early spring and again in May and July. Give extra iron if plants yellow or pale leaves appear.

Low and middle zones
Evergreen to partly deciduous
Soil: Improved loamy soil with good drainage.
Sun: Open shade, filtered or part sun. Avoid reflected sun.
Water: Moderate is preferred. Tolerates ample or spaced periodic irrigation.
Temperature: Hardy to 20°F (−7°C). Becomes all or partly deciduous below freezing.
Maintenance: Occasional.

Clytostoma callistegioides

COCCULUS laurifolius

Cocculus • Laurel-leaf cocculus • Laurel-leaf snailseed

Family: *Menispermaceae*

Cocculus is an exuberant, dark green shrub to small tree from southern Japan and the Himalayas. Bold laurel-like leaves densely cover the plant. It can be kept at 6 feet or will grow at a slow to moderate rate to become a small tree 15 feet high, occasionally to 25 feet. The crown makes a dense canopy of leaves 20 feet or more wide. Trunks bend and weave and usually lean to one side. Inconspicuous greenish flowers in spring produce a scattering of 1/4-inch black berries in summer. Plants drop leaves in spring just as new leaves appear.

Special design features: Luxuriant deep green creates a cool oasis or woodsy effect.

Low and middle zones
Evergreen
Soil: Prepared garden soil on the acid side. Needs good drainage. Avoid alkaline situations.
Sun: Open or filtered shade. Part to full sun.
Water: Moderate to ample, especially in summer.
Temperature: Hardy to 20°F (−7°C), lower if sheltered. May lose leaves below 20°F (−7°C) and can be badly injured at 15°F (−10°C). Recovers quickly in spring.
Maintenance: Little.

Uses: General purpose shrub or background plant. Wide unclipped screen or clipped hedge to 6 feet or higher. Espalier on cool walls. Trained as a patio tree, it creates a wide umbrella of dense shade, producing an effective screen.

Disadvantages: Iron chlorosis and leaf-tip burn in alkaline soils. Slow to develop good form.

Cocculus laurifolius

Planting and care: Plant from containers any time. Space 4 to 6 feet for a screen or hedge. To form a tree, stake young plants and gradually remove lower branches until the crown is as high as desired, then allow to spread. Once or twice a year give it a deep soaking to leach soil salts. Feed with iron sulfate if leaves turn yellow. See page 35 for details.

Cocculus laurifolius, Pittosporum tobira 'Variegata' (foreground)

CONVOLVULUS mauritanicus

Ground morning glory • Morocco glorybind

Family: *Convolvulaceae*

(C. sabitius). A low plant with a continual show of blue lavender morning glory flowers, the ground morning glory from northern Africa makes an undulating and somewhat irregular gray-green cover. Plants grow at a moderate rate up to 12 inches with a spread of 2 feet. Bloom is heaviest in spring and continues until weather turns cold. In mild winters there is sometimes an additional sprinkling of flowers.

Special design features: Loose informal plant.

Uses: Ground or bank cover in areas of no traffic. Combine with native plants or rock gardens. Does well in hot or windy places. Under plant beneath taller subjects. A good container plant.

Disadvantages: May die out in spots in cold wet soils. Subject to chlorosis. May have bare spots. Needs to be replaced every few years. Bermudagrass invasions.

Planting and care: Plant in cool of the year—fall, winter or spring. Place plants from flats 12 inches apart for cover. Gallon plants may be planted 18 inches apart for a fast cover. Add new plants where needed to fill after growth is established and to replant bare areas each fall. Pull or poison grass if it invades.

All zones
Evergreen perennial
Soil: Tolerates poor soil, but prefers improved light or loamy soils with good drainage.
Sun: Part, full or reflected sun.
Water: Moderate to little.
Temperature: May be damaged at 15°F (−10°C). Speedy recovery.
Maintenance: Little except invasions of grass or other plants.

Convolvulus cneorum, silver bush morning glory, has many characteristics in common with ground morning glory. Its white flowers and leaves are much larger, however, and its growth habit is mounding rather than flat. It seems quite tolerant of cold and it performs well during the cool of the year. It does have difficulties surviving intense summer heat especially if nights are hot. Culture is much the same as for *C. mauritanicus.*

Convolvulus mauritanicus

CORTADERIA selloana
Family: *Graminae*

Pampas grass • Selloa pampas grass

(C. argentea). A giant grass of the South American pampas, this dramatic landscape plant comes from Argentina, Brazil and Chile. It quickly forms a fountain of cascading leaves up to 10 feet high, spreading 7 or more feet wide. Ribbonlike leaves are 1/2 inch wide, bright green, with sharp cutting edges. A variegated effect is produced when older leaves dry to straw color. In late summer, several erect spikes 3 or 4 feet high rise in the center of foliage and sport fluffy plumes varying from white to pink to lavender, depending on the selection. Plumes last for months, and make striking interior decorations.

Special design feature: Large-scale grassy mound creates a tropical or waterside mood. Rustling sound.

Uses: Specimen or row for dramatic effect, or screen or windbreak. Dry area banks. Large areas including lawns. Waterside situations, median strips and parks. Not for the small garden.

Disadvantages: Sometimes messy and difficult to eradicate once established. Leaf edges are deceptively sharp. Older or neglected plants may not respond to rejuvenation attempts, have rank and weedy look.

Planting and care: Plant from containers in spring. Space 5 or 6 feet apart for row planting. Regular garden care produces a large and attractive plant in one season. Groom by removing old plumes in fall when they begin to lean or look unkempt. Always wear gloves when handling plants, and protect your eyes from the sharp leaves. Plants severely cut back often have a difficult time recovering to an attractive stage. If plants are battered by weather trim them lightly, feed in spring and water regularly until new growth appears.

Cortaderia selloana

> **All zones**
>
> **Evergreen perennial**
>
> **Soil:** Tolerant of a wide variety of soils from alkaline to acid.
> **Sun:** Part shade to full or reflected sun.
> **Water:** Established plants survive drought, especially if they get extra runoff. Looks best with moderate water.
> **Temperature:** Thrives in heat. May be battered by wind and cold so that leaf edges tatter and become straw colored. Fast recovery in spring, especially if given regular garden care.
> **Maintenance:** Needs some grooming.

COTONEASTER species
Family: *Rosaceae*

Cotoneaster

This group of shrubs from Europe and Asia includes plants that reach up to 20 feet in height as well as low-growing ground-covering plants. They tolerate minimum care and fairly dry conditions. Most have angled to arching thornless branches, gray-green leaves, white flowers and red fall berries. Cotoneasters are members of the rose family, closely related to the Pyracantha species. They even suffer from the same diseases, and are most susceptible to Texas root rot. A few kinds perform well in the warmer zones; most do best in cooler areas.

Cotoneaster glaucophyllus
Bright-bead cotoneaster

(C. glaucophylla). Bright-bead cotoneaster is a shrub reaching 6 feet high and as wide, but easily kept at any height. It is often seen at heights of 2 or 3 feet. Stiff angled to arching branches covered with gray-green leaves up to 2 inches in length are closely set on the branches. Small, pinkish white flowers in dense clusters appear in spring and 1/4-inch orange-red berries follow in fall for a long period. The best cotoneaster for the desert.

Special design features: Angular gray-green form.

Uses: Hedging or borders, foundation plant and space definer.

Disadvantages: Subject to fireblight, Texas root rot and iron chlorosis.

Planting and care: Plant from containers any time, but best in spring. Add iron occasionally to prevent iron chlorosis. Prune as desired in late winter.

> **All zones**
>
> **Evergreen to partly deciduous**
>
> **Soil:** Prefers loose, well-drained soil with some humus added.
> **Sun:** Part shade to full or reflected sun.
> **Water:** Deep, widely spaced irrigation to moderate watering.
> **Temperature:** Hardy to cold, but may become all or partly deciduous in cold winters. Tolerant of desert heat.
> **Maintenance:** Little to none.

Cotoneaster glaucophyllus

Cotoneaster lacteus
Red clusterberry _____

(C. lactea, C. parneyi). The red clusterberry is an informal arching shrub with the appearance of a large-leafed, but softer and thornless pyracantha. If unclipped, it grows rapidly 6 to 7 feet high and about as wide, sometimes wider. May be clipped and kept at any size. Loose open form bears leathery, 3-inch, round-tipped leaves, deeply veined and dull gray-green above, whitish and hairy beneath. Foliage reaches to the ground. In spring, pink-tinged buds open in clusters of small white flowers. Dull red berries follow in fall, hang on a long time. Native to western China.

Special design features: Informal. Winter color.

Uses: Foundation plant, espalier on cool walls. Border shrub or transitional plant. Handsome as thornless clipped hedge or in containers. Does well in areas somewhat neglected, or in hot sun, wind, poor soil and some drought.

Disadvantages: Subject to fireblight, Texas root rot, iron chlorosis and red spider mite.

Planting and care: Plant from containers in any season. Space 5 feet apart as an informal screen or mounding planting. Space 3 to 4 feet for a clipped hedge.

All zones

Evergreen

Soil: Accepts wide range of soils. Prefers well drained garden soil.
Sun: Part shade to full or reflected sun.
Water: Moderate to occasional deep irrigation.
Temperature: Hardy. Accepts heat.
Maintenance: Some to little.

Cotoneaster pannosus
Silverleaf cotoneaster _____

A large, rangy, informal shrub, silverleaf cotoneaster grows at a fast to moderate rate, forming a fountain of arching branches 6 to 10 feet high and as wide. Dense clusters of small, pinkish white blossoms in spring produce soft red berries in fall. Leaves to 1-1/4 inches in length are dull gray-green above, silvery and felty beneath. This native of China is widely tolerant of poor soil, heat, wind and cold.

Special design features: Rangy to graceful arching form.

Uses: Filler plant for large semi-neglected areas. Use as a wide screen, windbreak or foundation plant for large buildings. Transitional gardens. Can be clipped as a hedge, but loses its distinctive shape and character.

Disadvantages: Occasionally subject to fireblight, Texas root rot, iron chlorosis and red spider mite.

Planting and care: Plant from containers any time. Space 4 to 6 feet apart for a loose screen, windbreak or clipped hedge. Supply with iron if foliage becomes pale. This plant generally requires only as much care as you want to give.

All zones

Evergreen to partly deciduous

Soil: Prefers loose, well-drained soil with some humus added.
Sun: Part shade to full sun.
Water: Deep, widely spaced irrigation to moderate irrigation.
Temperature: Tolerant of heat and cold. Loses leaves in very cold areas.
Maintenance: Little to none.

Cotoneaster salicifolius
Willowleaf cotoneaster _____

(C. salicifolia) A vigorous grower to 15 feet, with upright arching branches spreading as wide or wider. Gray-green wrinkled leaves are narrow, to 3 inches long. Small white flowers in 2-inch clusters produce bright red berries 1/4 inch in diameter in fall. This plant is useful to fill large spaces as a wide, loose screen or space divider. Good in low maintenance areas or transitional gardens. Supply with some shade in the low and middle zones. See bottom of page 80 for cultural details.

Cotoneaster glaucophyllus

Cotoneaster lacteus

Cotoneaster lacteus

Cotoneaster pannosus

CUPRESSUS arizonica

Family: *Cupressaceae*

Arizona cypress • Rough-barked Arizona cypress

Arizona cypress is native to the Southwest and northern Mexico. Individual plants range from tall and pyramidal, to flat and broad, or globe-shaped. Colors may be silvery to dark or gray-green. Foliage reaches to the ground or can be trimmed several feet above ground to reveal a well-developed trunk. Trunks are covered with dark fissured or checkered bark. These scaly foliaged conifers grow at a moderate to fast rate, reaching 20 to 50 feet in height (depending on the cultivar) and spreading 15 to 25 feet in diameter. Inconspicuous flowers and small cones are unimportant. For a tree that will be consistent in form, size and color, seek a cutting-grown plant.

Cultivars: 'Gareei' has silvery blue-green foliage, is pyramidal in form and of medium height.

Special design features: Dense bold form. Some are strong verticals.

Uses: Large areas where a tall, dense tree is desired. Roadsides, windbreaks, visual barriers or background plantings.

Disadvantages: Overwatered trees may outgrow roots and blow over. Drought-stressed trees may be attacked by bark beetles that will hollow out twigs, causing twig blight. In an extended drought, they will kill the tree. Subject to infestations of red spider mite. Sometimes short-lived,

lasting only 30 years or so, possibly because older trees are more subject to bark beetles and other stresses. Subject to injury from nitrogen fertilizer.

Planting and care: Plant any time from containers. Space 15 to 30 feet for rows, depending on desired density. Give some irrigation until established, then minimum. Mulch with manure.

Cupressus glabra formerly was considered a variation of *C. arizonica*. It is similar in foliage and form but sheds its bark annually to reveal a thin, smooth, cherry-red inner bark that is most attractive on older trees. It is thought that some named cultivars of *C. arizonica* might really be *C. glabra*.

All zones

Evergreen

Soil: Tolerant of many soils. Prefers light, dry soils.
Sun: Full to reflected sun.
Water: Little to none when established with extra runoff, or where it receives rainfall above 15 inches annually. Best with deep, spaced irrigation in areas of low precipitation to encourage faster growth and good health.
Temperature: Tolerant of heat and cold.
Maintenance: None.

Cupressus arizonica

CUPRESSUS sempervirens

Family: *Cupressaceae*

Italian cypress

A stately tree of history and legend, Italian cypress does as well in the arid country of the Southwest as in its native lands of southern Europe to western Asia. The very dense, erect and columnar form is clothed nearly to the ground with scaly, dark, gray-green foliage, except for some cultivars possessing a blue or golden cast. The natural form is seldom found in the nursery trade, but is occasionally grown as a seedling tree. It is not columnar and compact, but open with rangy lateral branches. Most trees grow at a moderate to slow rate to 60 or even 80 feet, with a diameter at the widest part of no more than 10 feet, often much less. Inconspicuous flowers and tiny cones are unimportant design features. A durable tree, but too large for the average home residence.

Cultivars: The most commonly sought cultivar is 'Stricta' ('Pyramidalis') — extremely narrow and dense. Medium-sized 'Glauca', blue Italian cypress, is narrow, dense and columnar with blue-green foliage. Other cultivars have been popular at various times as fashions and architecture have changed.

Special design features: Strongest vertical in the plant kingdom. Heavy dark form. Formal contained effect, and strong accent. Dramatic silhouette. Stately, sober, subdued to elegant, or rich, de-

pending on the setting and size of plants.

Uses: Rows as emphasis plantings for roads, drives or boundaries. Tall, narrow, vertical windbreaks usually planted at distances will slow the wind effectively, but do not create a solid wall. A background plant as a spaced vertical giving cadence to a formal planting.

Disadvantages: Subject to red spider mite infestations.

Planting and care: Plant any time from containers. For architectural cadence and rhythm, space 3 feet apart. Place 6 inches on center to create a solid hedge-wall. To prevent tall pointy trees and to encourage trees as columns, some people top them at

an early age by removing the leader which contains their growth. This is a questionable practice as the plant never gives up trying to be a vertical spire. Tie errant branches to the column. To feed trees in poor soil, mulch basin with manure. Applications of nitrogen fertilizer near the roots can cause damage.

All zones

Evergreen

Soil: Widely tolerant. Prefers light dry soils.
Sun: Full to reflected sun.
Water: Regular irrigation until established, or continue to encourage rapid growth. Established trees may need deep, widely spaced irrigation.
Temperature: Hardy to approximately 10°F (−12°C). Tolerant of heat.
Maintenance: None necessary except for treatment of red spider mite. Treat with systemic pesticide.

Cupressus sempervirens

CYCAS revoluta

Family: *Cycadaceae*

Actually not a palm at all, the sago palm is a cycad that grows slowly to 5 to 10 feet. Sometimes plants, especially young ones, look like ferns. Finely cut, refined, glossy, palmlike leaves of dark green whorl in a spiral around a central cone at the trunk center. Plants may develop one or more heads, each 3 to 6 feet across. As plants grow, they develop a trunk 8 to 10 inches in diameter with tiny leaf stumps set closely together like a miniature palm. Offsets at the base may be left to form a multiple-trunked specimen or can be transplanted in warm weather. Remove most of the leaves before transplanting. New growth is rapid in spring, slowing during summer, nonexistent in winter. Plants are either male and female; central cones of females are larger than those of males. Native to southern Japan.

Special design features: Refined, elegant form and texture on a small scale. Oriental or tropical effect. A miniature for small gardens and close-up viewing.

Uses: Accent plant of great interest. Underplant for larger plants or as part of a miniature grouping, eventually becoming an overstory plant to small plants. In scale with entryways, atriums or in containers. Excellent in rock gardens or as bonsai.

Disadvantages: May be damaged by cold in middle and high zones, excessive sun in lower zone, or extreme soil alkalinity.

Planting and care: Plant in spring when frost danger is past or any time in protected areas. Trim old fronds in spring as they die and new growth appears.

Low and middle zones, protected microclimates in high zone

Evergreen

Soil: Adaptable. Prefers enriched garden soil with good drainage.
Sun: Full, open or filtered shade. Part to full sun. Should have afternoon shadow in low desert.
Water: Moderate to ample. Established plants have been known to survive long periods of neglect.
Temperature: Survives to 5°F (−15°C), but foliage is severely damaged between 15° and 20°F (−7°C). Recovery rate is usually fast in spring when new fronds emerge, but because leaves last more than one year, plant should not be used where frozen back every winter.
Maintenance: Little to none.

Cycas revoluta

CYPERUS alternifolius

Family: *Cyperaceae*

Similar to and a relative of the papyrus plant of ancient Egypt, the umbrella plant grows in or out of water. This unusual, grasslike sedge is native to Madagascar, the Reunion Islands and Mauritius where it grows naturally in marshes. In the ground or in a water garden, it grows fast to a height of 2 to 5 feet and spreads to an indefinite width by underground rhizomes that clump sideways. It is highly decorative with its 12-inch wide "umbrellas" formed by flat leaf blades that radiate from the center at the top of each stem. Tiny yellowish flowers wreath the center of each ribbed umbrella each spring. In arrangements, it is attractive dried or freshly cut.

Cultivars: 'Nanus' (sometimes sold as 'Gracilis') is a dwarf umbrella plant, smaller and more slender than the species. It seldom flowers.

Special design features: Waterside, Oriental or tropical effect. Sparse plantings among rocks reveal stems, giving a vertical emphasis.

Uses: Silhouette or accent. Containers or small sheltered spaces such as entryways or atrium gardens. Special effects for dry steam beds, ponds or gardens. Effective among rocks. Established plants tolerate short periods of drought.

Disadvantages: Unattractive in winter where there is frost. Messy and unkempt if allowed to overgrow. May reseed where it is not wanted.

Planting and care: Plant from containers when danger of frost is past, any time in warm-winter regions. Overgrown clumps may be divided in early spring. Discard overgrown center and replant strong side shoots. May be planted from seed in spring or fall, or started from cuttings in spring by taking a stem, trimming the umbrella to half its original size and setting it in wet sand. Groom plants by cutting off frozen, dry or broken stems. Feed as you would a grass lawn for rapid vigorous growth. For a water garden, plant in a pot and set the pot rim 6 to 8 inches below the water surface.

Low and middle zones

Evergreen perennial

Soil: Tolerant. Prefers moist garden soil.
Sun: Filtered or open shade. Part, full, or reflected sun.
Water: Constant and ample to occasional.
Temperature: Foliage is damaged at about 28°F (−2°C). Recovers quickly in spring.
Maintenance: Needs regular grooming to look its best.

Cyperus alternifolius

DASYLIRION wheeleri

Desert spoon • Sotol • Wheeler's sotol • Spoon flower

Family: *Agavaceae*

The desert spoon is a favorite plant of the Southwest and northern Mexico. Used by the Indians for food, fiber and to make an alcoholic beverage, caucasian settlers dismembered this plant to obtain the trunk ends of the leaves, displaying them as "desert spoons." In recent years, the desert spoon has become a favorite of plant rustlers who sell it without the required legal tag from the United States Commission of Agriculture.

Slender, toothed, gray-green leaf blades radiate from the center in all directions. Growth is moderate to 5 feet or more, spreading 6 feet. Older plants occasionally develop several heads and also a short trunk, especially in the home garden. From late spring through late summer some plants send up a bloom stalk 5 or 6 feet above the foliage, topped by a long plume of straw-colored flowers like a sheath of grain. Fortunately, bloom does not spell the end of the plant as it does with agaves. Plants may be obtained from some nurseries or grown from collected seed, but it is not always viable. There is speculation that the moth responsible for pollination is a victim of civilization and insecticides. It is possible seed collected farther from cities would be more viable.

Special design features: Bold or dramatic accent or silhouette plant.

Uses: Desert gardens, subtropical groupings, or in containers.

Disadvantages: Small 1-gallon plants have a poorer survival rate than larger specimens. Plants collected in the wild are slow to establish.

Planting and care: Plant any time; space 6 feet apart for massing. Remove old flower stalks and dried lower leaves to groom. Seek 5-gallon size plants at nurseries for a better survival rate.

All zones

Evergreen

Soil: Tolerant of many soils and alkaline conditions. Likes good drainage. Grows naturally in rocky soils.

Sun: Full to reflected sun. Accepts part shade.

Water: None once established, except in areas with less than 10 inches of annual rainfall where it needs supplemental irrigation. Grows fastest with deep, widely spaced waterings. Will accept moderate irrigation when used in a garden situation.

Temperature: Hardy.

Maintenance: None to very little.

Dasylirion wheeleri

DODONAEA viscosa

Hopbush • Hopseed bush • Switch-sorrel

Family: *Sapindaceae*

A tough, drought-resistant shrub of variable size and form, hopbush is native to a large area of the Western Hemisphere from Arizona to South America. It is widely distributed in warm areas of the Old World. A fast grower, hopbush can reach 15 feet and spread almost as wide if grown in an irrigated situation. In less favorable situations it is often much smaller and sparser in foliage. Foliage is variable in color and density. Lush growth is dense with slender, medium green leaves to 5 inches long. In difficult situations, shrubs look angular and sparse and leaves may be only 1-1/2 inches long. Plants are usually irregular in shape, but are sometimes symmetrical. Insignificant flowers at the end of winter produce clusters of decorative, flat, round seed in late spring. Plants may bloom and produce fruit at other times of year, such as after summer rains.

Cultivars: 'Purpurea' is more symmetrical in form and less hardy to cold. Its foliage has a bronze to purple cast, and is less dense. It is very erect in form with a strong central leader. 'Saratoga' is dependably purple. This is the form that is most often trained into a small tree in mild winter areas.

Special design features: Untrimmed plants give a willowy effect. Evokes the mood of the mountain canyon, upper grassland or chaparral belt. Purple-foliaged plants give color contrast.

Uses: Desert gardens and transitional areas. Irrigated areas or gardens. Large informal screen. Clipped hedge anywhere, especially the shrubbier forms such as 'Purpurea'. Foundation plant or espalier. Small trees or standards.

Disadvantages: Young plants, especially the purple-leafed forms, can be set back by a long, hard freeze. Strong odor at close proximity during certain times of year, especially when plant is producing pollen. Greedy roots take water from less aggressive plants nearby. Attracts whiteflies.

Planting and care: Plant from containers any time in warm areas when danger of frost is past in cooler areas. Space 6 to 8 feet for a wide screen or mass planting, 3 to 4 feet for a clipped hedge. Trim leaders to promote bushiness, control height.

Low and middle zones, warmer parts of high zone

Evergreen

Soil: Tolerant of alkaline, rocky or heavy soils. Does best in improved garden soil.

Sun: Part, full or reflected sun.

Water: Best with regular, widely spaced irrigation for large plants. Plants grow well with extra runoff in areas of 12 inches of rainfall.

Temperature: Tolerant of heat. Young plants are damaged at 20°F (−7°C), frozen to the ground around 15°F (−10°C).

Maintenance: None to regular according to use.

Dodonaea viscosa 'Purpurea'

ELAEAGNUS angustifolia

Oleaster • Russian wild olive • Silver berry tree

Family: *Elaeagnaceae*

(E. argentea, E. hortensis). This is a sturdy tree in cold and windy areas, where many plants have trouble growing. Growth is moderate to 20 feet high with a spread of 15 feet in favorable situations. In tough conditions it may remain a shrub. It can be kept to any size in the garden. The form, color and fruit are reminiscent of the true olive, although leaves are more willowlike. The angular often spiny structure is covered with shredding bark and supports slender gray-green leaves which are silvery beneath. Small, yellowish, fragrant flowers appear in spring, followed by small, inedible, olivelike fruit. This native of Europe and western Asia is tolerant to wind, drought, cold and poor soil.

Special design features: Informal. Nondescript at a distance, but trained trees in the home landscape are sculptural and picturesque.

Uses: Shade for the street or patio. Windscreens. Large deciduous hedges when planted close and kept to desired size.

Disadvantages: May be short-lived, especially in warmer areas where it may not defoliate in fall. The plant then grows continually which eventually saps its strength and aids its decline. This eventually shortens the lifespan.

Planting and care: Plant from seed in spring or from containers any time of year. Plant bare root in winter. Volunteers may be transplanted in winter or early spring. Space 4 to 6 feet apart for a hedge or screen, 10 to 12 feet apart for a windscreen, 15 feet or more for a street tree. Do necessary pruning in winter when dormant. Shear hedges any time.

> **High zone**
> **Deciduous**
>
> **Soil:** Tolerant. Grows best in a light, improved garden soil. Needs a heavy soil to anchor its roots if used as a windscreen.
> **Sun:** Part to full sun. Reflected sun in high zone.
> **Water:** Moderate for fastest growth. Accepts little to none where rainfall is 12 inches or more a year, but will probably remain small.
> **Temperature:** Hardy to cold.
> **Maintenance:** Occasional to none depending on situation.

Elaeagnus angustifolia

ELAEAGNUS ebbingei

Ebbing silverberry

Family: *Elaeagnaceae*

(E. macrophylla 'Ebbingei'). This hybrid was developed in Holland. A versatile shrub, it grows at a moderate to slow rate to 9 feet in height and nearly as wide. Its most fascinating feature is the foliage. Crinkly edged leaves to 4-1/2 inches are green above, nearly white beneath. The whole leaf is covered with scalelike silver flecks that shimmer in light or shade. A few brown flecks on the leaf undersides concentrate on the shrubby stems, giving it a bronze, metallic cast. Foliage densely covers the thornless branches to the ground. Plants are attractive at close range as well as from a distance, well behaved, easy to train and adaptable to many growing situations.

Special design features: Shimmering leaf form.

Uses: General purpose shrub for use as a specimen or massed as a background or wide, freely growing screen. Adapted to irrigated gardens or transitional gardens. Outstanding as an espaliered wall plant or against fences. Clips well as hedge, but loses the decorative quality of the leaves.

Disadvantages: Occasional scale problem.

Planting and care: May be planted any time, but best in spring. Some shaping or pruning can be done as required. Plants in transitional areas require no care other than some irrigation.

Elaeagnus ebbingei

> **All zones**
>
> **Evergreen in mild climates**
>
> **Soil:** Tolerant. Prefers improved garden soil.
> **Sun:** Open or filtered shade, part or full sun. Accepts reflected sun in cooler areas and in middle and lower zones if given more water.
> **Water:** Moderate. Needs ample in hottest weather. Accepts occasional in cool season.
> **Temperature:** Accepts heat and cold. May lose leaves in cold winters.
> **Maintenance:** Occasional, or as desired.

Elaeagnus pungens

Elaeagnus pungens, the thorny elaeagnus, also deserves mention. It is a sprawling, irregular, densely foliated shrub, brown or olive-green when viewed at a distance. The sometimes spiny branches support oval wavy-edged leaves 2 to 4 inches in length, heavily covered with flecks of silver and brown scales, especially on new growth. It grows at a moderate rate 6 to 12 feet in height and as wide if left untrimmed. With favorable conditions it can reach up to 15 feet. Unimportant flowers sometimes produce edible fall fruit in red, silver or brown.

ENCELIA farinosa
Family: *Compositae*

Brittlebush • Incienso • White brittlebush

Brittlebush is a shrubby perennial native to the Southwest and northern Mexico. It is an indicator plant: Where it grows naturally indicates a climate where citrus should do well. The gum that exudes from the stems of the brittlebush was chewed by the Indians and was also used as an ointment for pain. Early padres used the gum as a substitute for incense, which is how it got its Spanish name, *incienso*. The brittlebush is fast growing to a height of 2 to 3 feet with a spread of 4 feet or more.

Light gray-green, almost whitish green leaves to 3 inches in length are triangular in shape and densely cover the plant when grown under favorable conditions. Plants are sparse and open in cold areas or where they have to struggle to survive. Yellow daisylike flowers 1 inch in diameter rise above the plant in profusion in spring. Bloom may occur at other times of year following rains or supplemental irrigation. **Special design features:** Pale-colored form. Spring flowers.

Uses: Desert or natural gardens, rockeries or transitional areas. Foundation plant. Specimen, groupings or in combination with desert plants. Whitish color contrasts well with deep green or red foliage.
Disadvantages: Unsightly if frozen back. Older plants become woody and straggly and need to be replaced. Plants sometimes die for no apparent reason.
Planting and care: Plant from containers or from seed in spring. Space 3 feet apart for massing. Cut back in late winter to rejuvenate. If frozen back, wait until new growth starts to reveal the extent of damage before pruning. Prune occasionally to keep tidy. Shear after bloom to groom and rejuvenate.

Encelia farinosa

Low zone

Evergreen in mild climates

Soil: Prefers light gravelly soil with good drainage.
Sun: Full to reflected sun.
Water: Little to none in areas receiving 10 to 12 inches of annual rainfall. Best with occasional irrigation, but accepts moderate, becoming lush.
Temperature: Topkill occurs in the high 20's°F (−7°C). May die at 15°F (−10°C). Recovery may be rapid or only partial.
Maintenance: None to some.

ENSETE ventricosum
Family: *Musaceae*

Abyssinian banana

(Musa ensete, M. arnoldiana). An exuberant, colorful, bananalike plant from the Ethiopian highlands, the Abyssinian banana is appreciated as a tropical feature or bold accent plant. A rapid grower up to 40 feet in its native semitropics, it usually reaches 6 to 20 feet high in cultivation. The erect, purple to brown trunklike stem flairs at the base. Wide leaves are supported on long arching stalks, which continue as purple midribs through the leaf centers. A plant 20 feet tall may have a spread of 15 feet. When a plant is 2 to 5 years old, unimportant flowers are produced in a bronze to purple cylinder at the stem tip. The bloom signals the end of the plant, and it dies to the ground. There is no edible fruit, and no offsets are sent out. Plants are grown from seed.
Cultivars: 'Maurelli', the Ethiopian banana or red-leaved banana, is more compact, and produces offsets at its base that can be taken off to start new plants. It is more in scale with the residential garden, and perhaps more decorative with its red-tinged leaves.
Special design features: Bold, tropical or jungle effect. A strong vertical.
Uses: Bold accent or silhouette plant. Specimen, or group plants of various ages for a continuous planting. Containers, sheltered courtyards, atriums, entryways

Ensete ventricosum

or as an underplant beneath a tall tree.
Disadvantages: Leaves tatter in the wind. Looks run down in cold weather. Leafburn from hot sun. Plant dies every few years, necessitating replacement, which can create a sense of disappointment for some gardeners.
Planting and care: Plant from containers or from offsets when weather has warmed up. Plants may be grown from seed. Feeding helps it grow faster, but is not necessary. Remove stalk after bloom and replace with a young plant. Remove frost-damaged fronds in warm weather after new leaves have come out. Plants need shelter from wind to look attractive.

All zones

Evergreen

Soil: Prefers a light, rich garden soil.
Sun: Open to filtered shade. Accepts morning sun, but best with afternoon shadow, or shade from late morning on in the warmest areas.
Water: Prefers ample.
Temperature: Leaf tips are frosted at about 28°F (−2°C). With lower temperatures, the foliage dies to the trunk. The plant recovers rapidly in spring.
Maintenance: Only occasional grooming, except for the inevitable removal when it blooms and dies.

ERIOBOTRYA japonica

Loquat • Japanese plum • Japanese medlar

Family: *Rosaceae*

A small to medium-sized tree, the loquat is one of the most decorative and versatile plants around. The attraction is primarily the bold, dense foliage that grows in a decorative, rosette pattern at the growing tips. Young plants have wide, low, irregular crowns. Mature plants develop picturesque, single or multiple trunks with rounded umbrellalike crowns. Fuzzy, leathery, deeply veined leaves are dark gray-green above, whitish beneath. Fragrant, fuzzy white often incomplete flowers in fleshy woolly, cream-colored clusters form in the leaf rosettes, appear in fall and winter. They produce small,

All zones
Evergreen
Soil: Tolerant, but needs good drainage.
Sun: Open to filtered or part shade. Full sun in middle and high zones, likes afternoon shadow in lower zone.
Water: Moderate, but accepts ample, especially as fruit ripens and as new growth appears. Tolerates periods of drought when established.
Temperature: Hardy to 15°F (−10°C) or below. Flowers and fruit damaged at about 28°F (−2°C).
Maintenance: Little to much depending on the neatness desired.

delicious, pearlike fruits in clusters in late spring in warm-winter areas. Most plants are sold as seedlings, but to have fruit of a predictable quality, seek a grafted, named variety. In some areas, commercial trees are available that produce fruit for the market. Native to China and Japan.

Special design features: Tropical or Oriental effect. Bold foliage.

Uses: Lawn, patio or street tree. Espalier on cool walls or hedge on "stilts" to extend the screening effect of a wall.

Disadvantages: Very susceptible to fireblight. Sometimes attacked by red spider mites. Leaf drop is a nuisance, especially in spring, but occurs whenever tree is growing. Fruit is blighted by hard frosts.

Planting and care: May be grown from seed. Plant from containers any time, but best before summer heat. Shade young newly set plants from the sun during hot periods. Prune sparingly to train.

E. deflexa, the bronze loquat, is a smaller and more refined version of the above. It is usually grown as a shrub or small tree. This tree requires a more sheltered location because it is more sensitive to heat and wind. White flowers appear in late winter and spring, but no edible fruit is produced. Requires same basic culture as *E. japonica.*

Eriobotrya japonica

Eriobotrya japonica

EUCALYPTUS species

Eucalyptus

Family: *Myrtaceae*

Eucalyptus is the most commonly planted skyline tree in warmer parts of the Southwest and in similar climates throughout the world. More than 500 kinds are found in Australia and Tasmania, with some ranging to the Philippines, Java and New Guinea. Over 80 species were introduced into southern California before 1900; about 20 species are now commonly used. In the arid and colder parts of the West, however, the number of successful species is more limited.

A similarity in appearance and scent makes almost all eucalypts easily recognizable. The differences are interesting and essential when choosing plants for the landscape. Some are extra-fast growers and get too large for an average residence. Others have round or very narrow leaves or especially fragrant foliage. Color, form, rate of growth, eventual size, even the bark varies from species to species. Some are hardier to cold than others, or have especially beautiful flowers. The tallest reach over 100 feet high. Most eucalypts are drought and heat-tolerant, grow fast and provide a welcome shade in hot dry areas. Not to be forgotten are the *mallees,* the shrubby eucalypts, which are multiple-branched and lower growing.

Special design features: Handsome infor-mal silhouette. Fast shade. Mostly vertical form.

Uses: Specimen, row or grove. Excellent for shade, windbreaks or open areas.

Disadvantages: Some species produce a litter of bark, leaves, flower parts or seed capsules. Many have greedy invasive roots, so are not good choices for lawn areas. Larger trees sometimes break in the wind and drop branches. Iron chlorosis is a special problem in some places, especially after freezes. A symptom is new foliage that is bunchy and pale. If not corrected, trees may die from it. To give a tree immediate treatment, place iron chelates in common gelatin pill capsules. Follow package directions for dosage amounts. Insert them into holes of equal diameter drilled into the trunk in an ascending spiral. Plug holes with chewing gum or caulking. Once the iron pills have been inserted, administer iron sulfate´ at the base of the tree to the drip line, as described on page 35 and irrigate deeply. Foliar sprays are also effective, but are temporary and are difficult to administer to large trees.

Planting and care: Plant any time in an area large enough for eventual size of the tree unless you plan to remove it when it gets too big. Select vigorous plants with

Eucalyptus citriodora

Eucalyptus erythrocorys

Eucalyptus camaldulensis

roots that fill its container, but are not root-bound. Young established trees in 5-gallon containers are good buys. Be sure the species you select is suited to temperatures in your area or to a special microclimate or situation. Also be aware that many *Eucalyptus* species grow quite tall. Avoid planting the giants under structures, overhangs or powerlines, or you will be plagued with pruning troubles for the life of the tree. Most young trees need staking until a strong vertical trunk is developed. Prune young trees as stated under the individual plant description. Very tall or old trees may need to be headed back to prevent wind breakage, but this is best done by properly trained and equipped arborists.

Low and middle zones according to plant; *E. microtheca* is adapted to high zone

Evergreen

Soil: Widely tolerant. Deep soil and good drainage are important for best results. Trees are more likely to become chlorotic in alkaline or poorly drained soils.

Water: Give deep periodic irrigation, more often for young plants until they are established and for fastest growth. Established trees grown where they receive extra runoff or in areas receiving 10 to 12 inches annual rainfall will need little or no supplemental irrigation. Most trees benefit from monthly soakings through summer, especially in hot areas receiving no summer rains. Taper off irrigation to prepare trees for winter.

Temperature: Follow ranges for each species.

Maintenance: None to constant depending on the tree and where it is used.

Eucalyptus camaldulensis
**Red gum,
river red gum,
murray red gum**

(E. rostrata). Red gum is a tall, majestic tree with weeping branches and long, slender, medium green leaves. Its bark is mottled tan to light gray, and sheds in patches except on the lower trunks of older trees, where it becomes a consistent dark gray. Individuals vary widely in form and tolerance to cold, perhaps because of hybridization. Too large for the average residence except at distant property edges. Very fast growing; it may grow 10 to 15 feet a year, eventually reaching around 120 feet in height. Best used in open areas, parks, large public spaces or for roadsides or shelterbelts. Tolerates lawn watering. Space 15 feet apart for windbreak, 20 to 30 feet or more for row planting.

Damaged by cold at 15°F or 12°F (−10°C) for a brief period.

Eucalyptus cinerea
Spiral eucalyptus, mealy stringbark, argyle apple, ash gum

This is an irregular tree with leaning trunk and gray juvenile leaves which spiral around the twigs. Moderate to fast grower reaching 20 to 50 feet. Long, slender, mature leaves eventually take over unless tree is pruned. Usually used as an accent or silhouette plant, it is perhaps the most attractive when grown with multiple trunks. Handsome even in maturity, it maintains the gray foliage color.

Tolerates more shade and more water than some eucalypts and generally does well in lawns. Space 10 to 15 feet apart for a mass effect. If grown for form and

foliage, keep it pruned to 15 feet. Leaves are handsome in arrangements.

Hardy to 17° (−7°C).

Eucalyptus citrodora
Lemon-scented gum

Tall, elegant and graceful, the lemon-scented gum has a smooth, slender, pinkish white trunk, open spreading crown, and narrow, medium green leaves. When leaves are crushed, they give off a fragrance resembling lemons. Fast growing to a height of 50 to 75 feet, its silhouette is attractive against structures or the sky. Safe to grow near walls and walks because roots don't heave. Accepts large or small amounts of water. Trees should be sturdily staked when young. Thin and cut back until trunk becomes strong. Bark peels and drops annually all at once.

Tenderness to cold is the only drawback besides annual bark litter. Space 6 to 12 feet apart for grouping of tall slender trunks, 20 to 30 feet for grove. Trim off low branches to enhance the high trunk and silhouette effect.

Place trees in sheltered locations such as against south sides of tall buildings.

Hardy to about 23°F (−2°C).

Eucalyptus erythrocorys
Red-cap gum, illyarrie

Red-cap gum is a rapid grower reaching 10 to 20 feet or more, spreading 15 feet wide. Attractive at close range with smooth, light tan to whitish bark, irregular, somewhat open crown and very green 7-inch, lance-shaped leaves. Showy, yellow, brushlike flowers in clusters break from red-capped buds and are usually seen from fall to spring, but can appear any time.

Eucalyptus cinerea

Eucalyptus microtheca

This tree has one of the largest and most striking bud caps of the eucalyptus group—bright scarlet, square, with a raised cross. They shed as the flowers appear. Seed capsules are large with wavy margins—decorative in arrangements. Grow as a single or multiple-trunked tree. Roots are less greedy than other eucalyptus and it remains in scale with the average residence. Massed, it makes a handsome, irregular grove.

Plant any time of year from containers, but best set out in spring. Space trees 10 to 12 feet for a wide, loose screen or at random intervals for a grove. Cut back main shoots several times to form dense bushy growth or a multiple-trunked tree. Needs controlled staking and pruning to look good. Grows in lawns, but may need extra iron to prevent chlorosis. In its native Australia it grows in sandy or calcareous soils, so should tolerate them in the Southwest. This plant is hardy to 23°F (−5°C), and tolerates desert heat when given moderate amounts of water.

Eucalyptus leucoxylon
White ironbark

This is a tree of variable form to a height of 20 to 50 feet, usually slender and open with pendulous branches. Juvenile leaves are round and dark gray-green; mature leaves are the same color but long, slender and slightly curved to one side. Smooth whitish bark sheds in irregular flakes, revealing pinkish white inner bark. Widely tolerant of adverse conditions, including heat, wind, drought and heavy or rocky soils. Excellent as roadside or open-area tree. Often grown for timber or as an aid in honey production.

Cultivars: 'Rosea' is the most popular white ironbark for landscape use because of its stunning pink flowers. 'Purpurea' has purple flowers. Both are variable in shape and size, but usually are smaller than the species, reaching heights of 15 or 20 feet spreading 12, sometimes 20 feet.

All accept most soil conditions and are very ornamental. Select plants carefully, in bloom if possible, because they vary widely in form and color. Space the species 20 to 30 feet apart for roadsides, smaller forms at random distances 8 to 12 or 15 feet for small grove or informal screen.

Hardy to 15°F (−10°C).

Eucalyptus microtheca
Flooded boc
coolibah tree

One of the hardiest eucalyptus, this tree survives heat, drought and poor soil as well as cold. It is a picturesque, often leaning tree of fast to moderate growth, cleaner and less likely to break in the wind than many of the other eucalypts. Trunks of young trees are smooth and mottled white to gray. Older trees have wrinkled, cracked, fibrous bark on lower trunk. Leaves are light gray-green and ribbony on pendulous branches. Crown is bushy, vertical in youth, rounded in age. Trees may reach 40 feet in about 20 years. Young trees need staking to develop vertical trunks, but resist training and must be

Eucalyptus cinerea

Eucalyptus microtheca

stoutly tied. Space 10 to 15 feet apart for windscreen, 20 to 30 feet or more for a street tree. Also makes a handsome, multiple-trunked tree.

Foliage may burn in a hard freeze, but winter-hardened trees usually survive to 10° to 5°F (−15°C). This is one of the hardiest eucalyptus. Grows in all zones.

Eucalyptus nicholii
Nichol's willowleaf peppermint, narrow-leaved black peppermint _____

This is a graceful, fast-growing, erect tree reaching up to 40 feet in height. It has a vertical form in youth and a spreading crown when mature. Weeping branchlets support very narrow, willowlike, gray-green leaves which are sometimes purplish. Leaves smell like peppermint when crushed. Bark is soft, brown and fiberous. Because the bark does not shed as much as most eucalypts, it is a desirable garden or street tree. Flowers are small and inconspicuous. Space 15 to 20 feet apart for a graceful, willowy grove or screen, 30 feet or more for a row or street planting.

Hardy to 12°F (−10°C).

Eucalyptus polyanthemos
Silver-dollar gum, silver-dollar tree, red box _____

Silver-dollar gum is a medium-sized tree of moderately fast growth, picturesque and often asymetrical with an angular trunk. Individuals vary in foliage color and form. Round brown-green to gray-green juvenile leaves have a silvery cast, giving the tree an unusual informal texture; they are often used dried in indoor arrangements. Older trees develop more pointed, mature leaves, but that takes a number of years. Trunks are covered with fibrous to scaly reddish brown bark. Upper limbs are mottled and support pendulous branchlets. Often open and irregular in youth, trees develop a wide crown with age. Select plants carefully as individuals vary. The silver-dollar gum grows almost anywhere and under poor conditions, but do not place them in lawns or other well-watered locations. Good drainage is necessary. They make excellent windscreens. For a row, space trees 20 to 30 feet apart.

Range is to 14°F (−10°C), but trees have survived lower temperatures without damage.

Eucalyptus polyanthemos 'Polydan'
Polydan eucalyptus _____

More symmetrical and vigorous in appearance than the species without the pronounced, round, juvenile leaves. Vertical in form, its deep, gray-green leaves are shorter, slender and pointed. Crown is dense, the bark mottled and often fissured.

It does not take drought as well as other eucalypts, but does well in lawn situations.

Eucalyptus 'Polydan'

Many are short-lived, possibly because they are grown in dry conditions. A fairly recent introduction, the eventual height is not known. Most specimens seen are 15 to 25 feet tall. Space 30 to 40 feet apart for an avenue planting.

13°F (−11°C) is lowest known limit.

Eucalyptus pulverulenta
Silver mountain gum, money tree _____

A real conversation piece, this fast-growing tree is widely admired for its round, leathery, gray-green juvenile leaves that appear to be skewered on the branches. Attractive with a single trunk and perhaps even more so with a multiple-trunk, it benefits from pruning and shaping. Looks best when kept at 6 or 7 feet or a little higher. Sturdy and appealing, it has silky white bark which peels and sheds annually. If allowed to grow to 20 or 30 feet, the mature leaf form develops, which is slender and pointed—similar to other eucalypts. For a shrubby sapling grove, space plants 8 to 12 feet apart, for a row

Eucalyptus polyanthemos

space 15 to 20 feet on center. This is not a tree for windy areas, however.

Cold hardy to 15°F (−10°C), but it has survived lower temperatures without damage.

Eucalyptus rhodantha
Rose mallee

A hybrid with an erratic, dashing form and spectacular flowers, rose mallee is a moderate grower reaching 6 to 8 feet high sprawling 4 to 8 feet wide. Long slender stems bear round silvery leaves 2 to 4 inches in diameter that seem to be pierced by the stem and set at neat intervals along

Eucalyptus pulverulenta

Eucalyptus viminalis

it. Plant has a tumbling form that appears bluish or grayish at a distance. In addition to the bold leaves and informal habit, it has a continuous bloom of fringy, red, pomponlike flowers 3 to 5 inches across. A real conversation piece as a specimen providing welcome informal relief to a formal situation, or as foliage contrast to dark evergreens. It can be trained as an espalier, unique among eucalyptus. Branches and flower capsules are attractive in arrangements.

Supply with well-drained sandy soil and part to full sun. Accepts ample water with good drainage. Prefers moderate spaced irrigation. Prune and train as desired. Generally not as long-lived as other eucalypts.

Hardy to 8°F (−12°C).

Eucalyptus sideroxylon
Red ironbark, pink ironbark

(*E. sideroxylon* 'Rosea'). Red ironbark is a tree of variable form, height and color, growing rapidly to 20 feet, sometimes 80 feet in height in favorable situations. A striking tree—the slender trunk has deeply fissured, rusty, dark brown bark. In each fissure the red inner bark can be seen, which is very dramatic in appearance. Trees are narrow and open with gray-green foliage. Those grown in middle zone may be structurally distorted by occasional freezes. Flowers in pendulous clusters from fall to late spring are creamy pink to deep pinkish crimson. Usually the darker the foliage, the darker the flower color. Select individual plants carefully to get the characteristics you want.

Accepts high heat and poor or shallow soil. Best in light soil with good drainage and deep periodic irrigation. Use in a residential-sized garden as a wind screen. A roadside planting of tall dark trunks is very impressive. Space 15 to 20 feet apart for a grove, 30 feet or more along roadsides.

Hardy to 20°F (−7°C), an indicator of where citrus can be grown.

Eucalyptus spathulata
Narrow-leaved gimlet, swamp mallee

This small, erect plant reaches 6 feet, sometimes 20 feet in height, with single but usually multiple trunks. Form is slender, bushy and bright green. Narrow ribbonlike leaves are 2 to 3 inches in length. Bark is smooth and reddish brown. Tolerant of poor drainage, lawn situations and heat. Actually a shrub eucalyptus, or mallee, it can be used in smaller spaces than the full-sized eucalypts. Excellent as a tall garden hedge, wind or privacy screen. Place plants in south exposures or warmer microclimates where frosts are common. Space 6 to 8 feet apart for hedge or screen or randomly for small grove or filler planting.

Hardy to 22°F (−3°C).

Eucalyptus viminalis
Manna gum

This is potentially the tallest of the eucalypts. Manna gum often reaches 60 feet, but in favorable situations it grows to as much as 150 feet. Form is upright with long pendulous branches that eventually create a spreading, dome-shaped crown. Dense, weeping foliage is light gray-green with a shape typical of many eucalypts. Older gray bark peels to reveal shiny white under-bark on the trunk and branches, contrasting dramatically with foliage. Shedding bark makes a constant mess under the tree. Because of its large size and the constant litter problem, this is not a tree for the average residence. Best used as a significant silhouette or skyline tree for open areas, parks, highways and large scale boundary planting or windbreaks. To train as a smaller tree, cut a 5-gallon-sized tree back to 1 foot in height and allow it to grow into multiple trunks.

Grows best in deep soil, but tolerates poor soil and arid to moist conditions. Thrives on deep, widely spaced irrigation. It also has a wider tolerance to cold and heat than many eucalypts.

Hardy to 12°F (−11°C).

Eucalyptus sideroxylon

EUONYMUS fortunei

Family: *Celastraceae*

Common winter creeper

(E. radicans). This viny shrub deserves wider use. It sprawls over the ground or will climb a wall to 20 feet or more, clinging by rootlets. Thick leaves are dark rich green above, whitish beneath, 1 to 2-1/2 inches in length with scalloped edges. Branches are densely covered with tiny bumps and numerous leaves; some turn red in cold weather. Mature plants are shrubby and bear fruit. Plants on the ground sometimes root as they go. Native to Japan and South Korea.

Cultivars: 'Azusa' is a ground-covering prostrate grower which has small dark green leaves with light veins. In winter, leaves turn deep red underneath. 'Colorata', purpleleaf euonymus, is semi-prostrate and spreads widely. Deep green foliage turns a plum color in fall and winter. 'Kewensis' climbs or trails with 1/4-inch leaves that form a delicate tracery or dense ground cover. 'Silver Queen' is shrubby and spreading with white-edged leaves. It does not climb. *E. fortunei vegeta*, big-leaf winter creeper, is woody enough to support itself as a mound or train as a vine. It will cover an area of 15 to 20 square feet. Orange-seeded fruit come in fall. Spring growth is yellow-green. Grows in an irregular manner at first, sending out large branches.

Special design features: Leafy, deep green color.

Uses: A plant for sunny to partially shaded walls. Spills over containers. Ground cover. Porch posts or trellises.

Disadvantages: Foot traffic destroys it. Subject to Texas root rot, powdery mildew and sometimes root knot nematodes. Rabbits sometimes eat it.

Planting and care: Plant any time. Space 3 to 4 feet apart for a ground cover. Treat planting holes to prevent Texas root rot and never plant where soil is known to be infected. Apply a systemic or spray if

mildew occurs. Responds to the same bug and pest prevention program as roses.

All zones
Evergreen
Soil: Tolerant. Prefers improved garden soil.
Sun: Part shade to full sun. Afternoon shade in middle and low zones. Reflected sun burns plants.
Water: Moderate.
Temperature: Tolerant of heat and cold.
Maintenance: Occasional.

Euonymus fortunei (variegated form)

EUONYMUS japonica

Family: *Celastraceae*

Evergreen euonymus • Spindletree

The evergreen euonymus is amazingly drought and heat-resistant. It is also extremely durable, tolerating poor soil and cold to below zero. Densely foliated with shiny-toothed, leathery, very dark green leaves, it grows at a moderate rate 8 to 10 feet high with a spread of 6 feet or more. Older plants can be trimmed into small trees with dense, umbrella-shaped tops and curving trunks. As a shrub, the oval leaves cover plant completely to the ground. There are numerous selections including variegated forms, but variegations usually disappear with maturity.

Cultivars: 'Albo-marginata' has leaves edged with white. 'Aureo-marginata' has leaves edged in yellow. 'Aureo-variegata' has leaves with green edges and bright yellow blotches in the center. 'Grand-

ifolia' is compact and fast growing, good for shearing.

Special design features: Species forms a dense cover under conditions too tough for most green plants. Contained formal look.

Uses: Shears well as hedge, single or multiple-trunked tree or topiary plant. Untrimmed its a dense, specimen shrub or dense, wide screen when planted in rows. Leans against walls without attachment as an espalier. Mass as a background or large scale mound. Attractive in containers.

Disadvantages: Very susceptible to powdery mildew in shade or where air circulation is poor. Somewhat susceptible to root knot nematode and quite susceptible to Texas root rot. Stems sometimes deform and flatten, but this is not serious.

Planting and care: Plant in soil prepared to avoid Texas root rot; never plant in infested soil. Set out plants any time of year. Depending on the cultivar, space 3 to 4 feet apart for a clipped hedge, up to 5 feet for an unclipped screen. Place in sun on south or west sides or in the open where air circulation will help to prevent powdery mildew. Control powdery mildew by spraying or by giving the plant a systemic. (See page 39.) Avoid sprinkling plant, it is best to flood or use drip irrigation to prevent mildew. Disinfect hedge

trimmers if you trim plants infected with powdery mildew or it can spread.

Euonymus japonica 'Microphylla', the box-leaf euonymus, is so remarkably different in appearance from the species it deserves special mention. It grows at a moderate rate to only 1 to 2 feet in height, with tiny, rounded, dark green, evergreen leaves. Often used as a clipped box hedge in formal situations. Prefers some shade in the hottest areas, and needs more water than the species. Space plants 9 to 12 inches apart for a low hedge.

Euonymus japonica

All zones
Evergreen
Soil: Tolerant. Prefers good drainage.
Sun: Best in full or reflected sun.
Water: Little to moderate administered at plant root zone. Withstands periods of drought when established. Best with deep, widely spaced soakings.
Temperature: Hardy to cold. Tolerant of heat.
Maintenance: Some to practically none depending on use.

EUPHORBIA pulcherrima
Poinsettia • Christmas star •Christmas flower • Painted leaf

Family: *Euphorbiaceae*

(Poinsettia pulcherrima). Large, red, flamboyant flower bracts at Christmas make this plant from Mexico and Central America a holiday favorite. Planted outdoors poinsettias grow up to 10 feet in height with no frost, usually less in the colder middle zone. Leaves 4 to 7 inches in length clothe the plant. Plants may lose leaves after the bloom period, which means they are "resting." New foliage appears and grows rapidly in spring. As nights approach the 12-hour length in fall, buds start to form, especially if night temperatures are in the low 60's°F (17°C). Temperatures above or below may delay budding, as will too much artificial light at

night from house or street lamps. Large red bracts with small yellow centers, the true flowers, usually appear in midwinter and last for many weeks if the weather does not turn too cold. Most are red; there are selections with pink, white or yellowish flowers and some with double bracts.

Special design features: Winter color. A giant, temporary bouquet.

Uses: Feature plant for winter months when there is little other color.

Disadvantages: Sensitive to cold. Not an easy plant to get to perform well because of its many cultural requirements. Sometimes gets thrips or spider mites.

Planting and care: Place plants in a well lighted, draft-free location with a cool nighttime temperature reaching to about 50°F (10°C). Keep plants moist unless leaves drop and plants go dormant. If this happens, give occasional water until spring when frosts are over and plant outside next to a south or east-facing wall with an overhang. Here plants have sun, warmth and protection from cold. Fertilize three times a year: spring when the new growth starts, June, and again in fall when the bracts are forming. Prune back halfway in early spring and remove leggy growth when bloom is over and when danger from frost is past.

Low zone, protected locations in middle zone, indoor plant anywhere

Evergreen to deciduous

Soil: Rich organic soil with good drainage.
Sun: Full sun to part shade.
Water: Keep damp but not wet.
Temperature: Damaged by light frosts. Killed by severe freezes, but heavier wood of mature established plants usually survives. Best with a night temperature range in low 60's°F (17°C).
Maintenance: A lot for an attractive plant that will produce lots of blooms.

Euphorbia pulcherrima

FATSHEDERA lizei
Botanical wonder • Aralia ivy • Ivy tree

Family: *Araliaceae*

The inventive French are responsible for botanical wonder, a cross between aralia and ivy. Partly erect to partly sprawling or climbing. Bold, rich green leaves vary in size and form. Plants make an undulating and uneven ground cover or may be tied to a trellis, wall or other strong vertical support. Often used simultaneously as a wall and ground plant.

Special design features: Bold foliage for tropical effect. Rich leafy form.

Uses: Espalier or ground cover in shady places such as entryways, atriums, building recesses, stairwells and north sides of tall buildings. Container plant.

Disadvantages: Aphids, snails, slugs and scale can be problems. Subject to damping-off fungus in summer. Surprisingly

Fatshedera lizei

hardy, but can suffer foliage damage from sunburn, wind or frost. Only young leaves are usually damaged by cold.

Planting and care: Plant any time from containers in protected areas, but cool periods are best. Pinch tips to force side branches. Plants are not self-climbing and must be trained as they grow, which means attention through the year. Train plants before stems become too woody and brittle. Plants may be cut to the ground if they become old and woody or out of hand. They will grow back quickly. Give regular garden care. Bait for slugs and snails. Spray for aphids and scale.

All zones

Evergreen

Soil: Improved garden soil with good drainage.
Sun: Deep to filtered or open shade. Tolerates morning sun where it does not get too hot.
Water: Moderate to ample.
Temperature: Mature foliage withstands temperatures to 10°F (−12°C), perhaps lower. Unseasonal frost may nip new growth.
Maintenance: Regular maintenance required for best appearance and growth.

Fatshedera lizei

FATSIA japonica
Family: *Araliaceae*

Japanese aralia • Formosa rice tree • Paper tree

(Aralia sieboldii, A. japonica). A striking plant with extraordinarily large leaves, Japanese aralia grows at a moderate to fast rate to 20 feet in height. It is usually seen at 5 or 6 feet in the desert and warm interior areas, because of the difficult growing conditions and because it is often cut back to initiate new growth. Rich, green, fanlike leaves may reach 16 inches in diameter. Mature plants develop woody stems, forming a multiple-trunked plant. Decorative ball-like clusters of whitish flowers appear in fall and winter above foliage. Native to Japan.

Special design features: Bold foliage creates a tropical mood. Dramatic silhouette. Appearance is best in cool weather when other plants are past their peak.

Uses: Containers or planting beds in shaded places such as atriums, entryways, building recesses, against north sides or as an under-planting for tall high-crowned trees. Specimen or in combination with other tropical-looking plants for verdant effect. Houseplant.

Disadvantages: Subject to damping-off fungus, attack by snails, slugs, scale and aphids. Leaves may tatter in wind, yellow from iron chlorosis in alkaline soils or from exposure to sun.

Planting and care: Set out from containers any time. Shelter from sun and wind. Wash dust from leaves occasionally. Remove some leaves to reveal the slender stems. To rejuvenate, cut back a few of the oldest and tallest canes to the stump in late winter each year. Growing leaders may also be pruned to reduce size of plant.

Fatsia japonica

All zones

Evergreen

Soil: Tolerant. Prefers rich soil with good drainage.
Sun: Full, filtered or open shade. Tolerates morning sun in middle and high zones if there is no reflected heat.
Water: Moderate to ample.
Temperature: Hardy to cold although new growth is damaged at about 20°F (−7°C).
Maintenance: Regular.

FEIJOA sellowiana
Family: *Myrtaceae*

Pineapple guava • Feijoa

An attractive and versatile plant, pineapple guava does well in heat, cold and poor soil with little water. This South American native deserves wider use. Growth rate is moderate to 15 feet, but easily kept at any height or size by clipping. Leaves are 3 inches long, gray-green above and woolly white beneath, spaced at intervals along the twigs to give plant a distinctive character. A natural, somewhat open small tree with bending, weaving trunk or trunks, it accepts training to any form. Unusual, fleshy, pinkish white flowers with dark red stamen appear in spring. Petals are edible and may be used to liven up fruit salads. Green fruit 3 inches long ripen in fall—delicious eaten fresh or in jellies. Fruit that is still on the plant is not yet ripe. Wait until they fall to the ground. They will keep for about a week and can be gathered any time.

Special design features: Gray-green with patterned foliage.

Uses: Small patio tree or train as dense, erect, round-headed standard. Wide loose screen, clipped hedge or espalier. Adapted to large containers.

Disadvantages: Problems are minimal. In well-watered locations like lawns, plants can become chlorotic.

Planting and care: Plant from containers any time. Space 4 feet apart for clipped

Feijoa sellowiana

hedge, 5 to 6 feet apart for loose screen. Prune selectively in late winter to maintain shape and size, but shear as needed. Clipped plants lose their natural character, produce fewer flowers and less fruit.

All zones

Evergreen

Soil: Tolerant. Prefers good drainage.
Sun: Part shade to full or reflected.
Water: Best with occasional deep irrigation. Accepts moderate to ample. Established plants tolerate much drought and neglect.
Temperature: Hardy to about 15°F (−10°C), possibly lower. Tolerant of heat.
Maintenance: None to regular.

Feijoa sellowiana

FEROCACTUS species

Family: *Cactaceae*

The dramatic, cylindrical form of the barrel cactus is a stand-by in the cactus garden. There are about 25 species of this plant, native to the Southwest and Mexico. Those most commonly seen are about 1 foot in diameter. They have a fluted form and green, waxy skin. Clusters of spines grow on the outer ridges of the fluting, with several straight thorns and one heavy, curved spine like a fish hook. Mature plants vary in size according to the species, with some kinds growing up to 8 to 11 feet high. Most in nature are seen at

> **All zones to 5,000 feet elevation, depending on species**
> **Evergreen**
>
> **Soil:** Prefers sandy or gravelly soil with good drainage.
> **Sun:** Full or reflected sun to part shade except for plants collected in subtropical Mexico. They require more shelter from sun in middle and low desert interiors.
> **Water:** Occasional to none depending on local rainfall.
> **Temperature:** Species are hardy to where they are found naturally. Plants collected in warm, subtropical regions need winter protection in higher, cooler zones.
> **Maintenance:** None.

2 to 4 feet. Although the shape is similar to the saguaro, this cactus is heavier in form and has hooked thorns, which the saguaro lacks. Most plants found in nature lean toward the south. This is caused by faster plant growth on the north side. Yellow to orange to red waxy flowers appear at the top of the plant. Bloom may come in May to September, according to the species. Fruits like small greenish to yellow pineapples follow, filled with tiny, round, black seeds. There is no truth to the story that a decapitated cactus will produce a barrel of water for the wanderer dying of thirst. That is only a myth which has unfortunately contributed to the destruction of many cacti.

Special design features: Bold desert form. Desert and subtropical effect.
Uses: Emphasis plant for corners or entrances, but not too near walkways.
Disadvantages: Protected by law and illegal to obtain without tag. Slow to grow.
Planting and care: Plant any time. Buy tagged plants to ensure the cactus was acquired legally. Try to plant with same orientation to the south as the plant had originally. Irrigate occasionally during the first summer after planting, and anytime where the rainfall is less than 10 to 12 inches a year.

Ferocactus species

FICUS species

Family: *Moraceae*

Ficus species may not seem to be significant when you consider there are only one or two species that grow outdoors in the middle and high zones. But you might be familiar with these plants even if you live in a very cold area—many are grown as houseplants. *Ficus benjamina, Ficus elastica* and *Ficus lyrata* are a few examples. In areas that are nearly frost-free, ficus are important outdoor trees. The majority of the following plants are used as hedges and street trees in mild climates. They are truly tropical trees and are not dependable even in cooler parts of the low zone. Where they can be used they are extremely versatile and attractive landscape plants.

Ficus benjamina
Weeping fig, benjamin tree, weeping Chinese banyan tree

(F. nitida). This is a handsome, dark green tree with very shiny, wavy-edged, 3-inch leaves and an open irregular crown which broadens with age. Native to the Malay Archipelago, southeast Asia and north tropical Australia. A moderate grower, its ultimate size in the desert is not yet known, but probably 20 to 30 feet is maximum in frost-protected spots. It seems to enjoy desert heat and sun with little evi-

Ficus benjamina

dence of leaf burn or low humidity stress. Trunk is slender, erect and covered with light gray bark; side branches are weeping. Many tiny, hard, inedible but decorative orange figs are produced by mature plants.
Special design features: Handsome silhouette of great interest. Oriental effect. Weeping form. Mature trees are lush and tropical.

Uses: Street or patio tree. Clip into a hedge or formal tree shape. Large or small containers indoors if sunlight is adequate. Container plants develop into miniatures of trees growing in the ground. Excellent for confined spaces because it can be clipped and kept small.
Disadvantages: Tender to cold. Figs may be a litter problem. Periodic leaf drop is

disconcerting, but is usually nothing to worry about. Sometimes scale or mealy bugs infest plants grown indoors.

Planting and care: Plant in midspring after nights have warmed if grown outdoors. Provide sufficient room for root development or they can buckle nearby pavement. Prune only to shape or to create a formal tree. Do any final pruning in late summer so tree can harden off before winter. Fertilize outdoor trees in spring and early summer; indoor trees should be fed lightly with irrigation any time they seem to need a boost.

Warmest areas of the low zone, indoors anywhere
Evergreen
Soil: Any reasonably good garden soil supplemented with organic matter.
Sun: Open shade to full sun, even reflected sun if new plant is gradually exposed to it before planting. Once adapted to reflected heat, weeping figs seem to enjoy it.
Water: Ample to moderate.
Temperature: Hardy to about 30°F (−1°C).
Maintenance: General garden care to look best.

Ficus carica
Common fig, edible fig

This tree has been grown in the Mediterranean region for thousands of years for its pear-shaped fruit. Usually a sculptural, gray-barked form in winter, and a bold-leafed, rounded form in summer, its hairy, deeply lobed leaves to 8 inches across and 10 inches in length are lush in the landscape. A naturally low-branching tree, it must be pruned up continuously if it is to be walked under. Grows at a moderate rate 15 to 20 feet in height spreading 20 to 30 feet wide. First crop of figs sets with the new leaves in spring and ripens in June, sometimes later, depending on variety. A second crop may ripen later in summer or appear continuously into fall only to be interrupted by autumn chill when leaves turn yellow and drop.

Cultivars: There are many. 'Black Mission' or 'Mission' with dark purple figs is best for low and middle zones, and is the best landscape tree. 'Brown Turkey' is better for the higher zone, smaller, with brownish purple fruit. A nice garden tree, but it needs to be cut back severely to fruit well and to make harvesting easier.

Special design features: 'Black Mission' has attractive sculptural branches when bare in winter. Both cultivars have bold foliage in summer. Fruit attracts birds.

Uses: Grown mainly for fruit or for sentimental reasons rather than as landscape subjects. However, there are many handsome, mature fig trees in arid regions. Best planted in the farther reaches of the garden and alone with room to spread. They are most attractive when viewed from a distance.

Disadvantages: Fruit drop can be messy. Insects and litter from birds during fruiting season. Leaves are itchy and irritate the skin. Sap of green fruit is poisonous. Trees are often bothered by root knot nematodes. Plants are also subject to crown gall and Texas root rot and other fruit pests and diseases. Fig mosaic virus can be a problem.

Planting and care: Plant from containers

Ficus carica

any time or bare root in winter. Trees may be easily started from cuttings.

Space trees 20 to 30 feet apart depending on the variety. Prune only to shape, control size or to remove dead branches. Trees damaged by root knot nematodes may need severe pruning. Varieties have different pruning requirements. Feed in early spring.

Low zone, middle zone to about 2,500 feet, high zone in warm microclimates
Deciduous
Soil: Prefers poor gravelly soil with good drainage.
Sun: Full to reflected sun. Tolerates some shade.
Water: Give deep, widely spaced irrigation. For better fruit irrigate more often during growing season.
Temperature: Mature wood is fairly hardy. Succulent new growth may be killed by heavy winter freezes unless irrigation is tapered off from late summer onward to harden wood. Recovers from freeze rapidly.
Maintenance: Seasonal.

Ficus pumila (juvenile leaf form)

Ficus pumila (mature leaf form)

Ficus pumila
Creeping fig, climbing fig _____

(F. repens, F. stipulata). Creeping fig is prized for the delicate tracery formed by its juvenile leaves on masonry walls. A self-climbing vine, it attaches itself tightly to walls with aerial rootlets. A few inedible figs form among foliage of mature plants. Once established and when most of available wall space is covered, 2 to 4-inch mature leaves appear. They cover the plant densely, giving it a completely different appearance. Eventually the vine develops woody branches 2 feet or longer which stand out perpendicularly from the wall. Vines will cover a large building, and are very aggressive once they catch on. Never allow it to grow on wooden structures, because it will cause damage. Vines may be cut back occasionally to renew the charming juvenile leaves, which will start up the wall once again. Native to eastern Asia.

Cultivars: 'Minima' is smaller with leaves that appear to stay juvenile longer.

Special design features: Deep green woodsy feeling achieved with a minimum of spatial depth. Charming juvenile leaf form.

Uses: North or east masonry walls. Partly shaded pillars. Recesses of buildings. Hanging baskets.

Disadvantages: Sometimes slow to begin climbing. Greedy invasive roots. Filaments cling to masonry after the vine is removed. Vigorous vines may disrupt roof tiles or cover windows if not controlled. Subject to root knot nematode and Texas root rot.

Planting and care: Plant from gallon-sized containers any time, but best in spring. Space 10 feet or more apart. Cut to the ground so new growth will catch on and climb. Tying up growth does little good; it will climb when it is ready. To keep juvenile leaf and to control size, cut back all or part-way every few years.

All zones

Evergreen

Soil: Tolerant. Prefers improved garden soil.
Sun: Open or filtered shade to part sun.
Water: Moderate with a range on either side.
Temperature: Damaged at about 15°F (−10°C), but usually survives freezes well if protected by warmth radiated from supporting wall.
Maintenance: Trim branches occasionally to control growth.

Ficus microcarpa
Indian laurel fig, Chinese banyan, glossy-leaved fig _____

(F. retusa nitida). Recent botanical classification seems to have confused the proper scientific name of this plant and a definite variety formerly known as *F. retusa nitida*. Presently, it is unclear just where this variety belongs, so we are

Ficus microcarpa

covering it under Cultivars because of familiar past use. Originally from the Malay Peninsula to Borneo, this spreading tree has a dense, rounded crown to 30 feet or more with pendulous side branches and smooth gray bark. It provides welcome shade for large areas.

Cultivars: *F. microcarpa nitida*, Laurel de India, is the tree most often seen in Mexican parks and plazas. It is smaller than the species (to 20 feet) at least in youth, and is more upright and stiffer in form. The crown is more conical at top, and foliage is stiffer and more pointed. Symmetrical growth habit makes it easy to shear and train for formal effects.

Special design features: Formal and dense appearance. Rich tropical foliage. Clean tree near pools or in contained areas.

Uses: Street, park or patio tree for warm areas. Excellent for shearing or shaping or as a large container plant. Effective as a spreading overstory tree or for formal topiary gardens.

Disadvantages: Tender to frost and subject to foliage distortion when attacked by the laurel mite. Surface rooting and trunks which form buttresses—a problem near paving or lawns.

Planting and care: Plant from containers when frost danger is over. Large specimens are often available. Prepare a deepened plant pit where caliche or other soil problems exist. Shallow soil over impenetrable soil layers increases surface rooting problems. Trees tend to be low branching and will need training and staking if canopy shapes are desired.

Low zone, indoors anywhere

Evergreen

Soil: Tolerant, but prefers well-prepared garden soil.
Sun: Full sun to some shade. Appreciates afternoon shadow.
Water: Moderate to ample.
Temperature: Damaged at about 25°F (−4°C). Tolerant of high heat.
Maintenance: Little to regular depending on use.

Ficus microcarpa nitida

FOUQUIERIA splendens

Family: _Fouquieriaceae_

This dramatic plant is native to the Southwest and northern Mexico. It grows slowly to 12 feet, sometimes 18 feet high. A few to several woody, thorny, mostly unbranched canes rise upward and outward from its base. After several years plants may reach a diameter of 10 feet under favorable circumstances. Each spring and after summer rains canes suddenly sprout green leaves their whole length. Leaves remain for several weeks before turning yellow and dropping. Spiky flamelike clusters of red tubular flowers from new growth develop at cane tips each spring. They remain for several weeks before plant goes dormant. Ocotillo quick-ly responds to moisture, putting out leaves in 4 or 5 days. Conversely, plants go dormant equally fast when moisture dries up. Ocotillos are protected by the native plant laws of some states and cannot be sold or moved without a tag from the United States Commission of Agriculture.

Special design features: Stark and dramatic silhouette against masonry or sky. Bold emphasis. Groupings, especially "living fences," are more striking than single plants.

Uses: Dry areas, desert or natural gardens. Minimum landscaping with rocks or a few low plants with a mulch.

Disadvantages: Sometimes subject to rust or powdery mildew in cultivation.

Planting and care: Plant bare root from the nursery any time. Plants are most effective set with random spacing when used as a grouping. Plants root easily from cuttings, either long whips or short pieces, especially in spring and early summer. Spray for rust or powdery mildew.

All zones

Deciduous

Soil: Likes rocky soil with good drainage.
Sun: Full to reflected sun.
Water: Irrigate newly set-out bare root plants weekly until established. Later, allow soil to dry out in between. Supplemental water is needed only in dry years or where rainfall is under 6 or 7 inches annually. Constant or regular irrigation may kill plants.
Temperature: Hardy to about 10°F (−12°C).
Maintenance: None.

Fouquieria splendens

Fouquieria splendens

FRAXINUS uhdei

Family: _Oleaceae_

This upright tree grows 25 to 30 feet high in only 10 years. A native of Mexico, it is a favorite in the lower zone due to its fast growth. In 20 years, trees in favorable situations may reach 40 to 60 feet in height. Young trees are narrow but in time develop billowing crowns as wide as the tree is high. Trees are nearly evergreen, except for old foliage that thins or falls just before new, dark green, 4-inch leaflets appear in February. Some seasons new foliage appears before the last date of killing frost, damaging the tree. A young tree may lose branches as well as leaves from the cold, but recovers quickly. Unimportant blooms come out as leaves appear, winged fruit follows.

Cultivars: 'Majestic Beauty' is more dependably evergreen and has a better form and larger leaves.

Special design features: Vigorous, upright, billowing form. Fast shade.

Uses: Shade tree for large areas such as parks or large lawns or near tall buildings. Not for narrow spaces or areas near pavement; its buttressed trunk and surface roots will heave structures.

Disadvantages: Cold damage. Leaf burn from hot dry winds. Litter on paving from seeds and flowers. Limb breakage from high winds.

Planting and care: Plant from containers in spring when danger from frost is past. Space young trees 30 feet or more apart. Correct pruning helps form a strong tree. Cut back long side branches of young trees. Prune weak branches at deep V-shaped crotches. Leave only well placed, strong side branches to develop as the tree's scaffold—its basic framework. Correct buttress roots on old trees by cutting below the ground surface and forcing development of brace roots. Irrigate deeply to encourage deeper rooting. If tree shows signs of iron chlorosis, yellowing of leaves, apply an iron foliar spray for immediate treatment. Annual applications of iron sulfate will usually prevent chlorosis from occurring. See page 35.

Low zone, marginal in middle zone

Evergreen to briefly deciduous

Soil: Prefers deep soil.
Sun: Part shade to full sun.
Water: Give deep irrigation frequently, especially in summer.
Temperature: Branch and foliage damage occurs at approximately 22°F (−5°C). Serious damage occurs at 15°F (−10°C) or lower.
Maintenance: Regular garden care.

Fraxinus uhdei

FRAXINUS velutina
Family: *Oleaceae*

(F. toumeyi). This tree and its cultivars are some of the best and fastest-growing shade trees for the home garden to be found in these zones. Native to the Southwest, trees grow naturally in canyons and along water courses. Trees are erect and grow fast to 30 to 50 feet with a spread of 20 to 30 feet. Somewhat open irregular crown of strong scaffold branches supports medium green leaflets, duller and more velvety than some of the cultivars. Inconspicuous flowers appear in spring before foliage comes out and bear winged fruit among the leaves of female trees in late spring. Trunk is covered with a gray bark of even roughness.

Cultivars: 'Modesto' is vigorous, more refined and symmetrical than the species, usually smaller and more compact. Its dense crown of shinier, brighter green leaves is supported on scaffold branches. 'Rio Grande,' also known as fantex ash, has large, darker green leathery leaves resistant to burn from the hottest sun or wind. It is very vigorous, fast growing, highly resistant to drought and less subject to chlorosis caused by soil alkalinity than *F. velutina* or 'Modesto'.

Special design features: Handsome, upright shade tree. Yellow fall color.

Uses: Fast shade for lawn, patio, street or park. Excellent for summer-shade, winter-sun combination. Looks attractive in combination with evergreens. Specimen, row or grove.

Disadvantages: Iron chlorosis in poorly drained or heavily alkaline soils. Occasional mistletoe infestations. "V" crotch on 'Modesto' is subject to breakage by wind. Heart rot may infest pruned or damaged trees that have not been properly treated and sealed.

Planting and care: Plant from containers any time of year or bare root in winter. Space trees 25 feet or more apart for rows or groves. Carefully prune and shape young trees to remove weaker branches at each crotch to avoid wind damage. Treat cuts or wounds. If tree yellows, treat with iron foliar sprays and then feed iron sulfate once or twice a year on a regular basis.

All zones

Deciduous

Soil: Tolerant. Prefers loose soil with good drainage and some humus added.
Sun: Part to full or reflected sun.
Water: Best with widely spaced deep irrigation once established.
Temperature: Hardy. Tolerant of heat.
Maintenance: Occasional.

Fraxinus velutina

Fraxinus velutina

GARDENIA jasminoides
Family: *Rubiaceae*

(G. florida, G. grandiflora, G. radicans). Gardenias are plants to fuss over. The species is originally from China, but there are numerous attractive cultivars suitable for cultivation in arid zones. Although camelia is a doubtful choice in these zones for all but the most dedicated gardener, gardenia, often a companion plant, can be grown with great success. The species grows to 6 feet in height where conditions are favorable, but is usually seen at 4 feet in hot, dry areas. Leaves to 4 inches in length are a crisp, shiny, deep green. Its exceptionally fragrant double white flowers to 3 inches in diameter are seldom produced in these areas. The following cultivars are more successful.

Cultivars: 'Veitchii' is more compact, tougher and less temperamental than the species and has smaller leaves and flowers. Grow it in protected areas such as north or east sides or in sheltered patios. It will produce a profusion of fragrant, double white flowers to 2-1/2 inches for a period in spring before dry heat appears. 'Veitchii Improved' is said to be still more luxuriant, more compact and a better bloomer. Tiny 'Radicans' grows to only 6 to 12 inches, spreading 2 to 3 feet with smaller sometimes variegated leaves and 1-inch flowers in summer. Plant under shady trees, in shady planters or on north sides. It generally does not perform as well as 'Veitchii'.

Special design features: Lush green of the woodsy garden. Tropical fragrance. Refined elegant plant.

Uses: Small planting areas in protected gardens. Containers. Best with many other plants close by so that humidity is increased, especially in the desert.

Disadvantages: Temperamental. Very subject to iron chlorosis. Bud drop from stress. Caused by any sudden change such as a variation of temperature, a drop in humidity or if soil is too dry or soggy. Flowers damaged by hot dry air.

Planting and care: Plant in carefully prepared soil. Keep plant roots carefully mulched and avoid cultivating around them. Feed monthly with a commercial fertilizer especially prepared for gardenias. This is especially important in areas having water with a high pH. Feed extra iron if foliage yellows or pales.

All zones

Evergreen

Soil: Highly enriched, organic, gritty soil with good drainage.
Sun: Filtered to open shade to part sun. Needs afternoon shadow, especially in low zone.
Water: Ample.
Temperature: Hardy to 20°F (−7°C) or below.
Maintenance: Constant care.

Gardenia jasminoides 'Veitchii'

GAZANIA rigens
Family: *Compositae*

Gazanias hug the ground in leafy clumps and produce daisylike flowers 3 inches or more across on 6-inch stems rising above the foliage. Gray-green, sometimes lobed leaves are narrow, to 3 or 4 inches long, and radiate from the center of the clump. They bloom mostly in late winter and spring in low and middle zones before the heat of summer. In cooler summer areas they may continue to bloom throughout the growing season. Yellow and white clumping types tend to be everblooming.

Cultivars: 'Copper King' is generally considered to be the hardiest and blooms only in winter and spring. It has large orange flowers with wine-red markings toward the center, very dramatic as a mass planting. 'Gold Rush' is bright orange-yellow

---

All zones

Evergreen perennial

Soil: Prefers enriched soil with good aeration and good drainage.
Sun: Part to full sun but accepts filtered shade.
Water: Allow the ground to dry out slightly between waterings, especially in summer. This helps avoid damping off.
Temperature: Hardy to below 20°F (−7°C). Languishes in intense heat.
Maintenance: Occasional.

---

with brown spots at the base. Blooms are abundant and long lasting. 'Royal Gold' has bright double yellow blooms. The 'Colorama' strain comes in white, cream, yellow, gold, yellow-orange or pink with purple undersides.

Special design features: Brilliant color at ground level.
Uses: Borders, ground cover, under trees producing only filtered shade. Bedding plant or filler between young plants of a more permanent nature which are slow to mature.
Disadvantages: In heavy soils an unidentified root rot or damping off fungus may wipe out parts of plantings in summer. Plants fail to cover, become stunted or die out, leaving bare areas. They often decline during hot weather.
Planting and care: Plant from flats, root divisions, seed or containers in fall or spring. Space plants or thin seedlings to one plant 9 to 12 inches on center to cover. Refurbish bare spaces in spring and fall. Divide every 3 to 4 years.

Gazania rigens leucolaena, trailing yellow gazania, is not as hardy but makes a quicker and better cover. It is an evergreen perennial in the low zone, an annual in cold winter areas. Yellow or orange daisylike flowers to 2-1/2 inches across bloom all winter in milder areas, as well as during warm weather. A single plant spreads by trailing stems to fill an area 12 to 18 inches in diameter in a short time. Silvery gray foliage contrasts with flowers. Plant in masses or rock gardens or allow it to drape over walls or containers. Like the clumping gazania, use as a border, filler, foreground or bedding plant. Space plants 9 to 12 inches apart and follow cultural requirements for *G. rigens.*

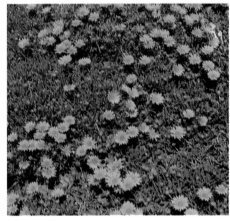

Gazania rigens leucolaena

GEIJERA parviflora
Family: *Rutaceae*

A graceful, weeping tree with medium green, fine-textured foliage, this distant relative of citrus looks like a refined, willow-leafed eucalyptus, and is just as hardy. Branches grow upwards with the branchlets holding narrow 3 to 6-inch leaves which hang down like a willow. A moderate grower to 15 to 25 feet high with a spread of under 20 feet. Bloom is insignificant. Although drought-resistant, it will grow in well-watered places such as lawns. This well behaved Australian native deserves wider use in warm climate areas.

Special design features: Willowlike weeping form, light shade, rustle of leaves.
Uses: Patio or street tree. Effective in grove or row plantings as a visual, wind or sun screen. Does well in neglected areas, on south sides or in lawns.
Disadvantages: Damaged by severe freezes.
Planting and care: Plant any time in low zone, in spring after frost is past in middle zone. Space 15 feet apart for a grouping or screen, more for a street tree. Only light pruning is necessary.

Low zone, protected places in middle zone

Evergreen

Soil: Tolerant. Needs good drainage.
Sun: Full to reflected sun. Tolerates part shade, but looks thin and spindly.
Water: Tolerates drought when established. To promote growth, supply moderate to ample water. Once it reaches the size you want, taper off and give only deep, widely spaced irrigation.
Temperature: Tree may defoliate if temperature reaches around 18°F (−8°C) and lose branches below that, but it recovers quickly.
Maintenance: Little.

Geijera parviflora

GELSEMIUM sempervirens
Carolina jasmine • Confederate jasmine • Yellow jessamine

Family: *Loganiaceae*

This is not a true jasmine, but its flowers resemble the Italian jasmine. It grows at a moderate rate to 20 feet, providing an uneven and tangled cover. Shiny pointed leaves are yellowish green and densely cover the billowing, twining growth. Clear, yellow, trumpet-shaped flowers to 1-1/2 inches long bloom in profusion along slender stems during late winter and early spring. Bloom is sometimes fragrant. Native to the southeastern United States.

Special design features: Vivid color when little else is in flower. Woodsy effect.

Uses: A vine which needs tying for porch posts, fences, walls and trellises. It can also be used as a ground or a bank cover, but may tangle and mound, covering unevenly. It is also effective spilling from containers or draped over walls.

Disadvantages: All parts of plant are poisonous. In time it becomes overgrown and top heavy with dead underbranches. Flowers can be a litter problem.

Planting and care: Plant from containers any time. Space 3 feet apart for a ground or bank cover, 4 to 8 or even 10 feet apart to make an attractive and interesting fence or wall plant. Tie to supports to train. For a ground cover, use U-shaped wires like giant hairpins to stake stems to the ground in the pattern desired. Prune overgrown plant severely in winter after the bloom is over; plants make a fast recovery.

All zones
Evergreen
Soil: Tolerant. Prefers improved garden soil.
Sun: Open to part shade, even full shade but blooms later in spring. Full and reflected sun for early bloom.
Water: Moderate. Established plants are quite drought-resistant.
Temperature: Loves heat. Hardy to 15°F (−10°C) or below. Recovers from cold damage quickly.
Maintenance: Occasional.

Gelsemium sempervirens

Gelsemium sempervirens

GLEDITSIA triacanthos inermis
Honey locust • Sweet locust • Honeyshuck

Family: *Leguminosae*

Thornless honey locust is very erect, growing at a moderate rate to 30 or even 75 feet high in the warm arid zones, spreading as much as 50 feet. Widely adaptable, needing little care. Train trees to have a tall trunk before scaffold branches spread out in a vase shape to form an eventual oblong crown. Also attractive as a multiple-trunked tree. Small leaflets range in color from bright spring green to deep green in summer, then turn yellow in fall. Unimportant greenish flowers produce large, flat, black pods to 12 inches in length that hang on the tree through winter. A long dormant period prevents new leaves from being damaged by late spring frosts. Get one of the cultivars of *G. t. inermis* if you want to avoid the sharp thorns of the species. There are a number of cultivars to choose from. All are excellent choices for summer-shade and winter-sun. Native to the eastern United States.

Cultivars: 'Sunburst' has golden green foliage and an irregular form. 'Moraine' is faster growing, more spreading in form. 'Imperial' is symmetrical with dense leaves that reach to about 35 feet at maturity. 'Shademaster' is less spreading and a rapid grower to 24 feet high and 16 feet wide in about 6 years, eventually reaching 40 to 50 feet.

Special design features: Erect vertical form. Handsome silhouette. Delicate summer foliage. Fall color.

Uses: Lawns, parks, boulevards, large patios. Deep roots do not heave paving. Foliage allows enough light to pass through to sustain lawns. Withstands city air pollution. Grow anywhere heat, wind, acid or alkaline soil are a problem.

Disadvantages: Occasional infestations of mistletoe. Sometimes seems to be slower growing in low deserts. Pods can be unsightly.

Planting and care: Plant from containers any time or bare root in winter. Easy to transplant large specimens during dormant period. Space 20 to 40 feet apart for boulevard plantings, 15 to 20 feet for groves. Irrigate deeply to encourage deep rooting and faster growth.

All zones
Deciduous
Soil: Tolerant except for very heavy soils.
Sun: Full sun.
Water: Deep widely spaced irrigation. Young trees need frequent watering during growing season.
Temperature: Endures heat and cold.
Maintenance: Little to none.

Gleditsia triacanthos inermis

GREVILLEA robusta

Silk oak

Family: *Proteaceae*

Erect and columnar to pyramidal in form, silk oak grows rapidly 50 to 60 feet in height with a spread of 25 feet when grown in deep soil. In desert areas it rarely reaches its maximum height of 150 feet. In shallow soils it may reach only 30 feet with a 15-foot spread. Medium to dark green fernlike foliage densely covers the tree. Mature trees become broad crowned and picturesque. Dense flat clusters of colorful yellow-orange flowers appear in spring on short leafless branches of old wood; they attract birds and bees. Small woody pods with black seed follow. Silk oak is considered evergreen, but can lose leaves from cold or become briefly deciduous in spring at bloom time just before new leaves come out. Native to Queensland and New South Wales.

Special design features: Strong vertical for a skyline silhouette.

Uses: Specimen or row in parks, by roadsides or in large public areas in warmer low zone. In the middle zone it is best on south exposures, protected from the full force of a cold winter.

Disadvantages: Very susceptible to Texas root rot. Neglected or older trees develop dead unsightly branches that require pruning. Susceptible to iron chlorosis, especially in lawns.

Planting and care: Can be grown from seed. When planted from containers set out in spring after frosts. Space 15 to 20 feet apart for massing or a grove, 25 to 30 feet or more for a row. Stake willowy young trees against the wind. Shorten side branches to balance framework. Give iron if trees become chlorotic.

Low and middle zones

Evergreen

Soil: Tolerant. Prefers deep soil. Good drainage is essential.
Sun: Part, full or reflected sun.
Water: Best with deep spaced irrigation. Accepts more water with good drainage. Tolerates drought when established but grows slowly and looks poorly.
Temperature: Young trees are more sensitive, damaged at about 24°F (−4°C). Older trees tolerate temperatures to approximately 15°F (−10°C) with minor foliage damage but recover quickly.
Maintenance: None to occasional.

Grevillea robusta

HEDERA canariensis

Algerian ivy • Canary ivy • Madeira ivy

Family: *Araliaceae*

Algerian ivy is a very appealing plant with large, deep green, lobed leaves which vary from 3 inches wide when young to 6 or 8 inches wide when mature. Foliage is supported on wine-red stems which add color to the dark green form. Mature growth is stiff and woody, producing eliptical leaves and greenish flowers, but these are rarely seen. Once established, Algerian ivy grows aggressively to cover a large area. Bolder and more casual in appearance than English ivy, but not as good a climber. Weight of mature leaves is likely to pull the vine away from its attachment. Vines are best tied to supports or grown on the ground. Native to North Africa and the Canary Islands.

Cultivars: 'Variegata' (variegated ivy, Hagenburger's ivy, Gloirede Marengo ivy), is an often-preferred form of Algerian ivy. Leaf edges are a creamy color and the center appears to be a grayer green. It is a bright contrast to dark wood or deep shade and dark green plants.

Special design features: Bold foliage. Tropical feeling.

Uses: Shaded places as a ground cover, especially beneath deciduous trees. Tie to porch posts, cool walls, fences or trellises where it will get afternoon shade.

Disadvantages: Subject to sudden death by a damping off fungus in summer. Can be affected by Texas root rot and snails and slugs. May be slow to start but then becomes very aggressive, overcoming weaker plants nearby unless restrained. Leaves burn in hot sun.

Planting and care: Plant any time of year. Thoroughly soak both plant and ground before planting. Space 3 to 4 feet apart for a fast ground cover. Do any severe pruning in late winter before spring growth begins. Trim lightly as needed. Feed heavily in early spring.

All zones

Evergreen

Soil: Tolerant. Prefers improved garden soil. Needs good drainage.
Sun: Morning sun. Full, open or filtered shade. Accepts more sun at higher elevations or in winter.
Water: Moderate to ample. Established plants withstand some drought, especially in winter. Water plants early in the day so foliage will have time to dry out to prevent fungus.
Temperature: Hardy to about 10°F (−12°C). New growth may be damaged below 20°F (−7°C).
Maintenance: Occasional control of aggressive vines.

Hedera canariensis

Hedera canariensis 'Variegata'

HEDERA helix

Family: *Araliaceae*

An old favorite, English ivy is a widely adaptable, self-climbing vine originally from Europe, Western Asia and North Africa. It covers densely, climbing walls, posts and tree trunks with tightly clinging aerial rootlets. Leaves are dark dull green, 2 to 3 inches long with 3 to 5 lobes and whitish veins.

Cultivars: 'Baltica' has small juvenile leaves which turn purplish in winter. It trains well as a pattern plant on walls. 'Pittsburgh' ('Hahn's Self-Branching') has lighter green leaves with dense branches. Juvenile leaves become larger on mature plants. Both take at least as much sun as *H. helix*, perhaps more. There are many other cultivars, some with miniature leaves, which make excellent houseplants.

Special design features: Dense cover of cool dark green.

Uses: North masonry walls, porch posts, ground covers, under stairwells or deciduous trees.

Disadvantages: Rootlets scar masonry walls. Aggressive and invasive once established, growing tips sometimes find crannies and enter the house. Sunburn where exposed to summer sun. Harbors slugs and snails. Subject to Texas root rot and summer fungus, especially in warm moist soils.

Planting and care: Plant any time of year from containers or flats. Moisten planting hole and soil around plant in container before planting. Space 12 to 18 inches apart from flats for a ground cover. To train on walls, one plant may be spaced 3 to 6 feet on center for cover. Do any heavy pruning in late winter before new growth starts. Trim lightly any time. At times ivy will overgrow and requires severe pruning to clear out old growth and to rejuvenate. Feed heavily in early spring.

Hedera helix

All zones
Evergreen
Soil: Tolerant. Best with some organic material added to top 6 inches of soil.
Sun: Full, open or filtered shade to morning sun. Accepts more sun in high zone.
Water: Moderate to ample.
Temperature: Hardy. Accepts heat with shade and ample water.
Maintenance: Occasional.

HESPERALOE parviflora

Red yucca ● Semandoque ● Red hesperaloe

Family: *Agavaceae*

(H. yuccifolia). Red yucca is a clumping plant with narrow, curving, straplike leaves which rise from its base to a height of 3 or 4 feet. Leaves are gray-green, stiff and fleshy with numerous fibrous threads along the edges. Plants slowly enlarge by clumping outward to form an irregular cluster of foliage 3 to 4 feet wide. Tall leaning spikes with scarlet to coral bell-shaped flowers appear late spring into early summer. Native to Texas and Mexico.

Cultivars: Variety *H. engelmannii* is a hardier form for the high zone.

Special design features: Desert effect. Spring color. Slender straplike leaves combine well with palms and yuccas.

Uses: Dry areas. Natural, desert or rock gardens. Transitional areas. Containers.

Disadvantages: Tough, but slow to develop. Sometimes eaten by rodents.

Planting and care: Plant any time from containers. Old clumps may be divided in late winter and reset to form new plants. Remove old flower spikes after bloom has passed.

Hesperaloe parviflora

All zones
Evergreen
Soil: Tolerant of many soils.
Sun: Full or reflected sun.
Water: Moderate to none. In areas with less than 10 inches of rainfall supply supplemental irrigation.
Temperature: Foliage damaged around 15°F (−10°C). *H. engelmannii* is hardy to 0°F (−20°C).
Maintenance: None except for removing old flower spikes.

HIBISCUS rosa-sinensis   Chinese hibiscus • Hawaiian hibiscus • Rose-of-China • China rose
Family: *Malvaceae*

(H. sinensis, H. chinensis, H. fulgens). This native of Asia imparts a sensuous, tropical feeling with its almost continual bloom of large bright flowers. The most common flower is single red, 4 to 8 inches across, with spreading petals and a yellow central stalk. Other colors are pink, white, yellow, orange, deep red and combinations of colors in the variegated forms. Flowers are available in double as well as single forms, but double flowers do not seem to hold up as well in desert heat. In the low deserts plants grow as erect shrubs, reaching only 4 to 6 feet high, often less in the middle zone where they are commonly nipped by winter cold. There are numerous cultivars, many with slightly different flowers, foliage and growth habits.
Special design features: Tropical color.
Uses: Feature plant for sheltered locations away from cold in the middle zone and from hot winds and intense afternoon sun in the lower zone. Best with southern and western locations against a wall and under an overhang. Excellent as an espalier or container plant.
Disadvantages: Susceptibile to cold, heat and wind damage. May also be attacked by aphids, bud weevil or Texas root rot.
Planting and care: Plant in spring when danger from frost is past. Space 4 to 5 feet apart for loose screens. Pinch stem tips during growing season to encourage bushiness and good form. Remove one-third of old wood on older plants each spring to keep vigorous and to encourage new growth. Feed monthly during growing season, more often in containers. In areas of frost, mound base of plant with soil in fall to protect lower stems.

Hibiscus rosa-sinensis

> **Low zone, protected locations in middle zone**
>
> **Evergreen**
>
> **Soil:** Good drainage is a must. Prefers light improved soil. Avoid cold, wet, heavy soils.
> **Sun:** Part shade to full sun. Needs afternoon shade in middle and low zones.
> **Water:** Soak deeply, allowing surface of ground to become dry to the touch before next irrigation. Best to water just before the plant wilts.
> **Temperature:** Foliage and tender stems are damaged between 28° and 25°F (−4° to −2°C). Plant freezes to the ground below that, but recovers quickly in spring with feeding and irrigation and will grow to 3 feet or more by the end of summer.
> **Maintenance:** Seasonal.

ILEX cornuta 'Burfordii'   Burford holly
Family: *Aquifoliaceae*

Although most hollies seem to languish in the warmer zones, this cultivar of Chinese holly does very well, especially if it gets an afternoon shadow in hot summer areas. It grows at a slow to moderate rate up to 6 feet high and 4 feet wide, but is easily maintained at a smaller size. Leaves are usually thornless, but retain the shiny crispness characteristic of holly. Berries are large, bright red, long lasting and need no pollinator. Berry production in the desert is usually sparse, especially on clipped plants.
Cultivars: 'Dwarf Burfordii' is smaller, slower growing and more compact.
Special design features: Crisp, bright to deep green. Woodsy effect.
Uses: Clipped hedges or screens. Pool areas or anywhere litter would be a problem. Espalier on cool walls. North and east exposures. Larger plants may be used as a loose screen, but have an uneven growth and shape which doesn't match the neatness of the foliage. Usually more attractive when clipped. The dwarf looks good close-up and can be used in small intimate gardens, such as entryways and atriums.
Disadvantages: Leaves may sunburn in hot sun. Subject to iron chlorosis in alkaline soils. Usually a problem-free plant, however.

Planting and care: Plant from containers any time. If pruning is necessary, do it in late winter before new growth starts. Mulch roots against summer heat. Do not cultivate around roots as it disturbs them. Flowers appear on old wood. Pruning and clipping lessens their number and the number of berries. Feed iron if plants become chlorotic.

> **All zones**
>
> **Evergreen**
>
> **Soil:** Tolerant. Best with improved garden soil.
> **Sun:** Open or filtered shade. Morning sun. Accepts full sun in cooler summer areas or the high zone.
> **Water:** Moderate. Accepts widely spaced soakings once established. Tolerates little to ample water.
> **Temperature:** Hardy to cold. Accepts heat with shade.
> **Maintenance:** Occasional to almost none depending on taste and use.

Ilex cornuta, the Chinese holly, is favored for its crisp, spiny, evergreen leaves and occasional red berries. It is very marginal except in the high zone and grows slowly to make a large shrub or a small tree, reaching an eventual height of 6 to 10 feet. Its open form bears leaves to 2 inches long with sharp spines at each of the four corners and at the tip. A plant for north sides and shaded places as a shrub, foundation plant, espalier or standard tree. Chinese holly is not as dependable or as widely used in these zones as the cultivar, 'Burfordii'. Besides a need for more shade and a greater susceptibility to sunburn, the cultural needs of Chinese holly are the same as 'Burfordii'.

Ilex cornuta 'Burfordii'

ILEX vomitoria

Family: *Aquifoliaceae*

Yaupon • Cassina

Yaupon is native to an area from Texas to Florida and southeastern Virginia, and does well in the deserts and warm interior valleys of the Southwest. Foliage and structure are different from Chinese and English hollies—denser, more finely twigged and with smaller oval leaves which have tiny thornless scallops at the edges. Dried leaves have long been used to brew a bitter tea. Tiny whitish flowers appear in small clusters in spring on wood from the previous year and produce numerous, small, scarlet berries in fall. No pollinator is needed. The species grows to 15 or 20 feet tall. There are also dwarf forms.

Cultivars: 'Pride of Houston' is medium size and erect with a loose, branching habit. 'Stokes', is very dwarf and compact, forming a spreading mound 12 inches high and 18 inches or more wide. 'Nana' is similar to 'Stokes,' with a mounding form that grows somewhat higher. Both are drought-resistant once established, and make excellent low-growing hedges.

Special design features: Dense dark green with a refined texture.

Uses: Larger species makes good topiary plants, columns or a large loose but dense screen or clipped hedge. 'Pride of Houston' can be used similarly, but will not get as large. Use dwarf forms where a low foreground plant is needed for close-up viewing. All are neat for pool patios, atriums, entryways or small patios. Or, use them in containers or confined spaces. They are easy to control, look nice when sheared and hold their form for a long time.

Disadvantages: Subject to occasional scale invasions.

Planting and care: Plant any time from containers. Space larger shrubs 4 to 8 feet apart for hedge or screen. Space dwarf forms 12 to 18 inches apart for continuous planting. Do any heavy pruning in late winter, but shear any time. Tolerant of neglect but looks best with regular care.

All zones

Evergreen

Soil: Tolerant. Prefers improved garden soil with good drainage.
Sun: Open to filtered shade. Part to full sun.
Water: Moderate to ample.
Temperature: Tolerant of heat and cold.
Maintenance: Little needed. Accepts as much as you care to give.

Ilex vomitoria

JACARANDA mimosifolia

Family: *Bignoniaceae*

Jacaranda • Green ebony

(*J. ovalifolia,* confused with and sold as *J. acutifolia*). This Brazilian native produces a breathtaking sight in mid to late spring when it produces large sprays of lavender-blue flowers. It grows at a fast to moderate rate reaching 25 to 50 feet high, spreading 25 to 40 feet wide. This tree revels in heat and requires it to flower. Flowers appear any time from April to September, but the majority are usually seen in May or June. Tree normally drops its leaves in February or March and stays bare until flowering time or sprouts new leaves soon after dropping old ones. Flowers are 2 inches in length and look like trumpets. Some cultivars have blooms of white or orchid-pink. White selections may blossom over a longer period, but with less profusion and with lusher foliage. All forms have dark, rounded 2-inch seed capsules.

Special design features: Tropical effect. Color during the warm season. Shower of petals on pavement is pleasant to some, a nuisance to others. Ferny foliage texture.

Uses: Specimen or grouping on south or west sides of tall buildings or at north end of courtyard where it receives southern sun and reflected heat. Plants can be located in the open in frost-free areas.

Disadvantages: Cold sensitivity may preclude bloom. Unattractive for a period in spring. Litter of leaves, blossoms and seed capsules.

Planting and care: Plant in spring from containers after weather warms up. Space 15 to 25 feet apart for a grouping. Stake young trees to form a single, erect trunk. Or, grow as a multiple-stemmed or multiple-trunked shrubby plant, especially where plants are frozen back regularly. Prune to produce form.

Low zone, middle zone if sheltered

Briefly deciduous in early spring

Soil: Tolerant. Prefers loose sandy soil.
Sun: Full to reflected sun.
Water: Give widely spaced, deep irrigation. Too much water in late summer and early fall produces succulent growth tender to frost. Taper off watering as fall approaches.
Temperature: Young trees damaged at about 25°F (−4°C). Older ones can tolerate temperatures a few degrees lower. When wood is damaged in hard freezes the next spring's flowers are destroyed.
Maintenance: Regular garden care.

Jacaranda mimosifolia

JASMINUM mesnyi
Primrose jasmine • Yellow jasmine • Japanese, Chinese jasmine
Family: *Oleaceae*

(J. primulinum). Primrose jasmine is a bold, sprawling shrub with branches trailing 6 to 10 feet in length. Untrained and untied, its natural form is a spilling fountain of rapid growth. It trains well into a wall plant and is one of the few plants for clipped hedges that will bloom profusely even when sheared. Medium green leaflets of three along the square stems are rather coarse in texture. Clear, yellow, single or double flowers bloom for a long period in late winter or early spring. Flowers are wide open and evenly spaced in pairs along the slender stems and look like popcorn after they dry out. No seeds

All zones

Evergreen

Soil: Tolerant. Appreciates improved garden soil with good drainage.
Sun: Part to full or reflected sun. Prefers moist but not soggy soil.
Water: Tolerates moderate to ample. Does well with widely spaced deep irrigation, but wilts quickly when the soil dries. It recovers rapidly when water becomes available.
Temperature: Hardy. Accepts heat of south and west sides.
Maintenance: Regular garden care for the most attractive plants.

are produced and there is no fragrance. Native to western China.

Special design features: A bold plant. Spring color. Informal.

Uses: Large scale ground cover, foundation plant or background plant. A wide screen or space divider. Excellent as a clipped hedge. Trains well over walls or high planters or banks.

Disadvantages: Rank growth unless cut back every few years or clipped regularly. Becomes woody with age. Develops iron chlorosis in overwatered situations, yet shows drought stress quickly.

Planting and care: Plant from containers any time. Space 6 feet apart for a mounding effect, 18 to 24 inches on centers for a low, clipped hedge, 3 feet apart for a higher one. Shear any time for a hedge. To rejuvenate, cut back overgrown plants nearly to the ground in spring after bloom has passed. Plants tolerate neglect but become woody and unattractive. If plants yellow, give extra iron.

Jasminum humile is another yellow-flowering jasmine that deserves mention. The single flowers are smaller but very fragrant. It is an evergreen shrub that grows to 6 feet high or more.

J. grandiflorum is also commonly planted in these zones. It is a self-climbing vine that will cover a fence or trellis with its evergreen foliage. Loose clusters of white, fragrant, tubular flowers bloom during the warm time of the year. Cultural requirements and adaptations for both of the above are very similar to *J. mesnyi*.

Jasminum mesnyi

JUNIPERUS species
Juniper
Family: *Cupressaceae*

Junipers are a numerous group of coniferous plants originating from parts of the Northern Hemisphere. Because they are easy to grow, dependable and tolerant of a range of conditions, they are widely planted in western gardens. Colors range from silvery to blue to bright green, deep green, yellow-green and gray-green, as well as variegated. Juniper forms are also many: prostrate or creeping, low spreaders to knee high and shrubs 6 to 8 feet high. Some kinds become small to medium trees after many years—others grow into tall trees seldom seen in home landscapes.

Form varies from horizontal or spreading to those with fountainlike, upward-pointing branches. There are upright columnar types and weeping forms. All are recognizable as junipers because of their dense, scaly, sometimes needlelike foliage similar to that of cypress.

Over the years junipers have been the subject of much manipulation and selection by horticulturalists. There are many, many varieties available, and a complete list would require many pages. Juniper admirers in every region have different opinions of which junipers are the best. The following includes a few selections in each general landscaping category.

Special design features: Dense evergreens with medium to fine-textured foliage. Woodsy feeling. Some produce an Oriental effect. Transitional when used with dry, desert-type plants, but also fit into the well-watered landscape.

Uses: Varied landscape uses according to plant.

Disadvantages: Large mature plants usually lose their form and charm when branches are cut back beyond the foliage. Foliage does not regrow quickly and may not come back at all if pruned severely. Foliage is somewhat prickly when handled. In hot areas the inner foliage on some types burns from the sun and

Juniperus horizontalis 'Wiltonii'

becomes straw-colored and unattractive. Some very low types burn out completely because the foliage does not shade the soil over their roots to keep them cool. Junipers are subject to infestations of spider mites. Control with malathion, kelthane or other contact spray. Aphids cause falling needles, sticky deposits and sooty mildew. Control is the same as for mites. Twig borers cause branch tips to brown and die back. Control by spraying from mid May to mid June with Sevin or diazinon or with copper sprays in mid and late summer. As with all ground covers, plantings of lower-growing junipers suffer from invasions of Bermudagrass or weeds. Plant only where Bermuda has been removed or destroyed to prevent this. Infestations of unwanted plants may be controlled by pulling or by careful application of an herbicide. Preemergents can be effective in controlling weeds.

Planting and care: Plant any time of year from containers. Avoid crowding and plant at recommended distances. Because most junipers are slow to moderate growers, space between plants can be filled with annuals, perennials, small shrubs or a mulch until junipers grow to fill the space. Do not plan to remove plants later in a closely set planting, which is sometimes suggested. Roots of individual plants usually intertwine so removal of any plant may injure others. Be sure to place plants at least 3 to 4 feet away from curbs or paved edges to prevent them from growing into an area where they will be in the way. In hot regions mulch roots of young plants to keep them cool until foliage spreads out so plant shades itself. Transplant junipers in fall or winter so they can regrow roots in cool weather. Never give heavy feedings of nitrogen fertilizers. As with all conifers, junipers are sensitive to all but the lightest touch of nitrogen and can be injured by heavy applications. The best fertilizer is a mulch of manure. As plants reach the desired size, trim them by removing the main leaders inside the foliage where the cut will not be visible. This leaves the secondary or side branches and helps maintain the plant's appearance while still containing it. Shearing and shaping destroys the natural form of junipers, but they can be successfully clipped and trained into dense hedges or bonsai. Clipping and shearing is an answer to overgrown plants if removal is the only alternative. Note that junipers placed closer together than their natural width will grow taller than normal. As they begin to crowd together, the only direction they can grow is up.

All zones

Evergreen

Soil: Good drainage is important. Plants are tolerant of many kinds of soil, but prefer enriched porous soils.

Sun: Full sun in high zone. Afternoon shadow in lower hot valleys is desirable. All accept part to full shade but growth may become sparse and rangy. The junipers listed accept full sun even in the lower zone.

Water: Best with deep periodic irrigation—allow soil to dry between waterings. Do not allow young plants to dry out completely until they have become established. Good-looking plantings are often seen adjoining lawns, but as a rule, avoid soggy soil because plants will often collapse from poor drainage.

Temperature: Hardy to 0°F (20°C). Those listed are tolerant of heat and low humidity. Most other junipers prefer some humidity or some shade if they are grown in hot places.

Maintenance: Plants tolerate neglect, but look best with occasional care. Weed and grass invasions of immature plantings are the chief maintenance problem.

Ground Covers

Juniperus horizontalis
Prostrate, creeping juniper _____

(J. prostrata, J. chinensis prostrata, J. virginiana prostrata). Flat heavy branches of this dark green juniper support dense short twigs. Space a minimum of 3 to 5 feet apart and 4 to 5 feet from any edge. Growth: Slow, to 18 inches high. Spreads 8 feet or more.

Juniperus horizontalis 'Wiltonii'
Blue carpet juniper _____

(J. h. 'Blue Rug'). This plant makes a dense, undulating cover of silver-blue, sometimes taking on a purple cast in cold weather. It is the flattest-growing juniper with long trailing branches and short side branchlets. Sometimes burns from reflected heat, especially new plantings in the low and middle deserts. Space a minimum of 3 to 4 feet apart and from any edge. Growth: Moderate to slow, 4 inches high, spreading 8 to 10 feet.

Juniperus sabina 'Buffalo'
Buffalo juniper _____

This juniper is bright green with soft feathery branches and foliage that spread out laterally. Massed, it gives a cross-hatched effect which makes a pleasing pattern. It is very hardy and prefers some humidity or a little shade in the low and middle zones. Space 4 to 6 feet apart and 3 to 4 feet from any edge. Growth: Slow to moderate, to 12 inches high, spreading to 8 feet.

Juniperus sabina 'Tamariscifolia'
Tam, tamarix juniper _____

(J. tamariscifolia). This juniper is dark blue-green and spreads to make a dense

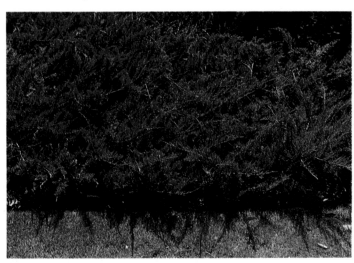

Juniperus sabina 'Buffalo'

Juniperus sabina 'Tamariscifolia'

symmetrical form. Varies from a low, spreading habit to one with more ascending branches which makes a higher plant. Pick plants with the habit you want. One disadvantage is that occasional plants in a mass planting fail for no known reason. Space 5 to 6 feet apart and 4 to 5 feet from use areas. Growth: Slow, to 18 to 36 inches high. Spreads 8 to 20 feet wide.

Shrubs

Juniperus sabina 'Arcadia'
Arcadia juniper

'Arcadia' is low growing with rich, green, lacy foliage that juts outward, forming an attractive tiered pattern. Plants open up in the center, making them less dense than other junipers, but they are very dependable growers. Space 3 to 4 feet apart and place about 5 feet from any edge. Growth: Slow to moderate, 2 to 4 feet high, spreading 4 to 5 feet wide.

Juniperus chinensis 'Pfitzerana Compacta'
Compact pfitzer juniper

(*J. pfitzerana nicksi, J. nicksi compacta*). Compact, densely branched and gray-green, this juniper has proved to be popular in the desert. It is low growing, but does not spread quite as much as a ground cover plant. Space 4 feet apart and 5 feet or more from any edge. Growth: Slow to moderate, 2 feet high, spreading 4 to 6 feet wide.

Juniperus chinensis 'Blue Vase'
Texas star juniper

The square, blocky form of this plant is covered with prickly blue foliage. Attractive as a shrub or barrier, space 10 feet apart for specimen plantings in hot areas, 3 feet apart in cooler regions for a continuous planting. Keep 6 feet away from edges or use areas. Growth: Slow to moderate, 3 to 4 feet high and spreading 3 to 4 feet but it gets much larger in hot areas.

Juniperus chinensis 'Mint Julep'
Mint julep juniper

Upward-jutting branches covered with bright green foliage give this juniper the effect of "taking off." Dependable green color for sunny places, hot or cold. Space 5 feet apart and from any edge. Growth: 2 to 4 feet or more high. Spreads to 6 feet wide.

Juniperus chinensis 'Pfitzerana'
Pfitzer juniper

This is one of the all-time durable plants for hot sunny spaces. Pfitzer has dense, deep, gray-green, feathery foliage on upward-sweeping branches which jut out at a 45-degree angle. Space 10 feet or more apart and 10 feet from any edge. Growth: Slow to moderate, 5 to 8 feet high, spreading to 15 feet wide.

Juniperus chinesis 'Mint Julep'

Juniperus chinensis 'Pfitzerana Compacta'

Juniperus chinensis 'Pfitzerana Glauca'

Juniperus chinensis 'Kaizuka' ('Torulosa')

Juniperus chinensis 'Pfitzerana Glauca'
Blue pfitzer _____

Similar to pfitzer in every way except its color is silvery blue. Space 10 feet or more apart and 10 feet from any edge. Growth: Slow to moderate to 5 to 8 feet high. Spreads to 15 feet wide. There is also a variegated form with yellow-tinged foliage, but most landscape plants variegated with yellow or gold look chlorotic in the desert.

Small Trees
Juniperus chinensis 'Kaizuka'
Hollywood juniper _____

(*J. c.* 'Torulosa'). This interesting, twisted, irregular and upright form makes a dramatic silhouette. Looks as if it was frozen in the wind. Deep, rich green in humid climates, dusty olive green in hot arid regions. Aside from the color difference, it grows well in these zones. Space 8 to 12 feet apart. Growth: Slow to moderate, to 15 feet or more high, spreading 6 to 8 feet wide.

Juniperus scopulorum 'Tolleson's Weeping'
Tolleson's weeping juniper _____

This is a small, weeping tree for a patio accent or Oriental effect. Blue or green form. Grows 2 feet a year. Growth: Fast to moderate, 15 feet or higher.

Juniperus scopulorum 'Tolleson's Weeping'

JUSTICIA brandegeana
Family: *Acanthaceae* _____

Shrimp plant • Rattlesnake plant • False hop

(*Beloperone guttata, B. tomentosa*). The shrimp plant presents a sprawling form with soft, medium green foliage and a continuous coppery bloom over the warm season. Bloom appears as a cluster of overlapping bracts to 6 inches in length, resembling a shrimp in form and color. True flowers are small, white, tubular forms among bracts. Leaves are slender ovals to 3 inches long. Once established, plants grow fast to 3 feet in height spreading 4 feet wide.

Cultivars: 'Yellow Queen' has chartreuse bracts.

Special design features: Neat color plant. Tropical to subtropical effect.

Uses: Warm microclimates in middle zone, anywhere in low zone. Effective for small patios, entryways, atriums or near swimming pools. Container plant. South or west sides under overhangs as foundation plant or small area ground cover. Color for public areas and large or small planters, especially where plants will drape and trail. Anywhere for close-up viewing.

Disadvantages: Brittle—may break if hose is dragged over it or if walked on. Sometimes slow to establish. Sometimes spindly. May look poor in winter when temperatures drop to the low 30's°F (0°C).

Planting and care: Plant in spring when danger from frost has passed. Space 30 to 36 inches apart for massing. Pinch young plant continually as it grows to develop a compact, multiple-branched form. Unpinched plants are more likely to be spindly. Keep plant attractive by never allowing soil to dry completely. Protect from cold. Apply a mulch around plant instead of cultivating so as not to disturb surface roots. Cut cold-damaged plants back in late winter to rejuvenate them.

Justicia brandegeana

> **Low and middle zones**
>
> **Evergreen perennial**
>
> **Soil:** Needs good drainage. Prefers average garden soil.
> **Sun:** Part to full sun.
> **Water:** Moderate with some leeway on either side.
> **Temperature:** May freeze to the ground when the temperature goes below 25°F (−4°C). Recovers quickly in spring.
> **Maintenance:** Occasional pruning and pinching.

JUSTICIA ghiesbreghtiana   Mexican honeysuckle • Desert honeysuckle • Firecracker plant
Family: *Acanthaceae*

(Jacobinia ghiesbreghtiana, Cyrtanthera ghiesbreghtiana, Sericographis ghiesbreghtiana). A wonderfully dependable nonwoody shrublike plant, Mexican honeysuckle is one of the most welcome introductions in recent years. It provides color more or less all year in mild winter climates. Medium green, soft, fuzzy-surfaced leaves are slender 3-inch ovals. Erect stems grow fast to 5 feet unless trimmed back; plants are easily kept at 2 or 3 feet with occasional trimming. A gallon-sized plant will spread to 2 feet or more at the base, with foliage it may be 3 feet or more wide. Clusters of red-orange tubular flowers loved by birds are heaviest in

spring, appearing after a growth spurt. Plant seems to cycle into growth, blooms and then rests.

Special design features: Lush, almost tropical effect. Some color nearly all year.
Uses: Containers. Small planting areas such as entryways, large planters, atriums and patios. Bedding or foundation plant. Does well in hot places or under trees with filtered shade, as a filler plant or an underplant. For clumps, masses or rows in irrigated, natural, desert or tropical gardens. Likes east, south or west walls.
Disadvantages: May become rank or look bad in cold weather. Occasionally yellows from chlorosis.

Planting and care: Plant when frost is past in middle zone, any time in low zone. Pinch tips to encourage bushiness. Feed to encourage fast growth, feed iron if it yellows. Trim back occasionally to keep neat and to rejuvenate.

Low and middle zones
Evergreen perennial
Soil: Average garden soil with good drainage.
Sun: Filtered shade to part shade to full or reflected sun.
Water: Moderate to ample. Tolerates less in cool weather.
Temperature: Freezes back at about 25°F (−4°C). Recovers rapidly in warm weather.
Maintenance: Some pruning.

Justicia ghiesbreghtiana

Justicia californica

Justicia californica, the chuparosa, is a similar plant deserving mention. It has a sparser appearance than Mexican honeysuckle, but responds well to rain or irrigation. Bright red tubular blossoms are quite showy and are loved by birds, especially hummingbirds. It is native to the low deserts so is not particularly hardy, damaged at about 28°F (−2°C). It can be grown in the middle zone and recovers quickly from winter cold setbacks.

LAGERSTROEMIA indica   Crape myrtle
Family: *Lythraceae*

(L. elegans). Crape myrtle grows at a slow to moderate rate to 15 feet high, sometimes to 20 feet, spreading 8 to 12 feet or more. Plants usually take on a vase shape, but are sometimes trained as small trees with a single trunk. Shiny leaves to 2 inches long are bright green, often tinged with bronze, especially when new. Depending on the weather they are golden, orange or red in fall before dropping to reveal a handsome outline. When bare it lends interest to the landscape with

rounded seed capsules that remain at the branch tips. Smooth, dappled gray to light brown bark may shed to reveal a pinkish, inner bark. Large clusters of delicate flowers in shades of red, white, purple and pink may bloom when hot weather arrives, sometimes as early as late spring. It loves heat and grows best when protected from wind and in deep soil. Originally from China.
Cultivars: 'Glendora White', a small tree with large, showy, white flowers. 'Near

East', formerly 'Shell Pink', a large shrub or small tree with shell pink flowers. 'Watermelon Red', a large shrub or small tree, with flowers of vivid red. Smaller plants include: 'Petites', a dwarf crape myrtle which grows to 5 to 7 feet high. 'Petite Embers', very dwarf, with rosy red flowers. 'Petite Orchid' has dark orchid blooms. 'Petite Red Imp' is very small with deep red flowers. 'Petite Snow', very dwarf with white flowers. Two outstanding new hybrids are 'Majestic Orchid', a small, heavy-stemmed tree with large clusters of orchid-colored flowers, and 'Peppermint Lace', medium size, upright and compact, with deep red flowers edged with white. The National Arboretum has recently introduced a group of crape myrtles called Indian Tribes, developed after years of research. 'Catawba' is a compact plant with dark purple flowers. 'Cherokee' is upright and spreading with red blooms. 'Seminole' is upright and spreading with medium pink flowers.
Special design features: A change-of-season plant—spring green and summer flower color, fall color, winter silhouette. Informal but luxuriant.
Uses: Color feature plant for summer. Combines well with plants in any type of garden. Transitional plant. Group, row or specimen.

Lagerstroemia indica

Disadvantages: May get leafburn or chlorosis in alkaline soils. Hot winds can also cause leafburn. In cool humid locations powdery mildew may be a problem; seek resistant plants.

Planting and care: Plant any time from containers. Space 6 feet apart for an eventual loose screen with continuous summer color. Prune old wood 12 to 18 inches back on large forms, twiggy growth and spent flowers on dwarf forms. Do this in the dormant season to increase next year's bloom. Give moderate feeding and leach the soil periodically (deep soaking) where water or soil are high in alkalinity. Spray for mildew just before bloom.

All zones

Deciduous

Soil: Prefers enriched deep soil with good drainage.
Sun: Tolerates part shade. Prefers full sun, even reflected sun.
Water: Best with deep spaced irrigation. Tolerates moderate watering or ample water when planted in lawns. Avoid overhead irrigation.
Temperature: Hardy to cold. Revels in heat.
Maintenance: Annual to occasional.

Lagerstroemia indica

LANTANA camara
Family: *Verbeneaceae*　　　　　　　　　Bush lantana • Shrub verbena • Yellow sage

This native of Tropical America provides color off and on all year as it cycles into bloom in frost-free areas. It grows to 4 feet high and as wide with sometimes prickly foliage and stems. Leaves are 3 inches or more in length, rounded at the tips with toothed edges and deeply veined fuzzy surfaces. They usually appear in pairs along stems. Flowers are multicolored: orange to yellow or orange changing to red, pink or white. There are many hybrids and cultivars of this plant which are widely available and often preferred.

Cultivars: Dwarf forms reach 2 to 4 feet high, spreading 3 to 4 feet. They are densely branched and more tender and like garden conditions as well as heat. They are: 'Dwarf Pink', bright pink flowers; 'Dwarf White', a white flowered form; 'Dwarf Yellow', corn yellow flowers; and 'Radiation', orange-red flowers. 'Christine', to 6 feet, cerise-pink flowers.

Special design features: Vibrant warm weather color. Informal.

Uses: Color accents for masses, containers, foundation planting, espalier, hedges or feature areas. Desert, transitional or natural gardens.

Disadvantages: Foliage has a pungent smell. Tiny black berries in clusters resembling blackberries are poisonous. Foliage and stems are prickly or abrasive when handled. Poor winter appearance, but less objectionable if combined with evergreens. Hybrids and cultivars seem more sensitive to cold and like the greater humidity of garden conditions.

Planting and care: Plant from containers in spring when danger from frost is past and nights are warm. Place 3 feet apart for massing shrub forms, 3 to 4 feet for spreaders. Overfeeding and too much water may diminish bloom and encourage foliage. Pinch tips to encourage fullness. Mulch plants in cold winter areas. Cut plants back in late winter or early spring after frost to remove unsightly or dead leaves and branches or to shape and encourage compact dense growth. Do this even in mild climate areas. In summer most plants are encouraged to bloom again more quickly after a bloom cycle if clipped or pinched lightly at the tips.

Low and middle zones, protected microclimates or annual in high zone

Evergreen perennial in frost-free areas to dormant with frost

Soil: Tolerant.
Sun: Part, full or reflected sun. Revels in heat of south sides.
Water: Moderate. Avoid giving plants too much water in winter.
Temperature: Foliage damaged at 28°F (−2°C). Most established plants recover quickly in spring.
Maintenance: Minimum amounts of pruning and grooming.

Lantana camara

Lantana camara

LANTANA montevidensis
Family: *Verbenaceae*

Trailing lantana • Weeping lantana • Lavender lantana

(L. delicata, L. delicatissima, L. sellowiana). Trailing lantana, a native of South America, has naturalized in warmer parts of the United States. It is a constant bloomer in warm weather with a profusion of 1-inch, verbenalike, lavender flowers. Leaves are small, prickly, gray-green and crinkly. Slender stems root along the ground so plants spread to an indefinite width and mound less than a foot high. Single branches may trail to 6 feet, find their way up fences, walls or other supports. When the soil dries out, plant becomes dormant. As soon as water is available it sprouts green leaves, begins to

grow and blooms again. This heat-loving plant will tolerate periods of neglect. It is a joy to the casual gardener who wants fantastic results with very little care.

Special design features: Trailing plant with constant bloom when given water in warm weather. Informal.

Uses: Ground cover or wall climber for warm weather or warm microclimates. Desert, wild, natural or rock gardens. Transitional areas. Spiller for banks or containers. Grows over hot pavement. Does well on south or west sides.

Disadvantages: Pungent smell. Prickly foliage. Very aggressive in flower beds and may take over. Unattractive winter appearance in cool areas.

Planting and care: Plant from containers or transplant rooted runners in warm weather. Space 3 feet apart for massing. Cut back any time to control or if plant looks poor. Best planted in a well defined area to prevent unwanted spreading. Prune back once a year to control its invasiveness. Mulch plants in colder areas in winter.

All zones, dormant perennial in coldest winter areas

Evergreen perennial

Soil: Tolerant.
Sun: Part, full or reflected sun. The more sun the better.
Water: Accepts spaced irrigation, but best with moderate water.
Temperature: Foliage is damaged at 25°F (−4°C), frozen to the ground at about 20°F (−7°C). Fast recovery in warm weather.
Maintenance: Occasional pruning to total neglect.

Lantana montevidensis

LARREA tridentata
Family: *Zygophyllaceae*

Creosote bush • Greasewood • Guamis

(L. divaricata, L. mexicana, Corvillea tridentata). This shrub is a personification of the desert plant; its adaptations allow survival in areas receiving as little as 3 inches of annual rainfall. Native to the Southwest and northern Mexico from sea level to 5,000 feet in elevation. Foliage is bright green with new growth in spring or the wet season, to yellow-green or dark olive green in the hot, dusty dry season. Plants also vary in density, dropping some foliage as water becomes scarce. Plants may stay the same size for years, or once established, grow rapidly with occasional irrigation. Individual shrubs have been seen up to 12 feet high and 8 feet or wider, but most reach about 4 to 6 feet in height. Single yellow flowers are numerous in spring, sometimes occur at other times of the year if conditions are right, but are less numerous. Small, pearly white, fuzzy seed balls that follow flowers were used medicinally by the Indians. Also, small quantities of lac, a resin found in tiny droplets on the branches, were used to fix arrow points and for mending pottery. There is a pungent fragrance to the foliage, especially when it rains. It is perhaps the most characteristic scent of Southwest deserts.

Special design features: An informal plant that adapts to its surroundings.

Uses: Natural or desert gardens. Very useful for screening areas. A handsome silhouette against structures. Espalier on warm south or west walls or for refurbishing. Combines well with other desert plants. Informal grouping or screen.

Disadvantages: May be difficult to obtain or slow to grow. Creosote is said to produce a substance which poisons the ground and prevents other plants from growing beneath it. Rain or irrigation leaches the soil and allows grass and other low plants to grow underneath seasonally.

Planting and care: Plant from containers any time. Small plants may be transplanted during the rainy season if the ground is soaked so most of the root system can be removed. Cut back about halfway after transplanting and keep soil damp until plant is established.

All zones

Evergreen

Soil: Widely tolerant of poor soils.
Sun: Part, full or reflected sun.
Water: Occasional deep soakings to none at all. Over irrigation may kill the plant, especially in heavy soils.
Temperature: Hardy to cold. Tolerant of great heat.
Maintenance: None required.

Larrea tridentata

Larrea tridentata

LAURUS nobilis

Family: *Lauraceae*

Grecian laurel ● Daphne ● Sweet bay

The bay leaf is familiar to most cooks. They are picked from this legendary tree native to the Mediterranean region. The modern term "rest on his laurels" came from the crowns of honor woven from this plant and worn by ancient celebrities. A slow to moderate grower to 15 or 20 feet high, sometimes higher, Grecian laurel usually takes on a conical form with several trunks that tend to sprout continually from the base. Foliage covers tree to the ground unless trimmed. Leaves are dark green, shiny, leathery and slender, to 4 inches long. Small yellowish flowers in spring produce inedible, dark purple berries in late summer or fall.

Planting and care: Plant from containers after weather warms in spring. Space 6 to 8 feet apart for massing. Do any heavy pruning and shaping in late winter. Shear any time; use hand clippers to avoid mutilating the leaves. To harvest leaves for cooking, pick them in the early morning and dry flat on a cloth indoors. When nearly dry, cover with another cloth or paper towels and press for a week to 10 days with a heavy board. Store in an airtight container.

Special design features: Compact foliage plant of many uses. Dark, rich, elegant quality possessed by few plants. Adapts easily to contained or formal gardens as a clipped plant.

Uses: Historically used as a formal hedge, topiary shape or standard tree. Looks nice, stays neat and holds its shape for a long time without shearing. Unclipped specimen or mass for a screen or background or patio tree.

Disadvantages: Sunburn in hot areas with reflected sun. Cold damage in exposed locations. Suckers at base of tree may be a nuisance. Subject to scale infestations.

Low and middle zones, warmer parts of upper zone

Evergreen

Soil: Needs good drainage, but is otherwise tolerant. Does best in enriched garden soil.
Sun: Part to full sun. Avoid reflected sun.
Water: Moderate to ample. Gradually decrease water in fall to harden off plant.
Temperature: Young plants and new growth may be severely damaged below 20°F (−7°C). Older plants and those hardened off withstand lower temperatures without damage.
Maintenance: Occasional light pruning.

Laurus nobilis

LEUCOPHYLLUM frutescens

Family: *Scrophulariaceae*

Texas ranger ● Ceniza ● Barometer bush

(*L. texanum*). Texas ranger is a striking plant when covered with its rosy lavender, bellshaped flowers. This native of Texas and Mexico grows slowly up to 8 feet tall. In some desert zones it may reach only 4 to 5 feet tall and as wide unless trimmed. Grown for its gray, feltlike foliage and open to dense rounded form, it blends with the desert or contrasts nicely with dense green plants. Foliage is denser in warm weather. Where moisture is available, it becomes luxuriant, almost succulent. Flowers need hot weather to bloom, often appear after summer showers, which is why it is sometimes called barometer bush. Branches look silvery purple to brownish contrasted against the foliage. Very tolerant of dry, adverse conditions, but will lose much of its foliage except at branch tips. Texas ranger and its natural adaptations to drought is also discussed on page 10.

Cultivars: 'Compactum' is smaller. A good, low foundation plant.

Special design features: Informal pearl-gray foliage. Desert effect. Sporadic warm season flowering can be spectacular.

Uses: Informal shrub with room to spread as a specimen, screen, clipped hedge or random planting. Foundation plant. Natural or desert gardens. One of the most important plants in the transitional garden. Blends well with a green garden on one side and the desert on the other. Foliage contrast. Use to vegetate disturbed areas.

Disadvantages: Will die out if overwatered. Can look sparse and scraggly in late winter or in long periods of drought.

Planting and care: Plant any time from containers. Irrigate until established and afterwards for faster growth or if in a low rainfall area. Clip or shape as desired.

All zones

Evergreen to partly deciduous

Soil: Tolerant.
Sun: Part to full or reflected sun.
Water: Best with occasional deep soakings. Tolerates none to moderate, the latter only with good drainage.
Temperature: Hardy to about 5°F (−15°C). Endures great heat.
Maintenance: None to regular according to desired effect.

Leucophyllum frutescens

LIGUSTRUM japonicum
Family: *Oleaceae*

Ligustrum japonicum

(L. texanum, L. kellermannii). An elegant subject in the right location, waxleaf privet is a handsome garden plant of many uses.

It grows at a moderate rate to 10 or 12 feet high with an equal spread if unclipped. Keep at almost any size by shearing and shaping. Habit is dense, compact and erect. Leathery, almost spongy leaves are shiny and crisp, dark green above, whitish beneath, forming wide-pointed ovals to 4 inches long. Foliage reaches to the ground if left untrimmed. Whitish, somewhat fragrant flowers appear in clusters at leaf tips of unclipped plants in summer. Small dark berries follow. Native to Japan and Korea.

Cultivars: 'Silver Star' is a new introduction—a variegated medium-size shrub with creamy silver leaf edges. A slow compact grower, good for shady gardens as a contrast to deep green plants.

Special design features: Formal architectural form. Rich foliage is pleasant close-up. Outstanding for clipping and topiary in formal or small gardens.

Uses: Clipped as a high or low hedge. Topiary plant, standard tree, column or globe. Feature plant for containers. Adaptable to narrow spaces because it clips so well. North or east sides. Swimming pool patios.

Disadvantages: Berries are poisonous. Pollen can be quite bothersome to those allergic to it. Foliage may sunburn in low or middle zones when exposed to full sun in summer. Subject to root knot nematodes and brittle leaf; the latter is easily cured with an application of ammonium sulphate. Very susceptible to Texas root rot.

Planting and care: Do not plant in soil infested with Texas root rot. Prepare planting hole to prevent it. See page 38. Plant from containers any time of year. Space 1 foot apart for a low hedge, 3 feet apart for a 6-foot hedge or screen. Shear any time but do any heavy pruning in late winter. Avoid intense, reflected heat.

All zones
Evergreen
Soil: Improved garden soil preferably with good drainage and a high percentage of humus.
Sun: Open to filtered shade and part sun. Avoid mid-day sun in warmer climates, especially in summer. Accepts full sun at higher elevations or in cooler places.
Water: Moderate, but accepts light to heavy.
Temperature: Hardy to 20°F (−7°C) or below. Tolerates heat with shade and water.
Maintenance: As desired, from little to much.

LIGUSTRUM lucidum
Family: *Oleaceae*

Glossy privet is one of the basic trees or shrubs in the desert landscape. This native from China and Korea is usually seen as a clipped hedge, but is often planted as a slow to develop round-headed tree. It reaches up to 30 feet in height with a spread of 20 feet at maturity. As a hedge it grows at a moderate to fast rate to 6 feet in height and may be coaxed to 10 feet or more. Plant is as wide as high unless clipped or trained. Leaves are crisp, dark green pointed ovals to 4 inches long and curve backwards. Privet may produce an abundant to sporadic bloom of whitish flowers in pyramidal clusters in late spring. Bloom is mildly fragrant, appearing at the branch tips and lasts for several weeks. Small black berries follow bloom.

Special design features: Formal feeling. A strong and richly foliated tree. Neat and adaptable to many uses.

Uses: Erect round-headed tree with one or more trunks for patios, rows, street plantings or formal settings. Clipped hedge at any height from 3 to 12 feet. Wide unclipped screens. Pool patios.

Ligustrum lucidum

Disadvantages: Berries are poisonous. Bloom is rather coarse and not too attractive at close range and fades to brown. Subject to root knot nematodes and privet weevil. Occasionally gets Texas root rot.

Planting and care: Plant any time from containers. Space 12 to 18 inches apart for a quick, low hedge, 3 feet for a 6-foot hedge, 3 to 4 feet for a 12-foot hedge. Space up to 10 feet for a wide, unclipped screen. Clip so bottom of hedge is wider than the top. This enables the sun to reach all parts of the plant to produce denser growth. Do any heavy pruning in late winter. Shear as needed to shape or keep neat. Trim blossoms when they fade or to prevent berries.

All zones
Evergreen
Soil: Tolerant. Prefers improved garden soil with good drainage.
Sun: Filtered to open shade or part to full sun.
Water: Prefers deep spaced irrigation. Accepts moderate or ample. Tolerates limited periods of drought when fully established.
Temperature: Hardy to 15°F (−10°C) or below. Tolerant of heat.
Maintenance: Little to much according to use and preference.

LIRIOPE muscari
Family: *Liliaceae*

(L. m. densiflora, L. graminiflolia densiflora). This appealing, stemless, grasslike plant grows to 18 inches high and as wide. Leaves are dark green, 3/4 inch wide and reach up to 18 inches in length. Tiny blue-violet flowers on spikes above the tufts of foliage appear in late summer or early fall and last for about two weeks. Native to Japan and China.

Cultivars: 'Silvery Sunproof' is a variegated form to 18 inches high with white to yellow stripes. It brightens shaded gardens, where it grows best.

Special design features: Oriental, tropical or woodsy garden effect, depending on the situation. Grasslike foliage. Effect of coolness and nearby water. Combines well with mondo grass, *Ophiopogon japonicus.*

Uses: Under planting or ground cover for small areas such as entryways, atriums, intimate patios, borders, rock gardens or in containers. Filler between slow-growing plants. Swimming pool patios. Use with pebbles and bamboo to give a waterside effect.

Disadvantages: Foliage yellows in the hot sun, tips burn from soil salts. Shelters snails and slugs.

Planting and care: May be planted from divisions in early spring. Plant from containers any time. Space smaller plants 12 to 18 inches apart for a dense mass, farther apart if you want each plant to show separately. Remove old flower stalks to groom. Clip ragged plants in early spring before new growth begins. Give occasional deep soaking to remove soil salts and prevent leafburn.

Liriope muscari

All zones

Evergreen perennial

Soil: Average garden soil with good drainage.
Sun: Open, filtered or part shade. Morning sun in low and middle zones, full sun in cool summer areas.
Water: Moderate to ample.
Temperature: Tolerant of heat and cold.
Maintenance: Appreciates regular garden care.

LONICERA japonica 'Halliana'
Family: *Caprifoliaceae*

Hall's honeysuckle

An old-fashioned garden vine, Hall's honeysuckle is probably the most commonly grown cultivar of *Lonicera japonica.* Although this eastern Asian native has become a pest in the Middle Atlantic States, it is welcome in desert regions for its adaptability and vigor where little else will grow easily. Soft, medium green leaves are oval, to 3 inches long. Flowers in pairs along the stems are white then turn golden as they age. Bloom is heavy and very fragrant in early spring and may continue in summer with less profusion. Plants are aggressive growers once established and can cover up to 1,200 square feet or more. Not a plant for close range, but very pleasant on the back fence or at the end of the patio where its shade and fragrance can be enjoyed. Tolerates some drought, hot winds, poor soil and hot sun.

Special design features: Fragrance in spring and summer. Quick shade on a support or trellis. Old-fashioned garden effect.

Uses: Erosion control on banks, although it may not spread evenly and may need some training. Fast shade or screening on trellises, fences, arbors or porches.

Disadvantages: Mature plants become woody underneath and need to be cut back. Runners spread and reroot and can be invasive. Avoid planting too near shrubs or trees, where it can climb and take over. Difficult to eradicate once established.

Planting and care: Plant in spring when the weather warms up. Set plants 2 to 3 feet apart for a ground cover. Plants are easy to propagate from divisions or cuttings. Space 8 to 10 feet apart for a wall, fence or trellis cover. Prune severely in late winter to give it a fresh start and to keep the plant from becoming woody. Train as desired.

Lonicera japonica 'Halliana'

All zones

Evergreen to all or partly deciduous

Soil: Tolerant.
Sun: Open shade to part, full or reflected sun. Revels in heat.
Water: Accepts a wide range from little to moderate to ample. Best with spaced soaking. Tolerates long periods of drought and neglect but looks mangy and sections will die out. Comes back quickly when water becomes available.
Temperature: All to partly deciduous in frosty winter areas, but hardy and leafs out quickly in warm weather.
Maintenance: Occasional to groom and control.

LYSILOMA thornberi

Family: *Leguminosae*

Feather bush

The feather bush from southern Arizona and northern Mexico lends a tropical mood to the garden or patio. It is a shrub or tree with large, finely cut, medium to dark green feathery foliage. A moderate to fast grower, it reaches 15 to 20 feet high in frost-free areas and spreads its canopy of ferny leaves as wide as high. In cooler locations or in unprotected areas of the middle zone, it is a summer foliage plant. When frosted back it grows to only 4 or 5 feet by late spring or midsummer as a delightful feathery shrub. Numerous, creamy white flowers like clusters of puff-balls appear in spring as new leaves come out. Flat brown pods 4 to 8 inches long follow bloom and hang from branches for a long period. Tolerant of drought and great heat.

Special design features: Luxuriant feathery to ferny foliage. Tropical to subtropical feeling. Filtered shade.

Uses: Summer feature plant with rocks or structural features. Silhouette plant. Textural contrast of foliage. Natural, desert, tropical or transitional gardens. Patio or garden tree in frost-free areas. Summer effect in colder areas. Planting with evergreens will mask its winter dormancy.
Disadvantages: Sensitive to cold. Occasional chlorosis from too much water. Beans can be unsightly and produce litter.
Planting and care: Plant from seed or from containers in spring when weather warms up. Plant in a sheltered spot in areas of frost. Prune selectively; only shape or remove dead branches. If frozen, cut plant to the ground.

Lysiloma thornberi

Lysiloma thornberi

Low and middle zones
Evergreen to deciduous in cold winters
Soil: Tolerant. Prefers good drainage. **Sun:** Part, full or reflected sun. **Water:** Best with deep spaced irrigation. Accepts moderate irrigation or none at all in areas of 10 to 12 inches of annual rainfall and in somewhat porous soil. It remains a shrub with meager irrigation, becomes a tree when given moderate water. **Temperature:** Hardy to 25°F (−4°C). Recovers quickly in spring. **Maintenance:** Occasional.

MACFADYENA unguis-cati

Family: *Bignoniaceae*

Cat's-claw vine • Yellow trumpet vine • Cat's-claw trumpet

(Bignonia unguis-cati, B. tweediana, Doxantha unguis-cati). A remarkable, self-attaching vine, cat's-claw is one of the best plants for hot walls. Once it is attached, it grows with increasing vigor to form tracery and heavy cover to the top of buildings, up to four stories high. Heaviest growth or cover is usually at the top of vine where it gets the most sun. Plants that climb pillars or posts will hang down and festoon the area beneath, seeking to attach themselves and creep across the connecting horizontal surfaces of a beam or roof. Three-clawed tendrils emerge from leafbases of new growth and cling to almost any material—plaster, brick, concrete, wood or stone. A few to many large yellow, trumpetlike flowers appear in spring on old garlands, providing impressive but short-lived color. Slender beans to 12 or 15 inches long may follow. This Central American native accepts periods of drought when established, made possible by underground tubers which hold reserve stores of water.

Special design features: Wall tracery in juvenile state. Informal festoons in mature form.
Uses: For hot walls, trellises, posts and fences as a durable self-climbing cover.
Disadvantages: Roots and upper parts of plant become very aggressive and invasive in old plants unless they are cut back severely and part of the fleshy underground roots are dug up. Greedy roots. Top-heavy plants may fall away from walls, especially in high winds. Difficult to eradicate established plants. Sometimes slow to begin clinging and climbing. A few tendril tips remain in the wall when plant is pulled down.

Planting and care: Plant from gallon-sized containers after frosts are over. Space 10 to 20 feet apart to cover wall as densely as possible. Place plants so they will start low on wall. To keep neat, remove old undergrowth and trim part of vine back each year after flowering. To preserve juvenile leaves and tracery, cut back plant drastically each year.

Macfadyena unguis-cati

All zones
Evergreen to all or partly deciduous
Soil: Tolerant. Prefers improved. **Sun:** Shady walls to walls in hot sun. **Water:** Tolerates a wide range from little to ample once established. Young plants need moderate irrigation. **Temperature:** Evergreen in warm winter areas. Turns purple with cold; somewhat deciduous in frosty winters. May lose all its leaves at 20°F (−7°C). Hardy to 10°F (−12°C). Fast recovery in spring. **Maintenance:** Occasional.

MAGNOLIA grandiflora
Family: *Magnoliaceae*

Southern magnolia • Bull bay

The southern magnolia, a native of the southeastern United States, is valued for its erect form, shiny bold foliage and large fragrant flowers. Once established, it grows at a moderate rate to as much as 40 feet in height with favorable circumstances. In the desert it is usually seen at 30 feet because of harsher conditions. Growth habit is conical in youth with leaves reaching to the ground. Older trees may develop a wide pyramidal crown and a sturdy trunk 6 feet up from the first branch. Individuals vary in size and form and the date of first bloom. Some plants take as long as 15 years to begin, so choose a cultivar which blooms earlier if

bloom is important to you. Leathery leaves to 8 inches long are glossy dark green above, fuzzy and usually rust colored beneath. Blooms are scattered among foliage from late spring to fall. Flowers are white, 8 to 10 inches across, turning buff as they age. They have a fragrance similar to gardenias. Rust-colored conelike fruits 4 inches in length expel bright red seeds on little threads.
Cultivars: 'Exonensis' to 20 or 30 feet high; 'Majestic Beauty', eventual size is as yet unknown, but probably quite large; 'St. Mary', to 20 feet high spreading 20 feet; 'Samuel Sommer', grows large and fast to an undetermined size.
Special design features: Cool oasis. Woodland or southeastern landscape style. Bold form and foliage. Tropical feel.
Uses: Patio or lawn tree as specimen, row or group. Courtyards, wall plants, container or roof garden plant. Neat plant for swimming pool patios.
Disadvantages: Leaves may burn or tree may get iron chlorosis in alkaline soils. Flowers brown in hot winds. Can be slow to establish.
Planting and care: Plant any time of year—early winter or fall is best in warmer zones. Be sure crown is planted above soil level; tree should be set no deeper in

ground than it was in the nursery container. Space 15 feet or more apart for a grove or row planting, closer for a solid screen. Protect trees from afternoon sun in very warm areas. Do not cultivate around the shallow roots and add a mulch to protect root zone. Feed sparingly. Soak deeply occasionally to leach out soil salts if leaves start to brown.

Magnolia grandiflora

> **All zones except where very windy**
>
> **Evergreen**
>
> **Soil:** Enriched garden soil, preferable with abundant humus added.
> **Sun:** Part to full sun. Supply afternoon shadow in low zone, at least for young trees. Avoid reflected sun.
> **Water:** Moderate to ample in sandy soils.
> **Temperature:** Hardy to about 10°F (−12°C).
> **Maintenance:** None to occasional. Leaves drop from previous year for a short period in late spring, creating some litter.

MAHONIA aquifolium
Family: *Berberidaceae*

Oregon grape • Mountain grape • Holly mahonia

(*Berberis aquifolium*). A charmer from the American Northwest, Oregon grape has something of interest to offer all year. Plants grow moderately to 6 feet high, spreading to 5 feet wide. Glossy, hollylike foliage is tinted bronze when new, with some red leaves remaining on the plant through summer. Foliage turns purple or bronze in winter, especially when grown in the sun in cooler zones. Showy yellow flowers in 2 to 3-inch clusters bloom for a long period in spring, followed in summer by round black berries with a blue blush. Berries can be made into a tasty jelly.

Cultivars: 'Compacta' grows to 2 or 3 feet high by 4 feet wide as a low shrub. 'Orange Flame' is somewhat larger to 5 feet high. New growth is bronze-red; older growth is glossy green. There are several other cultivars; some are hybrids.
Special design features: Interesting crisp form and foliage. Spring and autumn color. Woodsy feeling.
Uses: In high zone use as accent, foundation planting, garden shrub or screen. Grows on north sides or under deciduous trees in all zones. This is one plant for areas known to be infested with oak root fungus—it is quite resistant. Compact form is good in containers, planters, atriums or cool and shaded entry patios. Well-tended plants are attractive at close range.
Disadvantages: Sometimes gets iron chlorosis and occasionally attacked by whiteflies. Harbors slugs and snails. Occasionally difficult to get started in low and middle zone.
Planting and care: Plant from containers any time. Place larger forms 4 to 5 feet apart for a screen or hedge. Space compact plants 3 feet apart for a continuous planting. Cut old, tall, woody stems to the ground to control size and to rejuvenate. Feed with iron during growing season to prevent iron chlorosis.

Mahonia aquifolium

> **Best in high zone, shaded places and north sides in middle zone**
>
> **Evergreen**
>
> **Soil:** Tolerant, but prefers good drainage and slightly acid soils.
> **Sun:** Open, filtered or part shade in all zones. Full sun in high zone.
> **Water:** Moderate, with some tolerance on either side.
> **Temperature:** Hardy to cold, but languishes in heat.
> **Maintenance:** Occasional.

MELIA azedarach

Family: *Meliaceae*

(M. australis, M. japonica, M. semper-virens): The chinaberry from Asia is often overlooked in the nursery trade. It grows so easily from seed nurseries seldom carry it. A rapid grower in youth, it grows 20 to 30 feet in height, rarely reaching 50 feet, with an equal spread if not trimmed back. Growth slows as it approaches maturity. Crown can be dense, open, irregular or umbrellalike according to selection. Branches of the species are more loosely arranged than those of cultivars. Foliage is dense deep green, velvety in the wind, made up of compound leaves having numerous tooth-edged leaflets. Trees change with the seasons. Bouquets of small lavender to purple flowers appear in spring. Foliage is dense green in summer and golden in fall. Winter silhouette is decorated with clusters of yellow 1/2-inch berries that hang until spring. Usually erect with a single trunk, young trees are slender and tall, attractive when massed.
Cultivars: 'Umbraculifera,' Texas umbrella tree, has a very erect trunk with branches radiating in a regular pattern like ribs of an umbrella. Crown is very dense and domelike. Some gardeners prefer it due to its more regular symmetrical form.
Special design features: Interest all year. Dense shade.

Uses: Shade tree for hot places—patios, lawns and streets. Sapling groves along walls for silhouette and screening above walls. A good plant for alkaline soils.
Disadvantages: Susceptible to Texas root rot and heart rot. Branches break in strong winds, especially on older trees or trees infected with heart rot. Berries are said to be poisonous and have an unpleasant odor between ripening and the time they are dried out. Trees may sucker at base. May naturalize and become a pest.
Planting and care: Plant seed after berries have ripened. Young trees may be planted from containers any time, or bare root in winter. Select nursery stock that has a predictable form. Space 5 to 6 feet apart for a temporary sapling grove. Space 15 to 20 feet for a grouping or row. Control berry production and branch breakage by trim-

ming the tips of the branches in late winter. Trees with dense canopies may require inside branches to be pruned to allow wind to pass through.

All zones
Deciduous
Soil: Tolerant.
Sun: Part, full or reflected sun.
Water: Moderate to spaced irrigation. Tolerates lawn watering.
Temperature: Hardy. Accepts heat of low deserts.
Maintenance: Seasonal.

Melia azedarach

Melia azedarach

MORUS alba

Family: *Moraceae*

White mulberry is a fast-growing shade tree which tolerates a wide range of conditions. It is widely planted throughout the world, naturalizing in many places in North America. Native to China, where the leaves have been used for centuries to feed silk worms. Form is erect with a single trunk clothed in light gray bark. Saplings quickly develop a scaffold of spreading branches which radiate from the trunk like an umbrella. Trees may grow to 20 feet tall in 3 to 5 years with a dome-shaped crown of an equal spread. Mature trees reach 35 to 40 feet in height with a wider spread. Foliage is bright to deep green with large toothed leaves. Females of the species bear fruit in spring, eaten by

All zones
Deciduous
Soil: Tolerant. Prefers deep soil and good drainage.
Sun: Full sun.
Water: Ample to moderate. Accepts little where soil is deep. Size may be gained by giving ample to moderate irrigation to encourage growth and maintained with spaced irrigation when the tree has reached the size you want.
Temperature: Hardy to cold. Tolerant of heat.
Maintenance: Seasonal.

birds. Both fruit and birds make a mess, and children can stain their clothes as they nibble the fruit. Seek "fruitless" cultivars.
Cultivars: 'Fruitless' and 'Kingan' have heart-shaped leaves. 'Stribling' ('Maple-leaf') has lobed leaves like a maple. 'Chaparral' is fruitless, with a remarkable weeping form that evokes a strong, Oriental effect. It is grafted on a straight, 6-foot trunk, with drooping, woody branches that create a remarkable silhouette.
Special design features: Fast shade for hot summer areas. Golden fall color.
Uses: Street, patio or lawn tree. Row or specimen. Parks and other public areas. Excellent for fast shade in mobile home or RV parks. May be used to refurbish sanitary land fills.
Disadvantages: Often short lived, but trees given average garden care will last many years with no problems. Not considered a quality shade tree. Aggressive roots can cause problems in lawns. May grow vigorously for a few years and then suddenly develop root problems and begin to decline in vigor. The problem is often under-irrigation or root knot nematodes. Occasionally subject to a variety of other ills, such as sooty canker, heart rot and slime flux, all caused by untreated wounds or cuts.
Planting and care: Plant from containers

any time or bare root in winter. Space 15 to 20 feet apart for a grove for solid shade, up to 30 feet or more for a street tree. Stake and train young trees carefully. Select 3 to 5 well-spaced branches above head height to form scaffold branches for the crown. Shorten tips of very long branches that droop from their own weight. Protect bark with a latex paint until foliage provides shade.

Morus alba

MUSA paradisiaca

Family: *Musaceae*

Edible banana ● Plantain

(M. sapientum). Banana trees are the essence of the tropics and jungle—they grace the summer garden with their dramatic leaves. Edible bananas are made up of many hybrid plants and clones, the fruits of which are varieties of seedless bananas. Plants grow in a clump from a fleshy base which spreads outward to an indefinite width. Slender, treelike succulent trunks, like giant green onions, have no true woody parts. They are bundles of sheaths from which leaf stalks rise in a spiral fashion from the base. Thin, bright green leaves are 1 to 2 feet wide and up to 8 feet long on larger plants. Banana stalks grow rapidly in warm weather 10 to 12 feet high, sometimes to 20 feet in areas where there is no frost. Leaves and stalks seldom

reach over 8 to 10 feet where they are frozen back each year. Established plants send out new sprouts in spring which grow quickly to their leafy, summer form. Each new stem or stalk has a life of its own and is programmed to bloom and produce fruit after a certain number of leaves have been produced. Plants that bloom in June may have ripe fruit by September or October.

Special design features: Tropical or jungle effect. Bold foliage. Vertical form. Handsome silhouette. Pleasing light and sound effects from sun and wind. Looks best with flowering plants or in combination with other tropical-appearing plants.

Uses: Requires sheltered patio, entryway or atrium. Can be grown in large containers and moved inside in winter or out of sight during its unattractive periods. Effective when grown outside a garden wall where leaves can be enjoyed in summer and where bases will be obscured during the less attractive times of year.

Disadvantages: Leaves tatter badly in the wind. Needs considerable grooming to maintain attractive appearance. Looks poor during cooler parts of the year. Easily damaged by cold. Sap permanently stains clothing.

Planting and care: Divide established clumps and replant or set from containers

in spring after frost. Space new plantings at random distances for effect. Because plants require grooming to look attractive, keep clump small with two mature stalks, a few younger stalks and a few sprouts. Feed often for fast growth. High nitrogen liquid fertilizers are best. Mulch roots in summer and again in fall to protect from temperature extremes.

Musa paradisiaca

Low and middle zones, upper zone where protected
Evergreen perennial in warm areas
Soil: Highly enriched soil with good drainage.
Sun: Open, filtered or part shade. Best with afternoon shadow.
Water: Ample. Tolerates moderate.
Temperature: Foliage damaged at 30°F (−1°C). Mulched roots have survived much lower temperatures.
Maintenance: Frequent.

MYRTUS communis

Family: *Myrtaceae*

Classic myrtle ● True myrtle ● Greek myrtle ● Roman myrtle

(M. boetica, M. italica, M. latifolia, M. romana). Dependable, neat and easy to grow, myrtle always looks attractive whether clipped or not. It gives long service and requires little care. A moderate grower becoming a shrub 5 to 6 feet in height in a few years, with a spread of 4 to 5 feet. Mature plants reach 12 to 15 feet high and spread as wide unless clipped. Foliage is dark green, shiny and compact, densely covering the plant to the ground. Numerous, delicate, white to pinkish flowers 3/4-inch across appear in spring. Small blue-black berries may follow.

Cultivars: 'Variegata' has white-edged leaves. 'Compacta', compact myrtle, grows slowly 2 to 3 feet high and as wide. It has small, pointed, closely set leaves, used for low hedges or edges. Susceptible to root rot in summer if soil is too damp. If soil is well drained, there is usually no problem. 'Microphylla', dwarf myrtle, grows 1 to 2 feet high, densely covered with very small, overlapping leaves. Excellent as a very low, dense, clipped hedge, edge or space definer. It is very susceptible to summer root fungus in wet soils. 'Boetica', twisted myrtle, is quite different in appearance from the others. It is an angular, rugged, upward-pointing and informal deep green shrub, growing slowly up to 12 feet. It can be trained into a

gnarled tree when it reaches this size.

Special design features: Formal to informal appearance depending on use. Dense, deep green statement in the landscape. Aromatic foliage.

Uses: Wide informal specimen or screen. Clipped hedge or background plant. Excellent for clipping and shaping as topiary plants or standard trees. Older specimens may be clipped into small trees. It may be planted in areas infested with Texas root rot, as it is highly resistant.

Disadvantages: Chlorosis in poorly drained soil or from shallow watering.

Planting and care: Plant any time from containers. Space 1-1/2 to 3 feet apart for a clipped hedge, 4 feet or more for a wide unclipped screen or background planting. Shearing may be done any time.

Myrtus communis 'Compacta'

Myrtus communis 'Boetica'

All zones
Evergreen
Soil: Adaptable. Should have good drainage.
Sun: Part, full or reflected sun.
Water: Prefers deep, widely spaced irrigation. Established plants tolerate some drought.
Temperature: Hardy. Tolerates heat.
Maintenance: None to regular according to use.

NANDINA domestica

Family: *Berberidaceae*

Heavenly bamboo • Nandina • Sacred bamboo

This is a plant with grace and class which belies its toughness and drought resistance. Heavenly bamboo is not a bamboo at all, but looks a little like one. Delicate foliage grows outward from one or more vertical canelike stems. Individuals vary in size unless grown from cuttings of the same plant. Plants grow at a slow to moderate rate to as much as 8 feet in height, but are usually seen at 4 to 6 feet. Some plants spread by underground roots. Most keep their slender form and seldom get more than 4 feet wide. Easy to grow and very neat, heavenly bamboo adds interest to the landscape in each of the seasons. Sprays of tiny white flowers appear in spring and are occasionally followed in fall by long-lasting sprays of red berries. Foliage turns reddish or bronze in the winter sun; fresh green growth reappears in spring. Plants are highly resistant to pests, diseases and most soil problems, requiring little maintenance and tolerating periods of neglect.

Cultivars: 'Compacta' dependably reaches 4 to 5 feet high and about 3 feet wide. 'Nana Compacta' is a small mound, 12 to 18 inches high by 15 inches wide.

'Purpurea Dwarf', the same size, is often preferred for its deep red color. Both have wide, drooping, cupped leaves and are best used under south-facing overhangs in low and middle zones so they get summer shade and winter sun. 'Harbor Dwarf' is a recent introduction that grows to about 2 feet tall and spreads by underground runners to fill as a ground cover.

Special design features: Oriental effect. Refined and attractive at close range.

Uses: Specimen or accent plant. Foundation plant. Narrow spaces. Entryways, atriums or intimate patios. Container plant. Taller forms can be used as informal hedges or space definers, and are effective as background plants, especially in the flower garden.

Disadvantages: May yellow from iron chlorosis in alkaline soils. Sometimes stunted in hottest locations. Dwarf types may not survive intense sun of desert summer afternoons.

Planting and care: Plant from containers any time of year. Space larger forms 18 to 30 inches apart for a solid hedge or screen, farther for a more casual effect. Do not clip or shear as a hedge as it ruins the form and exposed canes look unattractive and awkward. Space dwarfs 12 to 18 inches apart; 'Harbor Dwarf' 18 to 24 inches.

> **All zones**
>
> **Evergreen**
>
> **Soil:** Average garden soil. Avoid highly alkaline conditions.
> **Sun:** Open, filtered or part shade to full sun. Dwarf forms need afternoon shade in middle or low zones; all forms require it in low deserts.
> **Water:** Tolerates a wide range of irrigation practices from ample to moderate to occasional deep soakings, to periods of neglect, from which it recovers quickly when water is again available.
> **Temperature:** Hardy, but can lose leaves at 10°F (−12°C). Recovers quickly.
> **Maintenance:** Little to none.

Nandina domestica

NERIUM oleander

Family: *Apocynaceae*

Oleander • Rosebay

(N. indicum, N. odurum). One of the more dependable plants for hot climates, oleander is a bright bouquet in the warm season. Growth rate varies from moderate to fast depending on conditions and plant type. Vigorous growers produce a dense hedge to 6 feet high in 3 years with favorable conditions. Mature hedges have been seen 20 feet high and more. Plants are about half as wide as high unless pruned. Erect in form, stems are almost canelike in youth, supporting narrow, dark, dull green leaves to 10 inches long that densely

cover to the ground. Older plants may be trimmed up as trees, revealing gnarled single or multiple trunks. Except as patio trees and possibly the dwarf varieties, oleanders are best used in farther reaches of the landscape. They are not for a refined intimate space. Flowers may be single or double and appear in profusion on the branch tips with the heaviest bloom in mid to late spring; some flowers bloom through summer until nights cool in fall. Blossoms are red, pink, salmon, soft yellow and white—some are scented. Seed pods may follow, releasing airborne seeds. Native to the Mediterranean across Asia to Japan.

Cultivars: Largest and most vigorous grower is white-flowering 'Sister Agnes'. Many other color selections grow nearly as large. 'Mrs. Roeding' will grow to a 6-foot shrub with double salmon pink flowers. Newer introductions include two intermediate growers from North Africa: 'Casablanca', with single white flowers, and 'Algiers', single red flowers. The more refined 'Petite' series is a fairly recent introduction. Most commonly chosen are 'Petite Pink', profusions of delicate pink flowers throughout the growing season, and 'Petite Salmon', salmon-pink flowers. There are also red, light yellow and deep pink selections. These

> **All zones**
>
> **Evergreen**
>
> **Soil:** Tolerant of a wide range of soils including heavy, poor and alkaline.
> **Sun:** Full to reflected sun. Tolerates part shade, but may be poor or leggy and have few flowers.
> **Water:** Best with occasional deep irrigation. Plants tolerate periods of drought but become unattractive.
> **Temperature:** Plants revel in heat. Damaged by cold below 20°F (−7°C), severely at 10°F (−12°C). They recover rapidly in spring.
> **Maintenance:** Usually little to occasional. Depends on location and taste.

Nerium oleander (dwarf)

tightly branched dwarf plants grow 3 to 5 feet high, occasionally larger in hot summer areas. They are effective in containers or for low unclipped borders, hedges or background plantings.

Special design features: Summer color. Dense foliage.

Uses: Best as unclipped hedges, screens and borders. Effective along roads or drives, at property edges or as background plantings. Oleanders may be clipped to any size, but yield fewer flowers. Larger forms make handsome single or multiple-trunked trees when trained. Trees are adapted to grow in patios, median strips or as street trees for color, character and shade.

Disadvantages: All parts of the plant are poisonous. Sap may irritate skin or eyes, even smoke from burning leaves is irritating. Airborne seeds may cause hayfever at close range. Plants are sometimes subject to yellow oleander aphids and scale. Warty growths and splitting branches as well as blackened deformed flowers may indicate bacterial gall. Very severe infections of this disease may follow a hard freeze which has split the bark on mature twigs, allowing bacteria to enter. Plants must be cut back beyond any galls that show on stems or trunks. Disinfect pruning cuts with a 50-50 bleach and water solution and treat pruners or saw between each cut to avoid spreading the disease. Yellowing and dropping of old leaves, especially in spring, may indicate drought stress.

Yellowing of new leaves sometimes occurs if plants are overwatered. Flowers litter pavements. Plants trained as trees sprout suckers from the base which must be pulled out rather than trimmed off.

Planting and care: Plant or transplant in any season, but best in spring. Space 2 to 2-1/2 feet apart for a fast hedge, 3 to 6 feet for an eventual tall hedge, up to 9 feet for a wide, loose screen. Prune in late winter or early spring to control size and form or to remove seed pods. To maintain desired height or shape, trim branch tips lightly or shear any time. Train tree forms carefully, supporting trunks with sturdy stakes. To rejuvenate older plants cut a few old stems to the ground each year. Spray as needed for aphids or scale.

Nerium oleander

Nerium oleander

OENOTHERA berlandieri

Mexican primrose ● Pajarito

Family: *Onagraceae*

(O. speciosa childsii, O. tetraptera childsii). Appreciated for the delightful pink color of its flowers, Mexican primrose is widely grown in the Southwest as a spring color plant. This native of Texas and Mexico grows rapidly to about 6 inches high with slender stems and narrow 1-1/2-inch leaves. Flowers to 1-1/2 inches are cup-shaped with pink petals and bloom in profusion for a long period in spring, although the foliage may stay green all year. Plants spread wide and thick by underground roots. A taller, similar plant reaches 12 to 18 inches high with paler pink flowers and is sold as a member of this species. Although there is some ques-

tion about its true identity, just ask for the taller form of Mexican primrose.

Special design features: Spring color.

Uses: Borders, bedding plant or under-plant. Along fences or walks. Containers and rock gardens. Transitional, natural, desert or wild landscapes. Sometimes used as a ground cover, but it is not that attractive all year in most situations. Best used for its seasonal color.

Disadvantages: Not especially attractive in winter.

Planting and care: Plant from containers any time or from divisions in spring. Space 2 feet apart for a cover in 2 years or less. Clip nearly to the ground after bloom.

All zones

Often evergreen perennial

Soil: Tolerant. Appreciates improved garden soil.

Sun: Full sun.

Water: Moderate during bloom. Less in winter.

Temperature: Hardy in warm areas where it can maintain some greenery all year. Killed to the ground in cold winters, but quickly recovers in spring and blooms by May.

Maintenance: Occasional.

Oenothera berlandieri

Oenothera berlandieri

OLEA europaea
Family: *Oleaceae*

Gracious yet picturesque, the billowing crown and gnarled trunk of the olive tree make it one of the most desirable plants for the arid landscape. The olive is believed to be native to Asia Minor, where it has been grown since prehistoric times. Greeks and Romans carried it to other parts of the Mediterranean where olive groves are a familiar sight. Some of the earliest trees in cultivation in the United States were seedlings planted in the 18th century by Franciscan padres at the San Diego Mission, now widely known as the mission olives. Olives live for hundreds of years, enduring drought, heat and poor soil.

Although older single-trunked specimens grown in deep soil have grown to 40 to 50 feet tall, the olive is usually a multiple-trunked, round-headed tree growing slowly to 15 to 30 feet with a spread almost as wide. Young trees may gain height quickly, but take time to develop substance. Leaves are stiff and leathery to 2 inches long, medium gray-green above, white to nearly silvery beneath. Gray-barked trunks become rough and gnarled with age. Clusters of tiny yellow-white flowers in spring produce fruit which ripens in fall. Fruit may be pickled or pressed for oil. Recipes are available from county agents or university extensions.

Cultivars: 'Fruitless' *does* produce some fruit, though not as much. 'Swan Hill' is said to be totally fruitless. Two varieties are commonly grown for landscape or commercial use: 'Manzanillo', round-headed, open and spreading form with excellent olives; and 'Mission', taller, more compact and hardier but with smaller fruit. Three kinds commonly grown in commercial groves also used as landscape trees are: 'Ascolano' with large fruit; 'Barouni', which thrives in heat; and 'Sevillano', which has very large fruit.

Special design features: Informal and picturesque. High quality tree. Filtered shade. Tolerant of desert conditions.

Uses: Patio, street or lawn tree. Specimen, row or grove. Large containers. Espalier on walls. Silhouette against structures. Looks at home in wild or natural gardens where it blends with other plants, especially gray-colored ones. Can be used as a canopy tree in a raked earth or paved patio, with plants in pots or as a specimen. Not for lawn areas, where it tends to get too much water.

Disadvantages: Basal suckering, flower and fruit litter and hay fever are objections. Fruiting can be prevented by hosing off flowers before they bear fruit. Verticillium wilt is another potential problem. It is a fungus present in some root stock and often appears after a cool spring followed by rapidly rising temperatures. It may show up in new growth any time of year, but appears most often in spring. Symptoms are new leaves that roll inward, becoming dull gray and brown. They usually remain on the tree. Flowers on affected trees may die but also remain attached. Occasionally a whole tree will die, but more often it is seen as a single affected branch on an otherwise healthy-looking tree. To treat, spread a 3-inch layer of sawdust in the tree basin to the drip line and rake into the soil lightly, taking care not to damage the roots. Add one cup of ammonium sulfate followed by a deep soaking every three weeks for 3 or 4 times. Be sure to spread the ammonium sulfate evenly so as not to burn the roots. Black scale is another problem that shows up in damp humid summers. Spray to treat. Olive knot—small woody knots—can appear after frost injury on twigs and branches. Cut off all infected parts and bury or burn them. Seal wounds and sterilize pruning shears or saws after each cut to avoid spreading the disease.

Planting and care: Container plants, carefully selected from healthy stock, may be planted at any time. Leave lower

Olea europaea

Olea europaea

Olea europaea

branches on spindly trees, but shorten to 12 inches. Once the trunk is the same size all the way up to what will be the permanent scaffold branches, remove the side-branching stubs. From the beginning, slender trunks should be carefully staked to grow at the desired angle. If a single trunk is desired, find a suitable single-trunked plant in a container, or prune any

side branches or canes below the point you want the branching to begin, so that only one cane is left. Tie to a stake for support. After tree is trained to the shape you want and the trunk is self-supporting, prune only to remove suckers and dead or errant branches. Larger trees transplanted from orchards sometimes have difficulties adjusting to changes in soil. Thin top

growth and be certain water is getting to the root ball and is not draining away into the surrounding soil. Although fertilization is not necessary, trees respond to an annual feeding of nitrogen and added potassium, boron and phosphorus in some soils. Do not cultivate under olives, because they develop a dense mat of feeder roots near the ground surface.

A relative of olive that deserves mention is *Olea verrucosa*, shrubby olive. This outstanding evergreen hedge and screen plant is becoming available from the United States Soil Conservation Service in Tucson, Arizona. Started from cuttings, it grows at a slow to moderate rate 12 to 13 feet in height and as wide. Branches are densely covered to the ground with small, narrow, dark gray-green leaves. It is practically trouble-free and can be irrigated or neglected in a manner similar to native plants in areas receiving 10 to 12 inches of rain annually. Sheared once a year in spring, it maintains a neat, dense form and screens out dust, wind and noise as well as objectionable views. Use where it has room to spread. Birds love the fruit. Also a choice plant for formal shaping. Cultural needs are similar to *Olea europaea*.

All zones

Evergreen

Soil: Prefers deep rich soil, but tolerate poor, stony, shallow and alkaline soils.
Sun: Full sun. Tolerant of part shade to reflected sun.
Water: Tolerant—moderate to no supplementary irrigation in areas with ample rainfall. Best with deep, widely spaced irrigation, especially in summer. May be neglected in winter.
Temperature: Damaged at 15°F (−10°C). Prefers temperatures above 20°F (−7°C). For good fruit production, olives prefer hot summers and a certain amount of winter chilling—about 12 to 15 weeks of temperatures fluctuating from 35°F (2°C) to 65°F (18°C) between day and night.
Maintenance: Trees over paved areas require more clean-up. Remove suckers from base of trunk several times a year.

Olea verrucosa

OLNEYA tesota
Family: *Leguminosae*

Desert ironwood ● Palo-de-hierro ● Tesota

Ironwood is a desert native found in warm-winter areas of California, Arizona and northern Mexico to 4,000 feet elevation. It grows slowly 12 to 20 feet high with a nearly equal spread. Gray trunk and lower branches are very thorny in youth, gradually absorbing the thorns as they enlarge to become rough and fissured at maturity. Trees naturally branch low, are often multiple-trunked and must be trained to become effective patio subjects. Foliage is gray-green and finely divided. At a distance, ironwood is similar in appearance to an olive tree. A profusion of pea-shaped orchid flowers in clusters appears in early summer, followed by 2-inch seed pods. Where it grows naturally indicates a climate belt compatible to citrus culture. Wood is hard, heavy and dense, valued for carving and as firewood. A tree to value and train if you have it on your property, but not the best choice if you are starting it from scratch. It is very heat and drought-resistant.

Special design features: Picturesque. Informal.

Uses: Hot areas. Natural, wild, transitional. Desert gardens or patios.

Disadvantages: Spiny branches. Slow to develop. Sensitive to cold. Blossom litter.

Planting and care: Plant from containers in spring. Prune only to shape and to

Olneya tesota

remove objectionable branches. Withstands neglect but grows fastest with some irrigation.

Low zones, warmer areas of the middle zone

Evergreen to deciduous in cold winters

Soil: Tolerant. Best with loose, sandy or gravelly soil with good drainage.
Sun: Full to reflected sun.
Water: Best with periodic irrigation.
Temperature: Tolerant of heat. Freezes at about 20°F (−7°C) and makes only moderate recovery.
Maintenance: None to some depending on location and taste.

OPHIOPOGON japonicus

Mondo grass • Dwarf lilyturf

Family: *Liliaceae*

(Mondo japonicum, Liriope japonica). This grasslike ground cover grows slowly to form a dense mound of narrow dark green leaves 8 inches in height, spreading up to 10 inches wide. Tiny, almost inconspicuous bluish to lavender flower spikes appear in summer among leaves. Plants occasionally spread by underground roots. Native to Korea and Japan.

Special design features: Oriental effect. Tufted, woodsy turf effect.

Cultivars: 'Nana' is identical to the species, except it grows to one-half the size.

Uses: Small-area ground cover or foreground plant. Plants grow at different rates and cover unevenly, creating a bumpy, casual look. Nice effect in containers. Place around stepping stones or to fill other odd spaces. Not a plant to walk on, however. Combines well with other low, woodland-type plants.

Disadvantages: Dry thatch on neglected or sunburned plants is unattractive.

Planting and care: Plant from flats, containers or divisions any time of year. Space 6 inches apart. When tufts become large

All zones
Evergreen
Soil: Improved garden soil. Avoid alkaline conditions.
Sun: Open, filtered or part shade. Accepts full sun in higher elevations or near coast.
Water: Moderate is best. Accepts ample in summer in the hottest areas. Tolerates periods of drought but looks poor.
Temperature: Roots are killed at about 10°F (−12°C).
Maintenance: Occasional.

they may be divided and replanted. Grows fastest and looks best with regular garden care.

A similar plant is *Ophiopogon jaburan* (*Liriope gigantea*), giant lily turf. It grows to 3 feet high and as wide, spreading by underground stems. Leaves are also dark green. Flowers are more white in color, produced in loose clusters. Uses and cultural requirements are the same as *O. japonicus*.

Ophiopogon japonicus

OPUNTIA species

Prickly pear and cholla

Family: *Cactaceae*

Opuntias are perhaps the most widely distributed cacti, with a multitude of species. Native to the Americas but planted around the world, they have become so successful at adapting they are sometimes a pest. As a group, opuntias are thorny succulent plants formed of jointed segments with large spines; or barely visible but sharp bristles called *flochids,* which appear in polkadot clusters on their waxy surfaces. Showy 2-inch wide, waxy flowers in late spring or early summer come in shades of yellow, orange, salmon, pink, red and in rare instances, white. Fruit in the form of a berry or pear ripens mid to late summer. They are green to shades of orange, red or purple, often changing color as they ripen. Many prickly pears produce large edible fruit which makes a tasty jelly.

The cholla group produces thorns which are exceptionally treacherous if they latch onto you. Certain types can be used effectively as barriers for bank plantings or at property edges. Chollas are harder to start, slow to grow, more difficult to transplant compared to the prickly pears. They are also often short-lived. They are generally not used as landscape subjects unless they already happen to be on the site.

The prickly pear group is quite a different matter. They are some of the

Opuntia phaeacantha

toughest and most tolerant plants in the desert, and whole groves can develop in a few years from segments set in the ground and more or less abandoned. They range in size from treelike plants 15 to 20 feet tall, to spreading prostrate forms less than 2 feet high. They are bold and sculptural in form, striking against architecture or with other bold desert plants such as agave. Tender new leaves of certain prickly pears are prepared and eaten as a vegetable in Mexico (*nopalitos*) and parts of the United States as well as in other places.

Special design features: Bold form. Desert effect.

Uses: Desert gardens. Cactus fences at property edges and around swimming pools. Barriers, combined with other thorny plants such as desert hackberry, agave or ocotillo. Bank cover for erosion control. Tall forms make bold silhouettes against structures. Low forms give a textural contrast against surfaces of different materials. Native plants are most effective for refurbishing disturbed areas and returning them to a natural state. Plant densely and at random for this purpose.

Disadvantages: Dangerous if planted close to walkways or similar use areas. Plants require special care in handling and can invade areas where they are not wanted. Large plantings of low-growing kinds may harbor pack rats which eat and destroy them. Rodents may eat tender new leaves.

Planting and care: Plant any time from containers. If you have plants and desire more, break off one or more continuous segments any time of year and place it part way in the ground. It will take root and grow, but might not put out new pads until spring if planted in fall or winter. For rapid development of new clumps, set three to five segments in a cluster at 12 to 18-inch intervals. To make a fence or impenetrable barrier, set segments in a trench at 12-inch intervals.

Opuntia species

Opuntia microdasys

Opuntia ficus-indica
Indian fig, spineless cactus _____

(*O. engelmannii, O. megacantha, O. occidentalis*). A tree-type to 12 feet or more, Indian fig develops a woody trunk and several branches of smooth flat segments. Segments are as much as 20 inches long, have no thorns but a few sharp glochyds in each polkadot. Yellow spring flowers produce edible fruits of yellow to red, prized in Mexico and parts of the United States. Effective as a silhouette plant, hedge or barrier. Tender to cold in high zone.

Opuntia microdasys
Rabbit ears, goldplush, prickly pear _____

Rabbit ears prickly pear may grow to 3 feet high or spread along the ground. Segments are deep to bright green or gray-green, 3 to 6 inches long and set with velvety yellow polkadots of sharp glochyds—decorative—but deadly to touch or handle. There are selections with white or rust-colored glochyds. All produce yellow flowers in spring or early summer. Effective as low fences or specimens. This native of Mexico is tender to cold in the high zone.

Opuntia phaeacantha
Engelmann prickly pear _____

(*O. engelmannii.*) This species covers a number of different prickly pears native to a wide area from California to Texas and south into Mexico, at 1,000 to 6,500 feet elevation, some to 7,500 feet elevation. They are well adapted to the desert and the most effective for refurbishing disturbed areas. Plants vary in growth habit, size of pad, number and length of thorns and to some extent the color of their flower. Fruit is green changing to red, then purple. Excellent for jelly.

Opuntia violacea santa-rita
Blue-blade, dollar cactus _____

(*O. santa-rita*). Striking because of its color, blue-blade prickly pear has segments tinged with purple or totally purple, a dramatic color contrast to other desert plants. Segments to 8 inches long may have 2-1/2 inch brownish to pink spines, or are almost completely spineless. Yellow flowers to 3-1/2 inches in diameter may turn red near the base inside and produce red to purple fruit. Excellent as color feature specimen or hedges of medium height.

Opuntia violacea santa-rita

Opuntia ficus-indica

OSMANTHUS fragrans

Sweet olive • Fragrant olive • Tea olive • Sweet osmanthus

Family: *Oleaceae*

Erect and elegant with refined foliage, sweet olive produces clusters of tiny, inconspicuous, whitish flowers off and on most of the year; the heaviest in spring. Flowers give off an elusive but pervading fragrance, usually followed by 1/2-inch long blue fruit. Believed to have originated from eastern Asia, it has been cultivated for centuries in China for its flowers used to scent tea. Sweet olive grows at a moderate rate up to 10 feet, but usually reaches about 6 feet in the desert. Mature specimens can be trained into small trees. Plant can be dense or open, with 4-inch long shiny, dark green leaves. Usually well behaved, it takes little shearing or shaping to maintain a refined and attractive form.

Special design features: Pleasant fragrance. Refined upright form.

Uses: Specimen in small patio or entryway for close-up viewing and fragrance. May be massed or planted in rows as background or foundation plant. Containers. Espalier on cool walls. Small upright tree under taller trees.

Disadvantages: Iron chlorosis. Foliage can be burned by sun in hot locations in low and middle zones.

Planting and care: Plant from containers any time. Space 5 to 6 feet apart for massing. Do any desired shaping in late winter or early spring. Cultural requirements are similar to those of gardenia. Feed regularly and mulch roots in summer.

All zones

Evergreen

Soil: Prefers rich soil on the acid side with good drainage.
Sun: Open to filtered or part shade with afternoon shadow in low and middle zones. Accepts full sun only in cooler areas. A good subject for cool northern exposures.
Water: Moderate to ample.
Temperature: Hardy to cold. Accepts desert heat with ample water.
Maintenance: Regular.

Osmanthus fragrans

OSTEOSPERMUM fruticosum

Trailing African daisy

Family: *Compositae*

This colorful plant from South Africa is favored for its masses of spectacular blossoms. Violet-colored daisylike flowers 2 inches in width are displayed above the foliage from November through March. Flowers change to a near-white by the second day. Plants grow fast 6 to 12 inches high, spreading 2 to 3 feet in a year. Branches root as they go; plants eventually form a solid cover. Lobed hairy leaves are almost succulent.

Cultivars: 'White Hybrid' has a more profuse bloom and accepts less cold than the species.

Special design features: Color in winter and spring. Appealing informal texture. Prefers locations away from extremes of heat, cold and sun.

Uses: Banks or ground cover. Container spiller. Borders. Permanent bedding plant.

Disadvantages: Older plantings need refurbishing or replacement every few years. Plants sometimes die out in spots from heat or cold. Sometimes becomes chlorotic and needs extra iron.

Planting and care: Plant in spring or fall so young plants will get a start before extreme temperatures of summer arrive. Space 2 feet apart for fast cover. Pinch young plants to encourage new growth and more bloom. Feed iron if plants yellow.

Middle and lower zones

Evergreen herbaceous perennial

Soil: Tolerant of all but heavy soils. Prefers well prepared garden soil.
Sun: Part to full sun. Avoid reflected sun.
Water: Moderate.
Temperature: Damaged at about 20°F (−10°C). 'White Hybrid' is damaged a few degrees above that.
Maintenance: Occasional.

Osteospermum fruticosum 'White Hybrid'

Osteospermum fruticosum

PARKINSONIA aculeata

Mexican palo verde • Jerusalem thorn • Ratama

Family: *Leguminosae*

Mexican palo verde is native to the southernmost tip of Arizona, Mexico and other warm parts of America, but has been widely cultivated and has naturalized in many areas. It is usually an erect, single-trunked and thorny tree, completely covered with smooth yellow-green bark. As trees age, the bark becomes rough and gray. Trees grow fast to 15 feet high and eventually reach 20 feet or more with a crown spread as wide or wider. During summer and all year in warm areas, the tree bears a narrow portion of its leaf, the midrib, which is 8 to 16 inches long. The entire leaf is made up of this midrib and tiny leaflets 1/8 inch long. In late spring or early summer there is a spectacular display of yellow flowers and the whole crown becomes a giant bouquet. Slender yellowish pods to 6 inches long follow. Shedding midribs creates a straw-colored thatch like pine needles.

Special design features: Spring or summer color. Bright yellow-green tree color. Can be used for tropical or desert effect.

Uses: Street, patio or lawn tree. Desert or tropical gardens. Transitional areas. Fast filtered shade.

Disadvantages: Palo verde beetle can be a problem. See page 37 for treatment. Litter can be bothersome. Seeds also litter and reseed profusely. Occasionally suffers mistletoe infestations and frost damage. Sharp thorns in spring.

Planting and care: Start from seed, or set out from containers in spring or summer. Space 20 feet or more apart for a street tree or row. Remove lower branches if necessary to form a high crown, although trees usually take on a handsome, high-crowned shape naturally.

Parkinsonia aculeata

Parkinsonia aculeata

Low and middle zones

Deciduous

Soil: Tolerant. Accepts alkali. Prefers sandy soils with good drainage.
Sun: Full to reflected sun.
Water: Will grow without irrigation in low zones with 10 to 12 inches of annual rainfall, but best with some supplemental irrigation or runoff. Tolerates ample watering, such as in a lawn.
Temperature: Thrives in heat. Young trees are badly damaged at about 18°F (−8°C). Older trees can tolerate slightly lower temperatures.
Maintenance: None to occasional depending on location and taste.

PARTHENOCISSUS quinquefolia

Virginia creeper • Woodbine • Five-leaved ivy

Family: *Vitaceae*

(Ampelopsis quinquefolia, Vitis quinquefolia.) Virginia creeper is a self-climbing vine, clinging with tendrils or aerial rootlets to cover a large space in a short time. Native to an area that includes northeastern United States to Florida, Texas and Mexico. A vigorous grower, runners root along the ground as they travel to create a dense cover. Leaves 2-1/2 to 6 inches across are formed of 5-tooth leaflets which fan out from center. Leaves turn gold and crimson in fall.

Cultivars: 'Engelmannii' has smaller leaves. Its flowers are unimportant features, but it produces colorful blue fruits that attract birds.

Special design features: Intimate woodland feeling. Garlands and festoons. Autumn color.

Uses: Fences, masonry walls or trellises as a fast summer cover. May be used as a random ground cover with other plants around the foundation of a structure or as a garland up a tree or post.

Disadvantages: Sometimes invasive—it can cover windows and other openings. Sometimes attacked in summer by grape-leaf skeletonizers. Occasionally bothered by root knot nematodes, powdery mildew and Texas root rot.

Planting and care: Plant any time from containers. Rooted runners can be dug up and transplanted in winter. Give late winter clipping to control size.

Parthenocissus tricuspidata, Boston ivy, is a very close relative to the above. Leaves on young plants vary, may be round to lobed like ivy, about 2 to 4 inches across. Mature leaves are deeply lobed, waxy and form an interesting shadow pattern where they cling densely and tightly to the wall. It will climb up and over walls in cooler climates, adding a tracery pattern of lush green. Sometimes grows successfully in low deserts on north walls receiving little or no sun. Marginal plant in the middle zone. Cultural requirements are similar to *P. quinquefolia.*

All zones

Deciduous

Soil: Average garden soil.
Sun: Part shade to full sun. Prefers afternoon shadow in hot areas.
Water: Moderate to ample.
Temperature: Hardy. Tolerates heat and exposure.
Maintenance: Occasional.

Parthenocissus quinquefolia

PASSIFLORA alatocaerulea

Family: *Passifloraceae*

(P. alato-caerulea, P. pfordtii). This tropical hybrid vine gets its name from remarkable 4-inch flowers thought to be symbolic of the passion of Christ. Although it grows rapidly to 20 or 30 feet and clings by means of tendrils and suction-cup aerial rootlets, vine needs some training and tying to supports to climb. Verdant and tropical in appearance, it produces a dense cover of usually 3-lobed leaves to 3 inches across. Flowers are fragrant, intricate in design, in shades from nearly white to lavender to deep purple, and are used in the perfume industry. Bloom is sporadic with scattered flowers from spring through summer. Thrives in about the same climate as citrus.

Special design features: Lush tropical feeling. Interesting bloom.

Uses: Anywhere on a trellis, fence or porch post in frost-free areas or on protected walls under an overhang in areas where there is light frost. Protected and warm exposures in middle zone. Use as an annual vine in cooler areas, but it may not bloom before frost. Grow in a container or tub and bring inside in winter.

Disadvantages: Sometimes attacked by caterpillars. Iron chlorosis.

Planting and care: Plant from containers in spring when danger from frost is past. Control growth with pruning, especially July to September.

Low zone, protected locations in middle zone

Evergreen to semievergreen

Soil: Improved garden soil.
Sun: Part shade to full sun. Leaves may burn in reflected sun.
Water: Moderate.
Temperature: Loses leaves at about 28°F (−2°C). May freeze to the ground around 20°F (−7°C). Roots die at around 10°F (−12°C). Lightly frosted vines recover rapidly in spring. Those frozen to the ground take longer.
Maintenance: Occasional.

Passiflora alatocaerulea

PENNISETUM setaceum

Family: *Gramineae*

(P. ruppelii). Fountain grass is a dramatic and durable plant from Africa. It forms large tufts or clumps 2 to 4 feet high and as wide. Narrow leaves and fuzzy, bristly seed plumes rise above foliage on slender stems. Seed plumes 4 to 6 inches long may be white or tinged pink or purple. Bright green in summer, plants turn straw colored in winter if frosted. Tolerant of drought, exposure to heat and wind and pest-free, fountain grass has naturalized in many areas where it gets a little extra surface runoff.

Cultivars: Recently available is 'Cupreum', (*P. cupreum*), with deep reddish foliage, wider leaves and coppery plumes. It is larger and coarser than the species. Other kinds sometimes available are 'Atrosanguineum', purple with purple seed spikes, and 'Rubrum', rose-colored foliage and rosy spikes.

Special design features: Desert, waterside or Oriental effect. Informal. Helps soften the landscape. Contrasts handsomely with various plant or material textures. Very decorative.

Uses: Almost anywhere as a foundation plant, specimen, space divider or border plant. Parkways and median strips. A few plants among rocks or mulches can carry the whole landscape. Visual break in row or mass plantings. Looks at home in rockeries, desert, transitional or natural gardens or in informal areas. Use to hold banks, perform as a low windbreak plant or for erosion control.

Disadvantages: In winter it may be a fire hazard in mass plantings unless cut back. Sometimes spreads and naturalizes where it is not wanted. 'Cupreum', however, does not set viable seed.

Planting and care: Plant from divisions or containers any time of year. Species plants may be started from seed planted in porous soil. Space 3 feet apart for a border or row planting, 18 inches on center for erosion control or a solid mass. Tufts should be cut back in the dormant season, but grow out quickly in spring. Older or established clumps may be divided.

All zones

Deciduous

Soil: Tolerant.
Sun: Full to reflected sun. Tolerates part shade.
Water: Little to moderate. With some runoff, plants survive and multiply in areas of 6 to 12 inches of annual rainfall.
Temperature: Established plants are root hardy to cold. New plantings may freeze out at 15°F (−10°C).
Maintenance: Prune plants in fall to clean up dried foliage.

Pennisetum setaceum 'Cupreum'

Pennisetum setaceum

PHILODENDRON selloum

Family: *Araceae*

Selloum philodendron

Selloum philodendron is a tropical, treelike plant possessing the usual lavishness of Brazilian natives. It grows at a moderate to rapid rate to 3 feet high, sometimes to 10 or 12 feet high in favored places. Width varies from 3 to 10 feet. Deeply lobed dark green leaves may reach 36 inches in length on larger plants, located at the ends of 4-foot herbaceous stalks which grow from the center of the plant. Mature plants develop attractively patterned trunks and woody aerial roots near its base.

Special design features: Tropical or jungle effect. Dramatic form. Bold foliage. Sculptural silhouette plant. Accent, emphasis or space filler.

Uses: Entryways, atriums, patios and under stairwells, either in the ground or in containers. Be sure plants have at least a 3-foot square area in which to develop. Purchase plants grown outdoors for outdoor plantings and indoor-grown plants for indoor use; transferred plants may suffer shock from which they recover slowly, if at all.

Disadvantages: Sometimes gets iron chlorosis or a mosaic virus. Exposed plants sometimes are damaged by extremes of heat or cold or from reflected sun.

Planting and care: Plant from containers when danger from frost is past. Space small plants 5 feet apart for an effective grouping. Remove damaged leaves in spring after new growth starts. Groom as needed. Grows faster with feeding and garden care. Plants stay smaller in containers. If aerial roots become unappealing, do not remove them. Plant a low ground cover around the base of the plant to cover.

Low and middle zones, indoors in winter in higher zone

Evergreen

Soil: Average garden soil with good drainage.
Sun: Filtered, open or part shade. Accepts full sun, but needs afternoon shadow in hot areas.
Water: Moderate.
Temperature: Tolerates light frosts. Freezes to ground or to leaf sheaths at its base below 20°F (−7°C). New plants not yet established may freeze out. Plants usually recover rapidly in spring.
Maintenance: Little to occasional.

Philodendron selloum

PHOENIX canariensis

Family: *Palmae*

Canary Island date palm • Pineapple palm

A massive-scale, lush, feather palm from the Canary Islands, Canary Island date palm develops slowly from an appealing pot-sized plant into a majestic giant 60 feet tall in about 80 years. A palm 12 feet high may have a crown spread of 20 to 30 feet and a trunk diameter of 3 feet. Fronds are dark green, orange at the base, formed of many shiny filaments which glisten in the sun. Palms bloom in spring. Males produce *spaths*, bracts or leaves surrounding flower clusters, and females produce broomlike structures which hold the developing inedible dates. Dates first appear orange then turn brown in the fall before they drop.

Special design features: Bold vertical form. Eventually a skyline tree. Tropical or desert oasis feeling.

Uses: Best in large areas, such as a park or boulevard tree or in large commercial, public or residential landscapes. Young trees are excellent container plants—durable and undemanding. Palms stay small a long time in containers. A clean tree for pool areas (except for ripening fruit from females).

Disadvantages: Whole season required for recovery from a hard frost. Mature trees are costly to groom. Leaves split in the wind. Ungroomed trees have messy fruit drop with many seedlings. Dry rot sometimes makes trunks and frond stubs unsightly. Subject to palm heart rot.

Planting and care: Can be grown from seed, but very slow. Plant from containers in spring after frosts are over. Large specimens can be transplanted during warm summer weather. Space 30 to 60 feet apart for a row, avenue planting or grove. Remove fruiting parts and old fronds in late spring and summer. Mulch roots, feed and irrigate deeply in summer to encourage faster growth. If leaf bases are skimmed from the trunk, be sure to leave about 2 feet of leaf stumps below crown. This helps support leaf fronds.

Low and middle zones

Evergreen

Soil: Tolerant. Prefers rich moist soil for fastest growth.
Sun: Part to full sun.
Water: Moderate to ample, although palms of all sizes withstand considerable drought, especially in the cool season.
Temperature: Heavily damaged at 20°F (−7°C).
Maintenance: Periodic grooming necessary for attractive plants.

Phoenix canariensis

Phoenix canariensis

PHOENIX dactylifera

Family: *Palmae*

Slender and more open and lighter in appearance than the Canary Island date palm, the true date palm is characterized by a rough gray trunk to 18 inches in diameter and a "feather-duster" crown of gray-green fronds. It grows slowly to 60 feet high with a crown to 25 feet in width. Both male and female trees are needed to produce edible fruit which ripens in fall. If no male trees are present, pollen may be obtained to fertilize females. Date palms produce many offshoots around the base and are multiple-trunked by nature. If left in place, the offshoots will form a thick clump, eventually growing outward to form a many-headed, angle-trunked grouping. To keep trees neat, limit to three or four offshoots. Offshoots are also a means of asexual reproduction, and replicas of the parent. For fruiting types do not waste your time on seedlings. Seek plants asexually reproduced from one of the many named varieties.

Cultivars: There are many fine date-producing trees. If in doubt about the date palm best for your needs, contact your county extension agent.

Special design features: Desert oasis feeling. Dramatic silhouette, accent or emphasis plant. Strong vertical. Eventual skyline tree.

Uses: Specimen and emphasis plant.

Disadvantages: Sometimes gets palm heart rot or fiber rot, causing trunk and frond ends to look moth-eaten. Sometimes gets bud rot, especially during humid, rainy, summer weather. Can get leaf spot in humid climates. Taller palms become difficult and costly to groom. Sharp spines at frond bases make leaf trimming difficult.

Planting and care: Planting and transplanting should be done in summer, preferably in June. Large specimens may be transplanted, but this should be performed by experts. Offshoots may be removed and planted when they reach 10 to 12 inches in diameter and weigh 35 pounds or more. Plant palms from containers any time, but warm weather is best. Space 30 feet apart for grove or row. Plants tolerate long periods of neglect, but respond to fertilization and irrigation with faster growth and better fruit. To groom, remove old hanging fronds and fruiting remnants each year.

Low and middle zones

Evergreen

Soil: Tolerant if used as an ornamental.
Sun: Full sun to part shade.
Water: Moderate to deep irrigation on a regular basis.
Temperature: Hardy to 18°F (−8°C).
Maintenance: Occasional to periodic depending on use.

Phoenix dactylifera

PHOTINIA fraseri

Family: *Rosaceae*

This hybrid is a popular plant in the landscape, adaptable to many uses. A moderate grower 10 to 12 feet high and as wide but easily kept smaller. Erect in form, it is covered to the base with toothed leaves 5 inches long, bright coppery red when new, becoming shiny dark green with whitish undersides at maturity. Scattered leaves are red-tinted in cold weather. Flat lacy clusters of white flowers appear in spring. A scattering of blooms may appear again in fall along with a few red berries.

Special design features: Color and interest throughout year. Dense and exceptionally attractive foliage.

Uses: Large specimen shrub or small patio tree. Massed or planted in rows as background, wide screen or clipped hedge. Does well in containers, parking strips or in other paved areas. Tolerates reflected heat. May be espaliered on walls, except extra-hot west walls. Transitional areas.

Disadvantages: Rank growth and bare branches in spots if not pruned or clipped occasionally. May get fireblight or Texas root rot. Plants sometimes gets chlorosis. If this happens, plants should be fed with iron, a complete mineral mix, as well as a balanced fertilizer. Occasionally attacked by aphids.

Planting and care: Set out from containers any time. Space 3 to 4 feet apart for a clipped hedge, 8 feet for wide screen or background. Do any major pruning in late winter, trim and shear any time. Tolerates neglect, but regular garden care produces more handsome plants.

All zones

Evergreen

Soil: Tolerant. Prefers average garden soil. Alkaline soils may produce chlorosis.
Water: Moderate to deep irrigation on a regular basis.
Temperature: Hardy to 5°F (−15°C). Thrives in hot sun.
Maintenance: Regular garden care.

Photinia fraseri

Photinia fraseri

PHOTINIA serrulata

Family: *Rosaceae*

This plant does best in the high zone where the climate is cooler and the sun less intense. It is slightly more hardy than *Photinia fraseri*. A shrub or small to medium tree with bold foliage, Chinese photinia, a native of China, can grow 35 to 40 feet in favorable locations. Usually seen as shrub 8 to 10 feet tall with crisp toothed leaves to 8 inches long. New growth is coppery in spring, turns deep green in summer with yellowish undersides. Some leaves turn bronzy or red in fall and winter. Small white flowers in flat 6-inch clusters appear in spring and sometimes produce a sprinkling of long-lasting red berries in fall.

All zones

Evergreen

Soil: Tolerant. Prefers average garden soil. Alkaline soils may produce chlorosis.
Water: Moderate to deep irrigation on a regular basis.
Temperature: Hardy to 5°F (−15°C). Thrives in hot sun.
Maintenance: Regular garden care.

Cultivars: 'Nova' grows to about 8 feet high and as wide. It is generally considered a better plant than the species. 'Aculeata' is a compact variegated form often sold as 'Nova'.

Special design features: Dense bold foliage. Woodland effect. Large scale for large open spaces. Interest all year.

Uses: Use as specimen in open places that have good air flow. Background or screen plant. Larger species makes a handsome tree in cooler areas. Smaller form makes a handsome small patio tree in low and middle zones when given good air circulation and some afternoon shade. Espalier on cool walls. Close hedge clipping ruins the bold leaf texture.

Disadvantages: See *Photinia fraseri.* Additionally, this plant is subject to mildew. Plants grown in the open are generally less susceptible.

Planting and care: Set out from containers any time. Space 3 to 4 feet apart for a clipped hedge, 8 feet for wide screen or background. Do any major pruning in late winter, trim and shear any time. Tolerates neglect, but regular garden care produces more handsome plants.

Photinia serrulata 'Nova'

PHYLLOSTACHYS aurea

Family: *Gramineae*

Golden bamboo ● Fishpole bamboo ● Yellow bamboo

(Bambusa aurea). This is a running bamboo, erect and rather stiff with hollow, yellowish canes commonly used for fishing poles. It reaches up to 10 feet, sometimes 20 feet high. Vigorous grower and spreader, it will form dense thickets or groves unless controlled. Leaves 2 to 6 inches long and 1/2 inch or more wide are light green beneath, darker above. Plants require time to establish and to gain height and width. Once established, they will send out runners to moist areas and can become an invasive pest. This is the bamboo often used for staking plants or for fences.

Special design features: Vertical. Properly used it gives a tropical, Oriental, jungle or waterside effect. Festive, light, airy and informal.

Uses: General landscape use as a hedge, specimen or screen. Attractive near ponds, but produces too much litter for swimming pool areas. A good container plant.

Disadvantages: May be slow to establish, but once established, it may become invasive. Occasionally gets crown rot. Litter of leaves and sheaths.

Planting and care: Plant from containers any time or from divisions in spring and fall. When purchasing a plant for containers, keep in mind that a root-bound plant will grow faster. If you plant in the ground, contain plants with an 18-inch deep metal or concrete edge. Burying plants in their containers also restricts spreading. Be certain water is getting to the root zone if you use this method. An area surrounded by a wide, dry space also prevents spreading. Control spreading by cutting 12 inches deep around plants with a spade to sever lateral shoots. To encourage fast growth, treat as a lawn grass—water generously and feed with a high-nitrogen fertilizer monthly. To form a screen or hedge, set new plants 12 to 18 inches or more apart. Or, make an irregular row by clumping plants at varying intervals to form stands.

All zones

Evergreen

Soil: Prefers moist garden soil.
Sun: Open shade to part, full or reflected sun.
Water: New plantings are especially sensitive to drying out during the first year, so make sure they receive enough moisture. Established plants tolerate moderate irrigation but prefer ample.
Temperature: Hardy. Accepts heat with irrigation.
Maintenance: Periodic.

P. nigra, the black bamboo, deserves mention here. It is a running type usually reaching 4 to 8 feet tall with some specimens reaching 10 to 15 feet high. Canes to 1-1/2 inches in diameter are green then become olive or brownish, speckled with black. Mature canes are often pure black with nodes edged with white below. Leaf sheaths are greenish to reddish buff, adding still more color and interest. Hardy to 5°F (−15°C). It appreciates shade on hot summer afternoons in low and middle zones.

Phyllostachys aurea

PINUS species

Family: *Pinaceae*

The pines are a group of single-trunk, cone-bearing trees with aromatic needlelike foliage. Native to various parts of the Northern Hemisphere, they are plants of great character and are usually dominant features in the landscape. Pines can be identified by their distinctive cones. Cones are either male or female and one tree may have different sizes because it takes up to two years for cones to mature. Many young pines have juvenile foliage, often similar to that of spruce—flatter and not in bundles—quite different from the needles they develop at maturity. This foliage change makes young trees difficult to identify. A swelling of the terminal buds at branch tips in spring indicates new growth. This temporarily changes the look of some trees such as the Canary Island pine, which looks as if it is bearing huge candelabras. Few pines have proven successful over a long period of time in warmer climate zones, but a number introduced recently have performed very well over the past few years and should probably be used more extensively.

Special design features: Pines suggest the coolness of forest or picturesque survivors of mountain crags.

Uses: Shade. Wind and dust screens. Around pools and water bodies as basically clean trees that filter out dust.

Disadvantages: For the most part, pines require little care, but are susceptible to some pests and diseases. Suspect aphids if honeydew or yellowing needles appear. Infestations of red spider mites cause the foliage to look dull and mangy. Both pests can be eliminated with systemic sprays. Pines with 5 needles to the bundle are subject to white pine blister rust, but this is not much of a problem in the desert. Pines with needles in bundles of 2 or 3 may be attacked by the European pine

shoot moth. So far this has happened mostly in the Pacific Northwest. Older Canary Island and Torrey pine trees weakened by drought and with less resistance are sometimes attacked by bark beetles which can kill the tree. Aleppo and sometimes others die back at twig tips; the cause is not known. To avoid this blight, give pines regular irrigation, especially during periods of drought.

Planting and care: Plant any time from containers. The best time for planting and only time for transplanting is late fall or early winter. This allows the tree to establish a good root system before warm weather and rapid growth begin. When transplanting small trees, do it just before or just as growth buds swell. Take a large ball of earth with the roots to help ease shock. It is difficult to move larger specimens and it's best to rely on an expert. Stake and tie young trees securely, especially in windy places. Staking helps ensure proper establishment of root systems, prevented if young trees are continuously rocked by wind. To maintain size of tree or to encourage bushiness, growing tips (terminal buds) of branches may be pinched off halfway or more. Do not break off terminal bud below needles or the whole branch will die back. For the most luxuriant growth and healthiest tree, place a 2-inch mulch of composted manure with a sprinkling of soil sulfur over root area basin, preferably in fall. This helps prevent blight and aids in correcting caliche, tight silty soil or other poor soil conditions. Keep area around tree base free of weeds and debris. Avoid nitrogen fertilizers—pines find them toxic. If fertilizing plants or lawn nearby, do not apply more than one pound of nitrogen fertilizer for every 50 square feet of space around tree to a little beyond drip line.

All zones

Evergreen

Soil: Tolerant of poor soils, but good drainage is essential.

Sun: Part shade to full sun.

Water: Do not allow young trees to dry out during the first year or two until roots are well established. Most trees do very well with deep periodic irrigation after that. Many tolerate lawn watering. Overwatered trees show a yellowing of older needles and poor appearance. Pines kept too dry are slow growing, have sparse foliage and a weakness to pests, blight and diseases.

Temperature: See specifics for each species.

Maintenance: Regular watering, raking needles and annual mulching are all that is required.

Pinus brutia eldarica
Eldarcia pine, Mondel pine

This recently introduced tree is similar to *Pinus halepensis,* but is generally more symmetrical, more hardy to cold and faster growing.

Low and middle zones. Damaged by cold at about 13°F (−10°C), but has survived temperatures to 5°F (−15°C).

Pinus canariensis
Canary Island pine

Native to the Canary Islands, this pine has a tiered vertical form. A rapid grower—to 10 feet in 5 years, to 18 feet in 10 years, to about 33 feet in 20 years. Mature trees may reach 60 to 80 feet in height and develop rounded forms. Dark, bluish green needles to 12 inches long occur in bundles of 3. Glossy brown cones are 4 to 9 inches long. Young trees are leggy with juvenile foliage resembling that of spruce,

Pinus pinea

Pinus edulis

Pinus halepensis

Pinus roxburghii

but soon take on a pyramidal form. They are quite handsome with long shaggy foliage but look unkempt unless old needles are shaken off. Otherwise, little care is needed. The fairly narrow form of this pine makes it appropriate for narrower spaces or as a lawn tree. Its handsome, open silhouette makes an interesting pattern against the sky.
Low and middle zones. Damaged by cold at about 20°F (−7°C).

Pinus cembroides
Mexican stone pine

This small, rustic tree or large shrub is found in the higher rocky foothills of Arizona and Mexico up to seven thousand feet elevation. It is very similar to piñon pine and used in the same way. Orange branchlets support needles 2-3/4 inches in length in bundles of 3, but sometimes 2, 4 or 5. Collected trees have the picturesque, windblown appearance of a mountain crag and are used to create that effect in landscapes tending to be slightly Oriental.
Middle and high zones.

Pinus coulteri
Big-cone pine, coulter pine

This native of the dry, rocky slopes of California and Baja California, is a moderate grower 30 to 80 feet high with wide spreading, lower branches and an open, shapely form. It is resistant to heat, drought and wind. Needles in bundles of 3 are dark bluish green reaching 12 inches in length. Decorative drooping, cylindrical cones reach up to 14 inches in length. A good subject for the high deserts. Make sure it has room to spread.
Middle and high zones.

Pinus edulis
Piñon, nut pine, two-leaved pine

(P. cembroides edulis). Very similar to *P. cembroides,* but with 2 needles in a bundle instead of varying numbers. Piñon is a slow grower, taking many years to reach 10 to 20 feet in height. Older specimens may reach greater heights, but this tree is usually small and irregular with a flat to round top and horizontal branches. Native to higher elevations of the Southwest from California to Texas and into Mexico, Indians gather its tiny, tasty pine nuts, prized as a delicacy. Piñon is symmetrical in youth, but trees collected from nature usually have a weatherbeaten appearance, unlike pruned cultivated plants. Plants grown in nurseries can be trained into picturesque forms by using standard bonsai methods. Needles are dark green and short, about 1-1/2 inches in length. Rounded light brown cones 2 inches in length are the source of the edible nuts.
Excellent for small gardens with an Oriental feel. Rock gardens, small spaces or in containers.
Middle and high zones. Tolerant of cold, but languishes in heat of the low deserts.

Pinus halepensis
Aleppo pine, Mediterranean pine, Jerusalem pine

The Greeks traditionally cut down an aleppo pine each year and decorated it with flowers and ribbons in honor of the dead god Attis. It is believed this custom was adopted by Europeans in honor of Christ, making aleppo the first Christmas tree. Aleppo is probably the best pine for the desert. It takes about a year to become es-

tablished and grows at a fairly rapid rate: 10 feet in 5 years, 20 feet in 10 years, eventually reaching 30 to 50 feet. Mature specimens have round to irregular billowing crowns. Needles, usually in bundles of 2, are 4 inches in length. Trees may be gray-green in dusty areas but are usually medium yellowish green. Cones are rounded, light brown, two inches in diameter. Most pines drop their lower branches so that the trunk is visible. Mature specimens are often seen with no branches below 10 to 15 feet, making them excellent overstory trees. Aleppo pine blight can be a bothersome nuisance. It is indicated by browning of growing tips. Cause is unknown, but trees mulched with manure and irrigated regularly seem to overcome this problem. Irrigation is most important during dry periods when tree is under stress from drought.

Because aleppo pines are fast growing, evergreen and adapted to tough growing conditions, they make excellent windbreaks and screens. In windy areas, young trees must be firmly staked to prevent trunks from bending or weaving. See page 33.
Low and middle zones. Hardy to about 13°F (−10°C).

Pinus monophylla
Singleleaf piñon pine

This is a small, hardy, drought-resistant pine native to an area from Idaho to Mexico in higher rocky foothills. It is similar to other piñon except for a more rugged appearance at maturity, slower growth and cylindrical needles held 1 to a bundle.
Middle and high zones. Cool sheltered spaces in low zone.

Pinus thunbergiana

Pinus canariensis

Pinus pinea
Italian stone pine,
Umbrella pine _____

Visitors to Spain and Italy know the form of the ubiquitous umbrella-shaped pines towering over all, with tall, naked trunks and wide flat crowns. What is not understood is that this shape is not natural for this Mediterranean native. Although trees develop some bare trunk like the aleppo, it is the wood gatherers who trim trees to such heights! Stout, bushy, globe-shaped young trees grow at a moderate to slow rate 40 to 80 feet in height. Thinning the tree's interior will cause more rapid growth. Mature trees have wide flat crowns but when they are unpruned, branches extend much farther down the trunk. Stiff, bright green needles to 6 inches long are in bundles of 2. Cones are

Pinus torreyana

brown and oval, 4 to 6 inches long and produce the edible pignolia nut of southern Europe. A large-scale tree that will eventually tower over the average residence, umbrella pine is best used for roadsides or in large open areas. A background tree or one of a cluster of eventually tall, overstory, skyline trees. It makes an excellent silhouette and is handsome at any age.
Low and middle zones.

Pinus roxburghii
Chir pine, emodi pine,
Indian longleaf pine _____

(*P. longifolia*). A towering tree from the Himalayan foothills, chir pine is similar in many ways to the Canary Island pine. It differs in being more cold hardy and has brighter green foliage and a less open silhouette. A moderate to sometimes fast grower, it reaches 60 to 80 feet in height, with some trees in favorable locations attaining heights up to 150 feet. Slender and pyramidal in youth with thick foliage and slightly drooping branchlets, it becomes a spreading, symmetrical, round-topped tree at maturity. Needles in bundles of 3 reach up to 12 inches in length; cones are 4 to 7 inches long. Use this tree anywhere you want a pine, but give it room to develop.
Middle and high zones.

Pinus sabiniana
Digger pine _____

More of an ornamental rather than a shade tree, this native of the dry slopes of California foothills has an almost transparent crown. Very drought and cold-resistant. It is sparse and open in nature; in cultivation it is denser and has a more regular shape. Needles 9 to 12 inches long are blue to gray-green, 3 to a bundle. Oval cones 6 to 10 inches long have edible seeds. Not available in much of the Southwest, but this decorative, drought-tolerant tree deserves wider use. Because it pro-

duces filtered shade, other plants will grow under it.
Middle and high zones.

Pinus thunbergiana
Japanese black pine _____

A moderate to fast grower in the desert to about 20 feet, this Japanese native can reach a towering 130 feet in areas where it grows naturally. Trees have spreading branches and are conical to irregular in form. Sharply pointed foliage is fresh dark green; needles in bundles of 2 are 4-1/2 inches in length. Brown oval cones to 3 inches long appear near branch ends. This pine is in scale with a small garden as a feature, mass or screen and is remarkably well-adapted to training. Use as a bonsai plant or shear to form a pyramidal Christmas tree. Prune into picturesque shapes or train downward and outward to make a cascade. Outstanding as a container plant.
Middle and high zones. Heat is a problem in low zone.

Pinus torreyana
Torrey pine,
soledad pine _____

This romantic and craggy pine is found growing naturally along the California coast near San Diego preserved in a state park, and on Santa Rosa Island. Strangely enough, it grows very well in the desert. It will probably be difficult to locate in a nursery, but is worth considering. As a landscape tree, this pine grows at a fast, sometimes moderate rate 40 to 60 feet high, but unless exposed to wind and weather it is much more regular in shape and less open than the wind-blown trees seen on the coast, especially if grown in heavy soils. Stiff, light gray-green to dark green needles are 8 to 13 inches in length in bundles of 5. Dark brown cones are 4 to 6 inches long. Give trees room to develop.
Middle and higher zones. Hardy to 12°F (−11°C).

PISTACIA atlantica

Family: *Anacardiaceae*

(*P. mutica*). This is a neat, clean tree, round-headed to pyramidal, with an erect trunk and compound, medium green leaves formed of narrow, 2-inch long leaflets. Originally from the Canary Islands and the Mediterranean region to the Caucasus and Pakistan, it now thrives in the Southwest. Growth is slow to moderate to 60 feet high. Loose clusters of tiny blue fruit on females ripen in fall. These are not the popular pistachio nuts—they come from females of *P. vera*. Trees stay green all year in mild climates and lose leaves late in fall in harsher areas. Valued commercially for its wood and for resin which oozes out to form tiny drop-lets on the branches. Established trees tolerate much heat, wind and drought. Because it is slow to develop, it stays in scale with the average residence for a long time.

Special design features: Handsome, refined shade tree.

Uses: Patio, street or lawn tree. Excellent for public places.

Disadvantages: Mostly trouble-free but occasionally subject to Texas root rot.

Planting and care: Plant from containers any time or bare root when dormant in winter. Stake young trees and prune to form a high crown. Little care is needed. *Pistacia lentiscus,* mastic tree, is another Mediterranean pistache that should be mentioned. It is a large evergreen shrub or small tree with foliage similar to *P. atlantica.* It is generally hardy and thrives in dry environments. It may be hard to find in some areas.

All zones
Semi-evergreen to deciduous
Soil: Tolerant. Best in deep enriched soil with good drainage. Tolerates alkaline soil conditions.
Sun: Part to full or reflected sun.
Water: Moderate until established, then deep periodic irrigation. Give plants deep soaking even if planted in lawn. Lawn watering does not penetrate deeply enough.
Temperature: Hardy to cold. Accepts heat.
Maintenance: None to occasional.

Pistacia atlantica

PISTACIA chinensis

Family: *Anacardiaceae*

Chinese pistache, native to China and Taiwan, is similar to Mt. Atlas pistache but larger and more common. It grows at a moderate, sometimes slow rate to 60 feet high with a rounded crown that spreads to 50 feet wide. Trees in shallow soils or other difficult situations may grow to only 30 feet. Leaflets are 2 inches long, pointed, often turning scarlet in fall. Clusters of 1/2-inch fruit on females turn deep red in early fall before foliage changes, later changing to purple. Fruit is not edible. Not as drought-resistant as *P. atlantica.*

Special design features: Large, dense, quality shade tree for summer shade and winter sun. Fall color.

Uses: Lawn, park or avenue tree. Good in public spaces. Should be used where it has room to spread.

Disadvantages: Sometimes gets Texas root rot and should not be planted in infected soil.

Planting and care: Plant from containers any time or bare root in winter. Stake young trees and prune to form a crown high enough to walk under. Little care is needed, but it is a good idea to apply prevention methods for Texas root rot annually. This not only helps prevent disease, it encourages growth.

Pistacia chinensis

All zones
Deciduous
Soil: Tolerates alkaline soils. Grows best with deep soil and good drainage.
Sun: Part to full or reflected sun.
Water: Moderate until estabished, then deep periodic soakings.
Temperature: Hardier to cold than the other *Pistacia* species. Tolerates heat.
Maintenance: None to occasional.

Pistacia chinensis

PITHECELLOBIUM flexicaule

Family: *Leguminosae* _____

Texas ebony

(Ebenopsis flexicaulis). This is a small, highly decorative tree of slow to moderate growth to 20 feet high and 15 feet wide. With favorable conditions this recent introduction from Texas and New Mexico can reach a height of 30 feet. Its most striking feature is an unusual twig structure that forms a wide, bushy crown supported by an erect trunk with smooth gray bark. Thorny-based leaflets are medium green. Fragrant, light yellow to creamy catkinlike flowers in dense clusters appear in spring, sometimes extending into summer. Brown woody pods 4 to 6 inches long follow and remain for a long period.

Low and middle zones, protected locations in high zone

Evergreen to partly deciduous

Soil: Tolerant. Grows best in deep soil.
Sun: Part, full or reflected sun.
Water: Supply deep periodic irrigation. Accepts moderate to ample and will grow faster. Once established it can withstand periods of drought but will not grow as well.
Temperature: Revels in heat. Hardy to cold, but becomes deciduous in the coldest winters.
Maintenance: Little.

Pithecellobium flexicaule

Special design features: Character tree. Silhouette. Rich dark green foliage, epecially striking in desert settings. Great for informal effect.
Uses: Natural, wild or desert gardens. Specimen for patio or garden. Transitional plant. Silhouette against structures. Screen or barrier when planted in row or grouping.
Disadvantages: Plant away from walks because of thorns. Slow to develop.
Planting and care: Plant from containers in spring or fall. May be grown from seed which has been scarified (scratched) but slow. Space 6 feet apart for screen or barrier, 20 feet or more for a row. Trim as desired. Do any major pruning in late winter.

Pithecellobium flexicaule

PITTOSPORUM phillyraeoides

Family: *Pittosporaceae* _____

Willow pittosporum • Narrow-leaved pittosporum

A slender, erect, willowlike tree with long trailing branches and narrow, light, gray-green, 4-inch leaves, willow pittosporum is a native of Australia. Trained as a single or multiple-trunk tree, it grows at a moderate to slow rate 15 to 20 feet tall with a spread of 10 to 15 feet. It can be grown in a narrow space and can be trained to have a narrower crown. Tiny yellow flowers appear along drooping branches in spring and produce round, 3/4-inch, deep yellow seed capsules in fall. One of the more decorative small-scale trees for the desert.
Special design features: Refined desert grove or evergreen waterside effect.

Low and middle zones, warmer spots in high zone

Evergreen

Soil: Tolerant. Prefers loose soil with good drainage.
Sun: Open to filtered or part shade. Full or reflected sun.
Water: Best with deep periodic irrigation. Accepts moderate water with good drainage.
Temperature: Revels in heat. Hardy to at least 15°F (−10°C). Lower limits are not yet known.
Maintenance: Occasional pruning or sweeping.

Decorative silhouette. Adapts remarkably well to a small space.
Uses: Specimen as emphasis or silhouette plant. Grove, or row as tall hedge or screen. Parkways or median strips. Clean enough to use near swimming pools or paved patios.
Disadvantages: Seed litter in abundant years, but not always a problem. Sometimes appears sparse and twiggy.
Planting and care: Plant from containers any time. To form tall, erect tree, encourage vertical leaders by pruning side shoots and staking one or more central shoots to grow as tall as possible. Maintain appearance by pruning dead or errant branches.

Pittosporum phillyraeoides

Pittosporum phillyraeoides

PITTOSPORUM tobira

Tobira • Australian laurel • Japanese mock orange

Family: *Pittosporaceae*

Tobira is a neat, large-scale shrub requiring little care. Its dense, dark green foliage is a welcome accent in hot arid climates. It grows at a slow to moderate rate to 6 to 8 feet high with an equal width; specimens may eventually reach 15 feet unless pruned. Foliage covers plant to the ground. Leathery sometimes glossy leaves grow in rosettes and have round tips. In spring, small, white, waxy flower clusters form in the rosettes. Flowers have the heady scent of orange blossoms. They are sometimes followed by 1/2-inch blue-green berries in fall which split open to reveal orange seeds. Native to China and Japan.

Cultivars: There are two notable cultivars. 'Variegata' is smaller than *P. tobira,* reaching only 5 feet high and as wide. Variegated leaves are gray-green edged in white, giving it a bright cheerful appearance. 'Wheeler's Dwarf ' is very compact, somewhat more refined, with richer foliage and a mounding form. It will grow to 28 inches high with a spread of 5 feet in 4 years. Give it some shade in middle and low zones.

Special design features: Dense rounded mound. Formal. Gives a feeling of mass.

Uses: Specimen, wide screen or space definer. Background or foundation plant.

Disadvantages: Young plants may show sunburn and chlorosis in sunny locations for the first two years. Plants are brittle and break easily. Sometimes gets Texas root rot or root knot nematodes. If a plant is in poor condition, look closely to see if it has cottony cushion scale. See page 37.

Planting and care: Plant from containers any time. Space species to 6 feet apart for solid row or mass. Space 'Variegata' 4 feet apart for a row. 'Wheeler's Dwarf' should be planted 2-1/2 feet on center. Plants naturally assume neat rounded forms which do not need pruning unless they become too large. Never shear plants or it will ruin the shape. Any corrective pruning should be done in late winter before spring growth starts.

All zones
Evergreen
Soil: Garden soil enriched with humus.
Sun: Filtered shade to part or full sun. Tolerates reflected sun but may burn in early summer. 'Wheeler's Dwarf' is the most susceptible to burning from reflected heat.
Water: Moderate, but tolerates ample watering such as lawns, to periodic drought. Once established give it deep periodic soakings.
Temperature: Usually hardy to cold, but may suffer some cold damage in high zone. Accepts heat.
Maintenance: Very little.

Pittosporum tobira

Pittosporum tobira 'Compacta'

PLATANUS acerifolia

London plane tree

Family: *Platanaceae*

This is a hybrid sycamore used in many parts of North America as a street tree. It is large and erect and grows moderate to fast to 40 to 70 feet high unless pruned. It has a symmetrical, pyramidal form and somewhat pendulous lower branches, with a crown spreading 30 to 40 feet wide. Short stout trunk is covered with decorative mottled bark, which is whitish and peels in patches. This gives tree a light-colored look which contrasts with darker grays and browns of most bare trees in fall and winter. Irregularly toothed leaves have 3 to 5 shallow lobes and are 4 to 10 inches across. Brief bloom of unimportant flowers in spring produces round, decorative, bristly, 1-inch seed balls that hang in clusters through fall into winter. Foliage turns golden brown in fall.

Cultivars: 'Pyramidalis' is more upright in form and lacks drooping lower branches common to the species.

Special design features: Large scale. Symmetrical. Formal. Erect. Beautiful structure and cheery winter form.

Uses: Street, avenue or park tree. Good for large lawn areas. Impressive as row, grove or arbor. Tolerates city conditions of smog, soot and dust. Tolerates poor soil, although not caliche.

Disadvantages: Subject to iron chlorosis in alkaline soils, especially in lawns.

Leaves may show marginal drying during hot periods in midsummer and later on in low and middle deserts, especially if planted in a very hot location. Sometimes attacked by red spider mites or scale.

Planting and care: Plant from containers any time or bare root in winter. Space 20 feet apart for an overhead arbor, to 40 feet or more for row. Prune during dormant season for any necessary shaping or allow to grow naturally. Feed with iron seasonally to prevent chlorosis.

All zones
Deciduous
Soil: Tolerant but best in deep rich soils. Does not do well where drainage and soil depth are limited by caliche. May become chlorotic in alkaline soils.
Sun: Part to full sun. Tolerates reflected sun but may suffer from leaf burn, especially if water is limited.
Water: Little in dormant season to moderate or ample in hottest weather. Supply deep periodic irrigation for best results.
Temperature: Hardy to cold. Tolerates heat.
Maintenance: Seasonal, but depends on use.

Platanus acerifolia

PLATANUS racemosa
Family: *Platanaceae*

This native of southern California and Baja, California is a vigorous, usually fast-growing tree. It will reach up to 50 feet or even 90 feet tall when grown in deep soil with ample irrigation. More irregular in appearance than the London plane tree, trunks often divide into leaning or spreading secondary trunks. Large velvety leaves are deeply lobed and turn golden brown in fall; some leaves may stay on tree all winter. Tree leafs out early in the season and new foliage is sometimes caught by late frost. Inconspicuous spring flowers produce decorative, bristly seed balls which hang in clusters.

Special design features: Bold tree with striking branch structure, especially in winter.

Uses: Large lawn areas, parks or roadsides. Informal or natural gardens with lots of space.

Disadvantages: May get chlorosis in lawn situations. Sometimes attacked by spider mites or leaf miners. In hot areas and without sufficient soil moisture, leaves sometime dry around the edges in midsummer. Sometimes a host to a foliage blight called *anthracnose* that causes intermittent leaf drop through the growing season.

Planting and care: Plant from containers any time or bare root in winter. Space at least 40 feet apart for row, closer for grove. Avoid planting in sand dune or dry windy areas. Prune and train young trees carefully to attain desired structure. Prune in dormant season. Apply iron sulfate if chlorosis develops. New plants may be started from branch cuttings.

Platanus racemosa

> **All zones to 5,000 feet**
>
> **Briefly deciduous**
>
> **Soil:** Tolerant. Best with deep, loose, gravelly soil.
> **Sun:** Full to reflected sun.
> **Water:** Accepts lawn watering. Requires deep periodic irrigation, more frequently in summer.
> **Temperature:** Hardy to about 10°F (−12°C). Early new growth is damaged at about 26°F (−3°C).
> **Maintenance:** Occasional.

Platanus wrightii, Arizona sycamore, is probably better adapted than the above to very hot, dry regions; its natural habitat is the canyons bordering the Arizona desert. Form is somewhat more irregular and spreading than *P. racemosa*, and it does not grow as large. Trees generally reach 40 feet high with an equal spread. Cultural requirements are the same as for California sycamore.

PLATYCLADUS orientalis
Family: *Cupressaceae*

(Thuja orientalis, Biota orientalis). A remarkably durable plant from China and Korea with many cultivars, Oriental arborvitae is widely used around the world. It is very popular in the Southwest because of its durability and low maintenance requirements. It requires careful use in the landscape; its bold heavy form is so dominant, and plants often outgrow the space allotted. Two tiny plants on either side of a door can soon become giants blocking the way unless a dwarf cultivar is chosen or plants are spaced far enough apart. Growth is usually moderate. The species grows to 40 feet high but is seldom seen.

Cultivars: Its many cultivars vary in color, shape and size. Ask your nurseryman about the mature size of plant you intend to buy. 'Aurea' is the form often seen in mature plantings. It reaches 12 to 18 feet high and nearly as wide at its base, but it can be hard to locate. More commonly sold is the compact form—'Nana Berkman's Dwarf'. It reaches 4 feet high with golden foliage tints. 'Aureus Nanus' grows to 6 feet high or more. 'Westmont' reaches 6 feet high, is slow growing and globe shaped, with golden foliage during the warm season. 'Fruitlandii' reaches 6 feet high, a cone-shaped mass of deep green foliage. 'Elegantissima' grows 12 to 15 feet high, is golden green in summer with a broad base.

Special design features: Heavy, dense and formal.

Uses: Specimen plant for emphasis. Large formal gardens or lawns. Dense formal screen or hedge. Windbreaks or sound barriers. Planters or containers. Corners of buildings to extend line, making the structure seem larger.

Disadvantages: Occasional infestations of spider mites.

Planting and care: Plant from containers any time. Space large forms 8 feet or more apart for hedge or screen; smaller plants 4 to 6 feet, depending on eventual size of plant. Place closer for a dense, continuous line. Give ragged plants an occasional trimming to keep neat.

Platycladus orientalis 'Aurea'

> **All zones**
>
> **Evergreen**
>
> **Soil:** Tolerant of most soils. Best with improved garden soil.
> **Sun:** Full sun.
> **Water:** Moderate until established, then deep periodic irrigation. Mature plants can withstand periods of drought.
> **Temperature:** Tolerant to about 10°F (−12°C) and to altitudes of at least 4,500 feet.
> **Maintenance:** Little to occasional.

Platycladus orientalis 'Aureus Nanus'

PLUMBAGO auriculata

Family: *Plumbaginaceae*

(*P. capensis*). This viny shrub revels in sun and is a delight to the eye, cycling into bloom off and on during warm weather. Clusters of pale blue, 3/4-inch flowers bring a coolness to the garden. Plants grow at a moderate to fast rate to form a mound 3 to 4 feet high, higher when given support. Native to South Africa.

Cultivars: 'Alba' has white flowers.

Special design features: Garden or warm-weather color plant.

Uses: Mounding shrub or foundation plant in moist soils for south or west sides with an overhang. Garden bedding plant. Espalier or train to drape over walls and fences. Bank cover in warm areas or nearly frost-free places. Tolerates reflected heat from buildings and other vertical walls, but reflected heat from pavement concentrated in a small space may damage plant.

Disadvantages: Iron chlorosis. Sensitive to cold.

Planting and care: Plant from containers in spring when danger of frost is past. Space 3 to 4 feet apart for a mounding mass, 4 feet for hedge or bank cover. Clip and shape as desired. If damaged by frost, withhold pruning until new spring growth appears to tell which parts have been damaged. Feed heavily with a balanced fertilizer in spring.

Plumbago auriculata

> **Low and middle zones, especially in warm microclimates**
>
> **Evergreen to deciduous in cold winters**
>
> **Soil:** Average garden soil.
> **Sun:** Part to full sun. Accepts reflected sun from wall on south, east or west sides, as long as plants are not placed too near adjacent paving.
> **Water:** Moderate. Accepts ample.
> **Temperature:** Foliage damaged at about 28°F (−2°C).
> **Maintenance:** Little to occasional depending on preference and growing space.

Plumbago auriculata

PODOCARPUS macrophyllus

Family: *Podocarpaceae*

(*P. longifolius*). This versatile shrub or small tree from Japan is related to the true pine. In its native habitat it grows to 45 feet high, but is seldom seen over 20 feet high in arid climate zones. Flat, narrow, dark green, 4-inch leaves are set in a spiral on bushy horizontal branches. An excellent plant for narrow spaces. especially if trimmed. Healthy trees in favorable locations bush out and take on an irregular shape. Catkin flowers are 1 to 1-1/2 inches long, but are seldom seen in these zones.

P. macrophyllus maki, shrubby yew pine, has dense foliage with narrow 3-inch leaves. It grows up 6 to 8 feet tall in 10 years. One of the best container plants for indoors or sheltered places outdoors.

Special design features: Oriental effect. Formal. Vertical. Rich refined foliage is attractive in small spaces and close-up.

Uses: Trained specimens make tall narrow espaliers on cool walls. Especially attractive for entryways, atriums, courtyards or small patios. Informal unclipped specimens produce an Oriental effect.

Disadvantages: Occasionally gets iron chlorosis. Burned and yellowed by hot sun.

Planting and care: Plant from containers any. time. For a hedge or screen, space 1-gallon plants 18 inches apart, 5-gallon plants 2 to 3 feet apart. Espalier by tying to a wall. Clip and train as desired, or allow it to grow naturally. Give yellowing plants iron or shade from sun.

Podocarpus gracilor, fern pine, is similar to the above, but with a softer, narrower leaf. Plant form is variable, from a tree to a hanging basket. This is a perfect plant where cleanliness is needed: in an entryway, patio or near a pool. Like the variety *maki,* this is also an excellent container subject. Its cultural requirements are similar to *P. macrophyllus,* but it is more tender to cold.

> **Middle and high zones, protected locations in low zone**
>
> **Evergreen**
>
> **Soil:** Rich organic soil with good drainage.
> **Sun:** Open, full or filtered shade to part sun. Accepts full sun in cooler areas, but needs afternoon shadow in middle and low zones.
> **Water:** Moderate.
> **Temperature:** Hardy to 10°F (−12°C). Accepts heat with shade and water.
> **Maintenance:** Little to occasional, depending on use.

Podocarpus macrophyllus

POPULUS species
Family: *Salicaceae*

Populus are some of the fastest-growing trees available. Their appearance is as striking as their growth rate, with attractive bright green leaves and light-colored bark. Poplars and cottonwoods are best planted in open areas because of their large size. Greedy surface roots are also problems if trees are grown in or near lawns, and also invade sewer or water lines in search of moisture. They are best used as temporary trees for fast screening or shade or in open locations.

Special design features: Informal but elegant country feel. Interest and brightness in the landscape all year. Evokes Mexican hacienda, western ranch and field. Striking silhouette.

Uses: Large lawns, golf courses, parks, country roads or field boundaries. Too large for average residence and in many cases unsuitable because of disadvantages noted. Effective specimen or as a row or grove tree.

Disadvantages: Greedy invasive roots get into septic tanks or pipes, and may develop buttress or surface rooting if grown in or near lawns. Roots will go under a walk or wall to adjacent areas in search of water and can eventually heave and crack them. Weakened branches may break in the wind. Susceptible to Texas root rot, cankers and heart rot from untreated wounds. Infestations of mistletoe and tent caterpillars. Often short-lived, but aged specimens are found near an ample water supply. Litter is sometimes a problem, especially from female trees when seeds are released in spring.

Planting and care: Start by planting large or small branch or twig cuttings from mature trees just before they leaf out in late winter. Plant from containers any time or bare root during the brief dormant season. Space 10 to 15 feet apart for sap-ling row or grove, 40 feet or more for roadsides. Saplings may need staking against the wind. Treat any malady immediately upon discovery to promote healthy attractive trees.

All zones

Deciduous

Soil: Prefers deep sandy or gravelly soils.
Sun: Full sun.
Water: Requires deep constant soil moisture. Can be given periodic deep soakings. This is not a drought-resistant tree. Leaves will turn yellow and drop if ground dries out. Must have ample or at least moderate irrigation.
Temperature: Hardy to cold. Accepts heat with water.
Maintenance: Occasional raking of leaves.

Populus fremontii
Fremont cottonwood

This striking tree of the open country and wide washes is native to water courses and canyons of the Southwest and Mexico below 6,000 feet elevation. Although there are many poplars and cottonwoods, Fremont is the one most commonly planted. Trees grow fast to 50 feet tall, sometimes to as much as 100 feet when supplied with ample moisture. Spread is about half as wide as high. Mature form is open-branched with a flat to billowing crown and rough gray bark. Bright green foliage emerges in late winter, deepens in tone during midsummer and turns golden in late fall before it drops. Leaves shimmer in the wind. Catkins come in late winter or early spring on males. Females produce cottony seeds which drift through the air, the "cotton" for which the tree is named.

Populus fremontii wislizenii
Valley cottonwood, Rio Grande cottonwood, wislizenus cottonwood

This tree is native to a wide area from southern Colorado and Utah to New Mexico, western Texas and northern Mexico. It is probably more widely grown in these areas than the species. A stout tree, it grows fast to 40 to 100 feet high with a flat, wide, spreading crown of large branches and triangular leaves to 4 inches long and as wide.

POPLARS

Poplars are other members of the *Populus* group. They are often grown at higher elevations. Trees are beset by many problems in middle and low zones so are seldom used there. Because of their bad rooting habits, they are best suited to open country, parks, large lawns, golf courses or along roadsides. They invade underground water structures, heave pavements and often sucker badly from root sprouts. Trees are messy and short-lived because of a variety of maladies, especially Texas root rot, cankers and heart rot. They are stronger and healthier in the high zone and do well in open country.

Soil: Tolerant. Prefers deep moist soil.
Sun: Full sun.
Water: Ample.
Temperature: Tolerant of cold. Accepts heat of middle zone with ample water.
Maintenance: Little to none in open country. Constant in home landscape.

Populus alba
White poplar, abele, silver-leaved poplar

This is a spreading, fast-growing tree to 40 to 60 feet high, higher in favorable locations. It has especially beautiful 5-inch leaves that are medium green on top, white and woolly underneath. Leaves produce a silvery shimmer in the slightest breeze. Native to Europe and Asia, it has naturalized in North America.

Cultivars: 'Pyramidalis', bolleana poplar, is narrow and columnar, quite different in form from the species. Often grown as a windbreak in the high zone where it gives a vertical effect and a formal appearance.

Populus nigra 'Italica'
Lombardy poplar

This is another columnar form often seen in rows along drives or field edges, especially in higher elevations. It is recognized by its narrow leaves and form, which is almost as vertical as the Italian cypress. Usually only male trees are planted because of their narrower form and lack of airborne seeds.

Populus fremontii

Populus species

Populus nigra 'Italica'

PROSOPIS species
Family: *Leguminosae*

Mesquite • Algarrobo

Mesquites are very adaptable and tolerant of adverse conditions and exposure. They are outstanding choices for the warm deserts and hot interior climates. These trees and shrubs are native to both North and South America and possibly other parts of the world. They have long supplied desert dwellers with shade and shelter, fuel for fires, building materials, food for livestock, and shredded bark for making baskets. Their beans are ground into food (the *pinole* of Mexico). Bees make an excellent honey from their flowers.

Because mesquite trees hybridize and look different in varying environments, there has been great confusion determining their botanical classification. Botanical names are important; knowing them allows you to choose the exact tree that is right for a particular landscape use. In the West, mesquites grown for the nursery trade are distributed under certain names. They are listed in the following descriptions under their *current* names; their former names or synonyms are shown in parentheses.

Mesquite trees usually have dark, sometimes rough bark, with a spreading crown composed of numerous, finely cut leaflets. They are often thorny in youth but thorns usually disappear with age as

growth slows. Flowers are yellowish catkins and appear in spring, followed by yellowish seed pods which may be straight, twisted or curved, depending on the species. Mesquites are not true desert plants in the same sense as cacti, saltbush or creosote bush; they do not store water or become dormant in drought. They will live on little water, but adverse conditions keep them small and cause them to develop twisted forms. With their long tap root and lateral roots they are able to penetrate to depths of 150 feet to search out water. Underlying ground water allows mesquites in low areas, washes or flood plains to grow to 30 feet or more. If the water table drops swiftly and trees have no time to send roots deeper, they will die.

Even if a mesquite is deprived of water in its youth, when given regular irrigation, it gains size but keeps its picturesque form. Ample irrigation will produce a large tree faster. All mesquites, twisted or not, have outstanding character and blend with a garden or natural landscape equally well. They are most attractive when trained as multiple-trunk trees, selectively pruned and shaped when young. Once a mesquite has obtained a desired size its growth will be slowed by reducing irrigation.

Special design features: Bold desert tree of great character. Feeling of the West.

Prosopis 'South American Hybrid'

Prosopis alba

Prosopis glandulosa glandulosa

Prosopis velutina

Ferny leaves and spreading form. Tropical or desert mood.

Uses: One of the most effective plants for summer-shade and winter-sun. Effective in controlling glare, cooling ground surfaces and filtering dust and breezes. Lawn, patio or street tree. Parks or roadsides. Can be used in difficult soil areas such as sanitary landfill sites or disturbed areas. In narrow spaces leaders may be trained upward and over rooftops or walls before they are allowed to send out laterals to form a crown.

Disadvantages: Invasive roots. Do not plant near septic tanks or leaching fields. Litter of catkins in spring, seed pods in summer and leaflets in fall. Occasional infestations of scale and mistletoe. Texas root rot is mainly a problem in Texas and is rarely seen in more western deserts. Dripping dark fluid from trees can stain patio floors. Young plants in open areas are nibbled by rabbits and their bark may be stripped unless protected.

Planting and care: Mesquites can be grown from seed, but take 2 or 3 years to equal trees you could purchase in a 5-gallon can. Tiny plants have long tap roots when they are only a few inches tall and may not transplant successfully. Plant from containers any time. Space 20 feet or more apart for a fast canopy or closer for a sapling grove. Stake young trees for support. To make a high crown or canopy, select the most prominent leaders and trim side branches to 12-inch nubs. Tie main leaders on long poles angled outward at the desired slant—2x2's are good for this. Continue to cut low side branches back to 12 inches until trunks are strong and permanent branches at the top are partially developed. When desired height is reached, remove stakes and cut off side branches. Prune only to train and be sure to treat and seal pruning cuts.

All zones

Deciduous to partly evergreen to evergreen

Soil: Widely tolerant. Best in deep soil with a high water table. For best results, make sure planting holes in caliche or clay soils have drainage.

Sun: Full to reflected sun.

Water: Widely tolerant once established, from little to none or ample. Best with deep periodic irrigation, especially in summer for faster growth.

Temperature: Hardiness varies with variety. Plants endure all but severest winters. It is said that mesquite leaves do not appear until after the last killing frost. This is generally true, but not always. Still, the leafing out of mesquite in spring is a good indication that spring has arrived.

Maintenance: Periodic to none depending on location, tree type and.

Prosopis alba
Argentine mesquite _____

Mesquites offered by nurserymen under this name come from a tree so labeled at the Desert Botanical Garden in Phoenix. Some authorities say it could be *P. nigra*, but until further identification is made, it is usually called *P. alba*. It is perhaps the fastest-growing mesquite and is quite vigorous. Rough dark trunks tend to grow more vertical than other trees. Crown is spreading, blue-green and dense. With irrigation, it can become a substantial tree in 5 or 6 years. Although nearly evergreen, it is not free of leaf drop. Old leaflets usually drop in spring just as new foliage emerges. This tree seldom hybridizes with other mesquites, because it flowers at a different time.

Prosopis chilensis
Chilean mesquite,
Algarrobo mesquite _____

There is confusion about the naming of this species. For a time trees questionably called Chilean mesquite have been sold by nurseries. The true Chilean mesquite may presently be available because several growers are importing the seed directly from Chile. True Chilean mesquite resembles the species often offered as *P. chilensis*, but is more upright and vigorous in appearance. It is rapid growing and needs 25 to 40 feet for proper development. Foliage is more open and ferny and the tree is more evergreen. Tree may be sensitive to cold at 10°F (−12°C) or below. In mild winter areas it keeps its leaves all winter, but will lose them with sharp cold. It bears fewer blooms and the few seed pods are curved and hang in clusters. Young trees sometimes have enormous white woody thorns. Crown and form are usually symmetrical.

Prosopis glandulosa glandulosa
Honey mesquite
Texas mesquite _____

This tree is usually associated with Texas, but also grows from Kansas to New Mexico and northern Mexico. It recently made its appearance in Arizona. It is believed seeds came west in manure from cattle being trucked west from Texas. Quite different from other mesquites, foliage is brighter green with large leaflets and a weeping form. It somewhat resembles the California pepper tree, *Schinus molle*. This species tends to be completely deciduous in winter. It develops a large, picturesque structure in favorable locations. In areas hostile to its development, it may remain a knee-high shrub, forming large thickets in some places. Not as widely used or known in the West as most of the other mesquites listed here. This tree should be used more.

Prosopis hybrid
South American hybrid, Misnamed Chilean mesquite

The second common name is best for locating this tree in most nurseries, but is botanically incorrect (see *Prosopis chilensis*). Due to the mix-up over naming mesquites, it will be quite some time before growers, nurseries and gardeners become accustomed to calling this tree by its correct name. This tree does have charac-

teristics of several South American mesquites. Trees offered in nurseries under the name *P. chilensis* are collected locally and are this tree, believed native to Arizona and the warm, Yuma-Imperial Valley region of California and into Mexico. Vigorous in appearance with a wide, spreading crown and deep green leaflets, it is deciduous and has more blossoms and seeds than the true Chilean mesquite. It looks more rugged in a desert setting.

Prosopis pubescens
Screwbean mesquite, tornillo

This is a small deciduous shrub or tree with a twisted bean pod. It is most-often used where it is growing naturally because it is not carried in nurseries. There is no confusion about its identity.

Prosopis velutina
Velvet mesquite, Arizona mesquite

(P. juliflora velutina). The true name of this tree is the hardest to figure out, which is unusual because it is very common and widely distributed. The common name velvet mesquite is appropriate—its foliage is soft and gray-green with a velvety appearance from a distance. Leaflets are fuzzy and small. Craggy, rough barked and picturesque, it has nubby twigs which are usually thornless. This is the mesquite surrounding populated areas of Arizona. It is the main source of firewood and for wood working and cabinet making. Although common in the wild and in new residential areas, it is seldom carried in nurseries. This is a shame, because it blends the best with the desert. It is deciduous and loses its leaves with the first cold spell.

Prosopis chilensis

Prosopis chilensis

PRUNUS caroliniana
Family: *Rosaceae*

Carolina laurel cherry • Laurel cherry • Mock orange • Wild orange

(Laurocerasus caroliniana). Carolina laurel cherry is a very refined small tree. Erect and elegant with shiny, dark green foliage, plants grow at a moderate rate to 15 to 20 feet high. Plants can be kept lower and used as a large shrub. A large plant will spread 15 feet unless cut back. Well-branched, dense and pyramidal in youth, it becomes looser and more spreading with age. Foliage covers plant to the ground unless trimmed. Tiny white flowers appear in dense clusters in spring, sometimes followed by shiny black fruit. Native to the southeast from North Carolina to Texas.

Cultivars: 'Compacta' and 'Bright 'n Tight' are compact cultivars with denser foliage than the species.

Special design features: Formal and refined. Air of coolness and control.

Uses: Patio or street tree. Near swimming pools. Large background shrub, clipped hedge or screen. Feature plant for small spaces and for close-up viewing. Small, formal standard tree or topiary plant. Espalier on cool walls. Container plant. Use in locations protected from the full force of sun and heat in low zone and to some extent in middle zone.

Disadvantages: Very prone to chlorosis in alkaline soils. Foliage yellows from reflected heat and burns from hot winds. Sometimes gets root rot or gummosis.

Planting and care: Plant from containers in cool season. Space 24 inches apart for clipped hedge, 6 to 8 feet for unclipped mass or screen. Feed with iron regularly to prevent iron chlorosis in alkaline soils.

Prunus caroliniana

Middle and high zones

Evergreen

Soil: Deep, highly enriched, well-drained soil is best.
Sun: Part shade to full sun.
Water: Moderate. May need ample in summer, but don't overwater.
Temperature: Hardy to about 10°F (−12°C). Leaves will often burn in hottest locations.
Maintenance: Little to occasional, depending on situation.

Prunus caroliniana

PRUNUS hybrids and cultivars

Family: *Rosaceae*

The flowering plums are cultivars or hybrids of *P. cerasifera* and *P. blireiana*. They are chief among spring-flowering trees. The original *P. cerasifera* was a native from central Asia to the Balkans. From there it has spread widely as root stock; the tree itself is not planted for ornament or fruit. The most common variety planted is *P. cerasifera* 'Atropurpurea' (*P.* 'Pissardii'), an ancestor of many hybrids described here. Trees of this group are erect with fairly vertical young branches and globe-shaped crowns in youth. Crowns become more spreading with age. Growth is usually moderate up to 18 to 40 feet high, depending on the cultivar and growing conditions. Most have foliage in some shade of purple. Rose, pink or white flowers appear in spring. Some have fruit that is edible. Larger types may be used as shade trees, but flowers and purple foliage are the main reasons for their use.

Cultivars: *P. blireiana*, a hybrid between *P. cerasefera* 'Atropurpurea' and *P. mume*, is usually considered the most desirable flowering plum. Grows at a moderate rate to 25 feet high with a spread of 20 feet. It has a graceful, branching growth habit. Reddish purple spring foliage turns greenish bronze in summer. Fragrant, semidouble, rose to pink flowers appear between February and April.

P. cerasifera 'Atropurpurea', purple-leaf plum, is an ancestor of many hybrids. A desirable tree, more rounded in form and faster growing than some, reaching 25 to 30 feet high. Foliage is coppery red when new, deepening to dark purple, then gradually becoming deep, bronzy green in late summer. It has white flowers and can set a heavy crop of small, red, edible plums. 'Hollywood' is a hybrid between 'Atropurpurea' and the Japanese plum 'Duerte'. It produces 2 to 2-1/2-inch red plums of eating quality. Growth is moderate with an upright form to 40 feet high, but it usually reaches 30 feet with a 25-foot spread. Leaves are dark green on top and red underneath. Light pink flowers appear in February or March. 'Krauter Vesuvius' is smaller than 'Atropurpurea', growing at a moderate rate to only 18 feet high with a 12-foot spread. Foliage is purple-black, darkest of the purple-leaf plums. It seldom produces fruit. Pale pink flowers appear in February or March. 'Thundercloud', another hybrid, sometimes sets red fruit. Its growth is moderate to 20 feet high and as wide. It has a more rounded form than 'Atropurpurea' and dark coppery leaves. Flowers are light pink to white.

Special design features: Spring bloom, purple foliage contrast.

Uses: Color accent for homes, parks, streets and median strips.

Disadvantages: Although some trees live to be old, many tend to be short-lived be-cause of a variety of maladies. They include Texas root rot, iron chlorosis, cytos-porea canker and other problems that plague the stone fruits.

Planting and care: Plant from containers any time or set out bare root in winter. Space 15 to 20 feet apart for massing as a grove. Thin trees to increase size and spread. Prune after bloom to encourage heavy bloom on new wood the next year. From the second year on, supply an annual feeding of nitrogen at least 6 weeks before flowering time.

All zones

Deciduous

Soil: Best in deep enriched soil with good drainage. Plums accept heavier soils than peaches which thrive only in sandy soils.
Sun: Full sun. Young trees accept part shade.
Water: Moderate.
Temperature: Hardy to cold. Accepts heat with irrigation.
Maintenance: Periodic.

Prunus cerasifera 'Atropurpurea'

Prunus cerasifera 'Thundercloud'

PRUNUS persica

Family: *Rosaceae*

These giant, floral bouquets are sometimes considered to be the "first breath of spring." Aside from their blossom color, they are not prime landscape plants. Trees are erect and grow fast to moderate to 15 feet high. Spread of trees vary according to pruning practices. Most varieties have double flowers. Flower color ranges from white to pink or deep red, sometimes variegated. Bloom appears early to late in the season, depending on variety. Planting several different cultivars prolongs the spring mood as first one tree, then another comes into bloom. Late-flowering types may have their blooming period shortened by unseasonably hot weather. A few produce fruit of varying quality.

Cultivars: Very early bloomers: 'Early Double Pink', pink flowers. 'Early Double Red', rosy red to deep purplish pink flowers. Early bloomers: 'Helen Borchers', large clear pink flowers. Mid-season bloomers: 'Double White', double white flowers. 'Peppermint Stick', many white flowers with tiny pink stripes, look pink from a distance. 'Weeping Double Pink', double pink flowers. Small tree with weeping branches. Stake and tie trees carefully. 'Weeping Double Red', same tree as above, but with double red flowers. Late-season bloomers: 'Burbank', double pink flowers. 'Camelliaeflora', double red flowers. 'Late Double Red', double red flowers. Plants that bear edible fruit as well as spectacular flowers: 'Bonanza', a dwarf, blooms and bears at 2 years of age when it is about 2 feet tall. It may reach 6 feet in time. Flowers are semi-double and rosy pink. Yellow-fleshed freestone-type fruit ripens early. Excellent in tubs or containers. 'Royal Red Leaf' is a late bloomer with deep pink flowers. It bears red peaches with white flesh. Decorative foliage is red when new, deepening to purple. Other fruit-producing peaches said to do well in arid zones are: 'Saturn', 'Early Elberta', 'Early Crawford', 'Blazing Gold' and 'Halehaven'. There are numerous others as well as dwarf types. Desert gardeners serious about fruit-producing trees should seek further information about culture, local problems, necessary pruning practices and prevalent pests and diseases from their agricultural extension agent.

Special design features: Masses of color in spring.

Uses: Massive color, best when planted in combination with other plants. Place at "bottom" of the garden or behind a wall where brief flowering period can be enjoyed, but where its nondescript appearance will be masked the rest of the year.

Disadvantages: Subject to all the problems of stone fruits. Trees are often short-lived. Requires some maintenance and pruning.

Planting and care: Plant from containers

Prunus persica (flowering peach)

Prunus persica 'Bonanza'

any time or bare root in winter. Space 15 feet apart for grove. For maximum flowering, prune branches to 6 inches soon after bloom is past. Many new branches will grow and produce abundant blossoms on new wood next spring. Light pruning is best for trees planted in patio areas. Bloom will be colorful but more scattered. Treat all pruning cuts to avoid cankers. Give a light feeding of nitrogen from the second year at least 6 weeks before flowering time. Increase amount as the tree grows. In spring, lay down a mulch of spoiled hay or composted manure.

Middle and high zones, marginal in the low zone
Deciduous
Soil: Light to sandy soils with good drainage
Sun: Part to full sun.
Water: Moderate.
Temperature: Hardy to cold. Tolerates heat, but needs a period of cold in winter. Early-blossoming types are sometimes nipped by frost.
Maintenance: General care to considerable, depending on use.

Prunus persica (flowering peach)

PUNICA granatum

Family: *Punicaceae*

The pomegranate is originally from southern Asia, but has been cultivated in the Mediterranean region since ancient times. Brought by the Spanish to the New World, it is one of the legendary plants of the Southwest and Mexico. It was planted by padres in mission gardens and copied by Indian silversmiths as the misnamed ''squash blossoms'' on traditional Navajo necklaces. Usually seen as a large shrub, pomegranate grows at a moderate to slow rate from 12 to 20 feet high and spreads to at least half as wide unless pruned back. In most gardens it reaches about 8 to 10 feet high. Its twiggy structure is covered with small glossy leaves that are bronze in spring, bright green in summer, golden in fall. Blossoms on most varieties are red. Many produce round, red to bronze fruit, 3 inches in diameter, filled with seeds surrounded by red to pale translucent pulp. The pulpy seeds are very sweet and can be chewed or used to make a delicious jelly.

Cultivars: 'Alba Plena' is a shrub 6 to 10 feet tall and does not produce fruit. Creamy-colored flowers bloom in summer. New growth is bright green. 'Double Red' has no fruit, double red flowers and reaches to 12 feet high with an arching form. 'Mme. Legrelle' has double creamy-colored flowers striped red, but bears no fruit. 'Nana', a dwarf form, is a very appealing landscape or container plant. It

Punica granatum

grows to only 3 feet high with tiny, single, orange-red flowers and small, red, inedible but decorative fruit. Nearly evergreen in warm places and blooms when it is only 12 inches tall.

Fruiting varieties include: 'Wonderful', thought to have been developed from original plants. It has especially good fruit. Single flowers to 4 inches across are orange-red. Plant grows as a 10-foot fountain-shaped shrub which makes an effective espalier or small tree if properly trained. It fruits only in warm areas. 'Sweet' is another popular fruit-producing pomegranate. Fruit is paler, less acid and milder in flavor than 'Wonderful'.

Special design features: Tough, durable and long lived. Pomegranate, especially original forms, tolerates hot winds, drought, salty soils, intense sun and cold winters. Resistant to Texas root rot and other fungus diseases, it is easy to grow and withstands neglect when established. An interesting bright green form with additional attraction of seasonal foliage and flower changes.

Uses: Specimen, hedge, screen or windbreak. Espalier on hot walls. Some can be trained into trees. Dwarf form is excellent as a low hedge, container or bonsai plant. Clipped hedges do not set much fruit. All are effective plants for transitional areas, desert or irrigated gardens.

Disadvantages: Twiggy winter form is not especially attractive. Fruit can be infected with a variety of diseases. A leaf-footed insect makes tiny holes in outer fruit cover and a fungus may enter. Internal black rot may enter through blossoms and decay fruit. Splitting of fruit can be caused by sunburn in the hottest areas or by a sudden increase in moisture.

Planting and care: Plant any time from containers or bare root in winter. Start cuttings in dormant season from pencil-sized twigs about 7 inches long. Space larger forms 3 to 4 feet apart for hedge or screen. Prune as desired in late winter; clean out twiggy growth in plant interior to maintain form. Shear any time, but shearing eliminates the arching form and most flowers and fruit. Keep irrigation steady when fruit is developing. A sudden increase can cause fruit to split.

Punica granatum

All zones
Deciduous
Soil: Tolerant. Accepts alkalinity.
Sun: Full or reflected sun. Accepts morning sun or open shade of the north side, but is more open in form.
Water: Tolerates any amount. Prefers deep occasional soakings. Once established, it tolerates periods of drought in areas receiving 10 inches or more of annual rainfall.
Temperature: Tolerant of a wide range.
Maintenance: Low, some pruning.

PYRACANTHA species

Family: *Rosaceae*

Pyracantha delights gardeners and birds alike with masses of tiny white flowers in spring and bright red or orange berries in fall. Slow to start, plants increase in vigor and grow rampantly once established. Plants grow from 4 to 12 feet or more high and as wide, depending on the selection and pruning practices. Form is angular with stiff branches that bear sharp thorns and dark green 1 to 1-1/2 inch leaves, slightly indented at the tip, with tiny teeth along the edges. Spring flowers are about 1/4-inch across and appear in large clusters along branches. They are followed by clusters of bright fruit which tend to become more colorful as the weather turns cold in fall. Some kinds hold fruit until the spring following bloom period, but berries are usually eaten by birds before then. Once established, pyracantha is a remarkably tough shrub, withstanding drought, hot winds, poor soil and beating sun.

Cultivars: *Pyracantha coccinea* 'Lalandei' is the hardiest of pyracanthas described here. It produces a profusion of bright orange berries in fall. Growth habit is upright.

Pyracantha coccinea 'Lowboy' is one of many dwarf types that can be used as a ground cover or on banks. It has orange berries.

Pyracantha fortuneana (P. crenatoserrata) is a vigorous plant with red berries. It grows to 15 feet tall and spreads to 10 feet wide. Its cultivars are better known than the species and are widely used in arid regions.

Pyracantha fortuneana 'Graberi' is upright in growth with large red berries produced midfall in huge clusters and lasting far into winter. One of the best for low desert areas.

Pyracantha fortuneana 'Rosedale' is one of the most vigorous plants. Upright in growth, its supple branches make it easy to espalier. Bright red berries are earliest to color, well distributed along branches and last long into winter and sometimes spring.

Pyracantha koidzumii is generally represented by its cultivars. It grows to 10 feet tall, spreading nearly as wide. Berries are orange-red, although some are darker red. *Pyracantha koidzumii* 'Santa Cruz Prostrata' is the prostrate form of *P. koidzumii*. A very satisfactory performer, but berries are often hidden by foliage unless plants are carefully pruned. It branches from the base and may easily be kept at or below 3 feet.

Pyracantha koidzumii 'Victory' is another pyracantha that has put in long and successful service in desert regions. Fast growth up to 10 feet high by 8 feet wide unless controlled. Showy, dark red berries. The hybrid 'Red Elf' is a compact dwarf with a mounding habit and closely intertwining branches and bright red berries. Excellent for containers, ground or bank cover.

Special design features: Dark green foliage and bright fall, winter and spring color.

Uses: Widely used as an espalier on walls and fences, even hot south and west walls, as a clipped hedge, specimen, formal topiary plant or standard tree. Pruned or unpruned, it is an effective traffic guide or privacy hedge or screen. Low forms can also be used as a ground cover.

Disadvantages: Subject to iron chlorosis. Sharp thorns make plants difficult to handle. Fireblight is its most common and serious affliction: Look for dying or dead twigs that are black or burned. Cut off twigs well below dead part and disinfect clippers with 30 percent bleach solution after each cut. Treat wounds by painting them with the bleach solution to prevent further infection. Infestations of red spider mites give plants a tired and dusty look and slows growth. Wash mites off plants with a garden hose or use a chemical spray. Sometimes susceptible to Texas root rot.

Planting and care: Plant from containers any time. Set larger forms 4 to 6 feet apart for hedge or screen. Ground cover types should be set 3 to 4 feet apart. Flowers and fruit are produced on 2-year-old wood. Established plants grow vigorously and need to be pruned often during the growing season. Pinch new growth to control size and form before they set thorns. Avoid trimming flowers that will produce winter berries. Any heavy pruning should be done in late winter before new growth starts and can be done by shortening long branches. Give iron regularly to prevent iron chlorosis.

All zones

Evergreen

Soil: Tolerant but needs good drainage. More likely to yellow in poor drainage situations or in alkaline soils.
Sun: Part to full or reflected sun.
Water: Deep periodic irrigation is best. Tolerates moderate, but may become chlorotic. Tolerates periods of drought, but looks unattractive.
Temperature: *P. fortuneana* and varieties are hardy to 10°F (−12°C). *P. coccinea* 'Lalandei' is hardy to 0°F (−20°C).
Maintenance: Constant in summer for clipped plants.

Pyracantha koidzumii cultivar

Pyracantha (espalier)

PYRUS kawakamii

Family: *Rosaceae*

Always graceful and civilized, evergreen pear produces a breathtaking display of white blossoms in late winter before other plants begin to bloom. Usually trained as an upright, single or multiple-trunk tree, this native of Taiwan is a sprawling shrub in its natural form. Tree forms have pendulous branchlets supporting bright green, glossy foliage which turns golden or orange and drop just before blossoms appear. New leaves emerge as bloom ends, but branches may be bare for a brief period, especially if weather turns cold. Trees will wait for a break in the weather before putting out blossoms or leaves. Growth is moderate to 25 feet high with a crown spread as wide in rich deep soils. Plants usually reach 15 to 20 feet high in the desert. If pears form at all they are small, hard and inedible.

Special design features: Refined and elegant. Spring bloom. Oriental elegance.

Uses: Street, patio, courtyard or garden tree. Attractive as specimen or grouping. Effective espalier for narrow spaces. Untrained plants make excellent background shrubs or wide screens.

Disadvantages: Subject to iron chlorosis and occasionally zinc deficiency. May get fireblight, crown gall, root knot nematode or Texas root rot.

Planting and care: Plant from containers any time. Set in soil prepared to prevent Texas root rot and never place in soil known to be infected. Space 10 to 15 feet apart for massing, 20 feet or more for street or drive. For a tree form, obtain pruned and staked plants. Keep staked until trunks are self-supporting. Once trained, little shaping is needed. Feed with a general fertilizer and extra iron or zinc as needed.

All zones

Nearly evergreen—briefly deciduous before flowers appear

Soil: Rich garden soil with good drainage.

Sun: Filtered, open or part shade to full sun. Best with afternoon shadow in low deserts.

Water: Moderate.

Temperature: Hardy to cold, but can become deciduous in coldest winters.

Maintenance: Little, once plants have become established.

Pyrus kawakamii

Pyrus kawakamii

QUERCUS ilex

Family: *Fagaceae*

An erect and symmetrical tree from the Mediterranean region, the Greeks associated holly oak with the god Zeus because its height attracted lightning, thunderbolt of Zeus. Plant remains a shrub in adverse conditions, but will grow at a moderate rate to as much as 60 feet high when planted in deep soil and supplied with adequate moisture. Pyramidal in youth, it develops a heavy crown as wide as it is high at maturity. Foliage appears gray-green from a distance. Leaves are variable in size and form—from 1 to 3 inches long and eliptical to lance-shaped. They are smooth dark green on top and feltlike with a silvery or yellowish cast underneath. Leaf edges may be smooth or spiny. Bark is fairly smooth. Adapts well to most locations and is resistant to pests and diseases which attack other oaks.

Special design features: Quality tree. Strong vertical. Formal and contained.

Uses: Tree for large areas such as streets, expansive lawns, parks, golf courses and large patios. Plant close together as a tall screen or clipped hedge. Takes clipping well enough to be used as a large topiary plant or standard tree.

Disadvantages: None known.

Planting and care: Plant from containers any time. Space 8 to 10 feet apart for substantial hedge or screen, 15 to 20 feet for grove or row, 20 to 30 feet or more for street tree. Stake tree carefully until it supports itself. Any pruning should be done sparingly in late winter.

Quercus agrifolia, California live oak, is a similar tree that deserves mention. It is slow growing in dry climates to 20 feet high, occasionally to 50 feet after many years. Stiff dark, leathery leaves to 3 inches are eliptical and hollylike with a sharp point at each tip. Dark gray bark, smooth for an oak, clothes the trunk which becomes gnarled with age. The California live oak is appreciated for its picturesque form and for the slow growth that keeps it in scale with the average home for many years. Cultural requirements are similar to *Quercus ilex.*

All zones

Evergreen

Soil: Tolerant.

Sun: Full sun.

Water: Fastest growth with deep irrigation at well-spaced intervals. One of the few evergreen oaks that will tolerate lawn watering. Withstands periods of drought when established.

Temperature: Hardy to about 10°F (−12°C). Tolerant of heat.

Maintenance: Little unless kept sheared as a hedge, standard or topiary.

Quercus ilex

QUERCUS suber

Family: *Fagaceae*

Cork oak produces cork from which many commercial products are made. The cork is harvested in sheets from the outer bark of trunks of trees growing in forests and commercial groves in southern Europe, Spain and North Africa, where it is native. Young trees are narrow and conical in form. They grow at a slow, sometimes moderate rate 30 to 50 feet high or higher with an equal width at maturity. Growth rate and eventual size depend on soil depth and moisture. Mature trees have a tall, erect trunk covered with deeply furrowed, tan to gray resilient bark. Branches are somewhat pendulous. Foliage appears gray-green from a distance. Leaves to 3 inches long with toothed edges are shiny dark green on top and grayish and feltlike underneath.

Special design features: Quality tree of great character. Well worth the time it takes to grow.

Uses: Lawn, street or park tree or other large space. Good in dry areas or near large structures. It will eventually tower over a house as a canopy.

Disadvantages: May yellow in alkaline soils. Soft cork bark invites mutilation and graffiti in public areas.

Planting and care: Plant from containers any time. Little care is needed, but some corrective pruning may be desirable and should be done in late winter. If leaves yellow give iron to correct chlorosis.

Another oak with interesting bark is *Quercus emoryi,* a blackjack oak. At maturity, its bark is thick and black and divided into plates, giving it an alligator-skin look. This southwestern tree grows slowly to 20 feet high, sometimes reaching 50 feet with optimum conditions. Its form is open to dense, with bronze-green, leathery leaves to 2 inches in length. Where space is restricted, it is more vertical and narrow. This tree may be difficult to find, but worth the search. Cultural requirements are similar to the other oaks.

> **All zones**
>
> **Evergreen**
>
> **Soil:** Best in deep soils. Needs good drainage.
> **Sun:** Full sun. Tolerates reflected heat.
> **Water:** Best with deep spaced irrigation. Tolerates lawn situation only with good drainage.
> **Temperature:** Hardy to about 5°F (−15°C). Tolerates heat.
> **Maintenance:** Very little.

Quercus suber

QUERCUS virginiana

Family: *Fagaceae*

Southern live oak, native to the southeastern United States, is a sentimental tree to many. It is the fastest growing oak tree mentioned here and the most attractive for hot interior valleys. Crown is thick and twiggy in youth with pronounced lateral branches, giving it a distinctive silhouette. In its native habitat a mature specimen is a great, spreading tree with nearly horizontal branches, dense foliage and dark fissured bark on a short, massive, gnarled trunk. It is known to reach 50 to 70 feet tall in cultivation with a crown spread that is twice its height. It is not yet known how large this tree will grow in the West, because of the rigorous conditions and because there are not any very old specimens. Trees grow at a moderate rate, faster in deep moist soil. Leaves to 3 inches long are smooth-edged, dark shiny green on top and fuzzy white underneath. It is reported southern live oak is resistant to pests and diseases which attack western oaks.

Special design features: A quality shade tree that should be used more often.

Uses: Large gardens and patios. Avenues, parks, golf courses and other public areas where a large tree is needed. It is the best oak for wet or well-irrigated places, but withstands periods of drought after it has become established.

Disadvantages: Leaf drop in spring just before new leaves come out causes trees to look unattractive. Slow growing.

Planting and care: Plant from containers any time. Bare root plants may be transplanted in winter and early spring. Space 30 to 40 feet apart for a spreading canopy or avenue planting. Do any necessary corrective pruning in late winter.

With this and other oaks, watch for the fungus disease *armillaria*, or oak root fungus. It is encouraged by excessive moisture at the plant's crown. Native California oaks are especially susceptible. The disease can be transmitted from oaks to other plants nearby. See page 39 for symptoms and treatment.

> **All zones**
>
> **Evergreen**
>
> **Soil:** Prefers deep moist, enriched soil. Tolerates most soils, but may not grow as fast or as large.
> **Sun:** Full sun.
> **Water:** Moderate to ample. Tolerates widely spaced irrigation once established, or well-irrigated lawns.
> **Temperature:** Loses leaves at about 15°F (−10°C). Hardy to 0°F (−20°C). Tolerates heat with irrigation.
> **Maintenance:** Little.

Quercus virginiana

RAPHIOLEPIS indica
Family: *Rosaceae*   Indian hawthorn

Indian hawthorn is a delightful shrub with shiny, dark green, leathery foliage. This native of China is attractive all year, with an interesting leaf pattern and breathtaking display of spring flowers. Growth is moderate; its size ranges (depending upon the cultivar) from 24 inches to over 6 feet high, with a spread as wide or wider. Maximum size is dependable and plants will not overgrow the space allotted. Toothed 2 to 3-inch leaves are set on bronzy stems and often have a reddish or bronzy cast when new. Small flowers in various shades of pink and sometimes white appear in clusters in spring over a long period. Small, dark blue, berrylike fruits may follow in summer. The species has off-white flowers and is often rangy so cultivars are preferred.

Special design features: Shrubby green forms for cultivated gardens. Refreshing late winter and early spring bloom. Dependable size. Woodsy or eastern garden feeling. Oriental quality with other plants of an Oriental nature.

Uses: Larger types make dense background plantings, specimens or small trees. All plants are effective as foundation plantings. Informal hedges, edges or dividers. Large scale ground covers.

Foreground plants for small patios. Container plants.

Disadvantages: Iron chlorosis. Leafburn from intense or reflected sun or soil salts. Unpinched plants are often open and rangy looking.

Planting and care: Plant any time from containers. Space larger forms 2 to 3 feet apart for massing. Pinch at least once after bloom to keep growth compact. Do not shear or clip because it ruins the character of the plant. To encourage vertical growth, pinch side shoots. Give plants an occasional deep soaking to leach out soil salts.

All zones
Evergreen
Soil: Best with enriched garden soil and good drainage.
Sun: Open or part shade. Accepts full sun in middle and high zones. Needs afternoon shade in low zone or where there are reflective surfaces such as concrete walks or white walls.
Water: Moderate to ample.
Temperature: Hardy to about 10°F (−12°C) or below.
Maintenance: Regular garden care.

Raphiolepis indica

Raphiolepis indica

RHUS lancea
Family: *Anacardiaceae*   African sumac

This tree is native to the arid lands of South Africa and is well adapted to the southwestern part of the United States. Trees create a large, dense screen to 15 feet or more high at maturity. Plants vary in growth rate from slow to fast depending on conditions. Maximum height is usually 20 feet high with a 30-foot spread. Crowns of trees 12 to 15 feet high are flat-topped and spreading. As trees gain height, crown becomes more dome shaped. Foliage appears dark green and fine textured at a distance. Leaves are formed of 3 narrow pointed lobes 2 or 3 inches long. Unimportant greenish flowers in late winter produce red or yellow pea-sized berries in clusters on females. Trees have naturalized in the Southwest.

Special design features: Ornamental and picturesque shade tree of bright to deep dark green. Dense shade canopy.

Uses: Shade tree for paved or unpaved areas, patios, courtyards or lawns—where drainage is excellent. Set close together as a mass planting to create a continuous canopy. Street tree or in public spaces. Unpruned plants in row makes a screen. Single plant makes a bold emphasis point.

Disadvantages: Susceptible to Texas root rot, especially in well watered places with poor drainage. May get butt rot or iron chlorosis in poorly drained soils.

Planting and care: Plant from containers any time, but best in spring and summer. Space 15 to 25 feet apart for a canopy, 20 to 30 feet or more for street tree. Place 10 to 15 feet apart for a dense screen. Corrective pruning may be done any time. Never plant in an area known to be infested with Texas root rot and for best results, prepare soil before planting with the recommended mix to prevent it. Trim young trees as they grow and encourage high scaffold branches so that you can walk under the tree.

Middle and low zones
Evergreen
Soil: Tolerant of a wide range of soils but needs drainage. Tolerates less than perfect drainage if given little water, but growth may be slow.
Sun: Part to full or reflected sun.
Water: Little to moderate. Space irrigation so soil dries out between waterings. Accepts ample with good drainage.
Temperature: Hardy to 15°F (−10°C) but may burn and look unkempt below 20°F (−7°C).
Maintenance: Little, once plants are trained as desired.

Rhus lancea

RHUS ovata

Family: *Anacardiaceae*

A shrub native to the chaparral belts of California, Arizona and Mexico, this broad, slow-growing plant may reach 10 feet, rarely 15 feet high after many years. It is surprisingly drought tolerant and gives the look of being well watered, with a rounded form and crisp, leathery, bright to deep green foliage. Leaves to 3 inches long are often folded at midrib and curve backward. They have reddish, often wavy margins and are supported on red stems.

Branches and trunk have smooth, light gray bark. Deep red buds on 1-inch spikes form in fall and remain for a long period and are actually more decorative than the small, creamy flowers which open in spring. Deep red hairy fruits occasionally follow. These plants sometimes naturalize in high deserts where they seem best adapted.

Special design features: A very decorative plant for dry areas, it gives a lush green look to the landscape. Dense, cheerful, leafy mound, wide shrub or small tree.

Uses: Specimen or emphasis plant. Wide screen. Desert or dry natural gardens. Pool patios.

Disadvantages: Damping off fungus in summer is an occasional problem.

Planting and care: Plant from containers in fall, winter or early spring. Space 5 to 6 feet apart for large screen, 5 to 10 feet for informal mounds or massing. Accepts some pruning, but most attractive if allowed to grow naturally. Because of its regular form, it rarely needs training of any sort, unless you want to train a tall plant into a tree.

A discussion of *Rhus* species would not be complete without mentioning *Rhus choriophylla* and *Rhus virens*. Both of these large, shrubby sumacs are native to higher elevations of southwestern deserts. They grow well at lower elevations with a little supplemental irrigation. They are very much like *Rhus ovata* in character, but are not as troubled with damping off and may actually be the best *Rhus* species for desert regions. Both have compound leaves with a midrib and small leaflets. Leaves are deep green with a rich waxy appearance. Size and cultural requirements for both plants are the same as *Rhus ovata*.

All zones

Evergreen

Soil: Good drainage is a must, but it is otherwise tolerant. Seems to prefer loose rocky fill and likes to grow among boulders.
Sun: Part to full or reflected sun.
Water: Prefers deep irrigation in winter and spaced but light irrigation in summer. Overwatering (especially in summer) and poor drainage can kill plant overnight.
Temperature: Hardy to approximately 15°F (−10°C) or below. Remains attractive after freezes that often destroy many plants in these zones.
Maintenance: None, except occasional leaf raking.

Rhus ovata

RICINUS communis

Family: *Euphorbiaceae*

Castor bean • Palma Christi • Wonder tree

This native of tropical Africa will grow from 4 to 8 feet high in a single season, depending on the cultivar. Where not damaged or killed by frost it can become a tree to 20 feet or more high. Castor bean is erect in form with a soft woody trunk, spreading semi-woody branches and large lobed leaves. Its fast growth is welcome in the desert when shade or a splash of green is needed to soften a barren landscape. New foliage is very large, with stiff lobed leaves reaching to 18 inches across. Leaves on mature plants may be only 9 to 12 inches across. Unimportant flowers on tall branched stalks are followed by round 1-inch fruit covered with soft spines.

Seeds, from which castor oil is made, are deadly poisonous if chewed. They are so offensive to rodents that an old-fashioned repellent was to lay seeds in rodent paths to discourage them—a natural rodent control. Foliage comes in shades of red, bronze or green. Pods vary in color according to cultivar from red, yellow and brown to blue-green. There are dwarf or variegated cultivars; one has dark red metalic leaves. Check the selections available at your local nursery. Catalogs from national seed companies may be a source of unusual varieties.

Special design features: Tropical effect.

Vertical form. Bold and decorative foliage. Some varieties have colorful foliage or seed clusters.

Uses: Very fast green. Summer screening. One of the best plants to make a new bare lot liveable for the first summer or two.

Disadvantages: Seeds are poisonous if eaten. Removing flowers as they form eliminates seeds and plants grow faster. Plant has a coarse look at close range.

Planting and care: Plant from seed in spring and give regular irrigation. If grown near a south wall or other warm location, damaged plants in areas of light frost will quickly sprout and grow in spring.

All zones as an annual, perennial in low zone

Evergreen in frost-free areas

Soil: Widely tolerant.
Sun: Part to full and reflected sun.
Water: Moderate to ample.
Temperature: Freezes at 28°F (−2°). Revels in heat.
Maintenance: Very low. Occasional pruning to control.

Ricinus communis

ROBINIA pseudoacacia

Family: *Leguminosae*

Black locust • False acacia

This tree grows fast to moderate to 40 or 50 feet high, with a spread as wide in open areas. In deep soils and with ample moisture it may reach 75 feet high, in shallow soils it remains much smaller. Structure is open with brown furrowed bark and thorny branchlets. Medium green leaves 1 to 2 inches long are divided into many rounded leaflets. White, fragrant, pea-shaped flowers in 4 to 8-inch clusters appear in spring, followed by 4-inch brown pods which hang on tree all winter. Handsome when pruned and trained in its early years. This tree is native to the eastern and central United States and widely grown in the temperate zones. It has naturalized in parts of the West.

Cultivars: There are numerous cultivars available. 'Idahoensis' is one of the most popular in the Southwest. It is fast growing and erect with pink spring flowers in clusters. Grows to about 20 feet high in 3 years in deep soil, and can reach an eventual 40 feet with open arching branches. Of the selections of this cultivar, 'Idahoensis Purple Robe', with purple-pink flowers, is preferred.

Special design features: Attractive at close range and from beneath. Interesting all year as it goes through seasons. Woodsy mood. Old-fashioned garden feeling.

Uses: Garden, lawn or street tree. Anywhere soil is deep and there is enough space for it to grow. Deciduous tree for windbreaks.

Robinia pseudoacacia 'Idahoensis'

Disadvantages: Untrained trees can be unattractive. Suckers from roots are a problem around tree base. Lawn mowing keeps suckers in check. Subject to borers, iron chlorosis, root rot and mistletoe invasion. Bean litter may be objectionable. Tree may be short lived in warmer zones with shallow and alkaline soils. Leaves may burn if soil salts build up.

Planting and care: Prepare planting pit carefully. It should be as large as feasibly possible with good drainage and soil prepared to prevent Texas root rot and to encourage growth. Plant from containers any time or bare root in winter. Space 20 to 30 feet apart for grove or roadside planting. Stake and prune young trees carefully to form handsome structure.

All zones, best in cooler areas

Deciduous

Soil: Tolerant. Best in deep moist soils.
Sun: Full sun.
Water: Prefers moderate to ample. Tolerates some drought when established. Needs occasional deep soakings to leach out soil salts, even when planted in lawns.
Temperature: Hardy to cold.
Maintenance: Give regular garden care for the first few years.

ROSA banksiae

Family: *Rosaceae*

Banksia rose • Lady Bank's rose

The famous rose in Tombstone, Arizona, reported to be the largest rosebush in the world, is a banksia rose. Plants grow vigorously, sending out long slender branches which arch and sprawl, to make a large, deep green, informal mound on the ground or a rooftop trellis cover. Given support, it can be trained as a tree or on a wall, fence or arbor. It may be kept close and dense as a wall plant by shearing. Untrained and unsheared, it is best for large-scale areas. There are two kinds—

yellow-flowering and white-flowering. Flowers consist of many crinkled petals that appear all at once in spring along the branches. From a distance they look like popcorn. They have little if any fragrance.

Cultivars: 'Albo Plena', the white-flowering form, has thorny branches and is completely evergreen. 'Lutea', with yellow flowers, does not have thorns and is less evergreen.

Special design features: Profuse spring bloom. Large green mound. Rampant

growth. Wild informal look unless kept pruned and sheared.

Uses: Use only where you are willing to give it the time and care necessary for training and control. Often used as bank plant along highways at interchanges or in median strips where it will tangle with out-of-control cars. White-flowering form makes a thorny barrier. Not a plant for close-up viewing unless given constant trimming.

Disadvantages: Rampant growth can take over. Occasional iron chlorosis. In cold winters yellow form may lose some or all of its leaves.

Planting and care: Plant from containers any time. Space 8 feet apart for a bank cover. Prune and shear as needed to control size and appearance in limited spaces.

Rosa banksiae (white)

Rosa banksiae (yellow)

All zones, warm spots only in high zone

Evergreen to deciduous in colder areas

Soil: Tolerant. Prefers well prepared garden soil.
Sun: Part to full sun.
Water: Moderate.
Temperature: Hardy, but may lose leaves in coldest winters.
Maintenance: Occasional as an informal bank or trellis cover. Constant where it needs control during growing season.

ROSMARINUS officinalis
Family: *Labiatae*

Rosemary

Rosemary is one of the best low-growing plants for arid lands. It is a stringy-rooted plant that does well in poor or shallow soils. Valued for its durability as a landscape plant, especially prostrate forms. It has dark, gray-green, almost needlelike foliage, and a picturesque form 4 to 6 feet high. Foliage densely covers plant to the ground and may appear lighter gray in dusty areas or gain a yellowish cast during dry hot summers. Numerous, small, pale blue flowers appear along branches from winter into spring and sometimes bloom again in fall when the weather cools.

Cultivars: 'Prostratus' is the popular low-growing form with trailing branches mounding to 2 feet high and spreading 3 to 6 feet or more wide. 'Lockwood de Forest' is similar to 'Prostratus' but has bluer flowers and lighter-colored foliage. 'Collingwood Ingram' (*R. ingramii*) has gracefully curving branches, blue-violet flowers and spreads to 4 feet wide. 'Tuscan Blue' is upright to 5 feet tall with rigid branches, rich green foliage and blue-violet flowers.

Special design features: Informal aromatic plant for tough situations. Transitional, desert or wild gardens. Birds love its tiny seeds. Fall, winter and early spring color. Rabbits will not eat it.

Uses: Upright forms may be used as shrubs which will not get too tall. Prostrate forms are excellent as ground or bank covers, edging or cascading from planters.

Disadvantages: Older plants become woody. Sometimes attacked by a fungus if drainage is poor. Bermudagrass invasions create a tangled mess in large plantings. Flowers attract bees.

Planting and care: Plant any time from flats or containers. Place prostrate forms from flats 18 to 24 inches apart, gallon-can plants 24 to 36 inches apart (on centers) as a ground cover or edging. Plant shrub types 3 feet apart. Pinch tips of young plants to encourage fullness and bushiness and to prevent long leaders from becoming woody. Trim tips lightly after bloom to groom. Shearing tops will encourage side branches to spread. Hedge plants may be sheared or left to grow as picturesque forms.

> **All zones**
>
> **Evergreen**
>
> **Soil:** Widely tolerant but needs good drainage.
> **Sun:** Part to full and reflected sun.
> **Water:** Moderate to little. Drought-resistant when established.
> **Temperature:** Generally hardy, but may be damaged if a hard freeze follows a springlike fall which has encouraged new growth. Tolerant of great heat, but may yellow slightly if kept too dry.
> **Maintenance:** Some to practically none in wild or natural areas. Sheared plants require regular trimming.

Rosmarinus officinalis

Rosmarinus officinalis

SALIX babylonica
Family: *Salicaceae*

Weeping willow

(*S. elegantissima, S. pendula*). The prominent feature of the weeping willow is its long, weeping branches. Trees grow fast to as much as 30 feet high in the desert, higher in less rigorous places and with a spread equal to its height. Branches may be green, brown, even golden depending on the selection, covered with luxuriant, slender, 3 to 6-inch, medium green leaves. Foliage emerges early in spring, although leaves may stay immature until warm weather arrives. They are bright fresh green at first before deepening in color. They stay on the tree well into fall before the autumn chill turns them to a golden hue. Bloom of unimportant catkins along branches appears as leaves come out and produce inconspicuous capsules.

Special design features: Strongest weeping effect of all plants. Delicate appearance. Waterside effect. Oriental feeling.

Uses: Near ponds or in well watered situations such as large lawns, parks or golf courses. Some like it as a garden tree for the home.

Disadvantages: Greedy roots. Trees are structurally weak because the wood is brittle and soft. Subject to a number of problems such as borers, root rot, cankers and crown gall. Often short lived, but many consider it worth the problems.

Planting and care: Mature branches may be taken in fall and planted directly in the ground where they will root. Twigs and branches will root any season in a bucket of water. If you plant young trees, stake trunk securely and high for a long time until leaders reach 12 to 18 feet high. Do not let side scaffold branches bud out until leaders are the height you want them. Shorten side branches and remove them as they age and lose their vigor.

> **All zones**
>
> **Deciduous**
>
> **Soil:** Tolerant.
> **Sun:** Full sun.
> **Water:** Moderate to ample.
> **Temperature:** Hardy to cold. Accepts heat with ample water.
> **Maintenance:** Much at first, then after establishment, occasional.

Salix matsudana, the globe or Navajo willow, is sometimes planted in the high desert. It is an interesting garden or backyard tree which develops a broad, dome-shaped crown. It does not have a weeping growth habit, and makes a very good shade tree. Cultural requirements are similar to *Salix babylonica,* although it is more manageable for residental use.

Salix matsudana

Salix babylonica

SAMBUCUS mexicana

Mexican elderberry

Family: Caprifoliaceae

(S. caerulea arizonica, S. c. mexicana). A somewhat weedy tree or shrub, this plant is valued because it reverses the seasons— it becomes green and lush in the cool of the year, deciduous (or nearly so) in summer. Plants grow naturally along washes and ditches from western Texas to southern California and Mexico. When irrigated, growth is rapid to 20 feet or more high with a 15 to 20-foot spread. Form is irregular with light brittle wood and rough, furrowed, light-colored bark. Bright green, toothed leaflets 1 to 3 inches long densely cover tree from fall until hot weather arrives. Crown is covered with flat clusters of tiny white flowers for several months from winter into late spring or summer. Birds love the summer-ripening berries. Blossoms and berries are edible; berries make a fine jelly or wine.

Special design features: Lush green in winter months.

Uses: Wild gardens or large areas where lush green is desired in midwinter when other plants are bare. Field edges. Along streams, ditches or washes. With other trees as a windbreak. Bird gardens.

Disadvantages: Brittle wood. Weedy growth unless severely pruned. Uneven performance as a landscape plant. Reseeds profusely in favorable situations.

Planting and care: Plant from containers any time, or grow easily from seed. Space 10 feet apart for hedge, row or windbreak, 20 feet or more for row of trees. Do any heavy pruning in late summer or early fall before growth spurt begins. Light trimming or shearing may be done any time.

Sambucus mexicana

> **All zones 1,000 to 4,000 feet elevation**
>
> **Deciduous**
>
> **Soil:** Prefers deep soil kept on the moist side during growing season. Avoid rocky or dry shallow soils.
> **Sun:** Part to full sun.
> **Water:** Moderate. Accepts lawn watering. Tolerates ample. Needs less water when it is summer dormant.
> **Temperature:** Hardy to cold.
> **Maintenance:** Periodic grooming in a garden situation.

SANTOLINA chamaecyparissus

Lavender cotton

Family: Compositae

(S. incana, S. tomentosa). Lavender cotton is a striking plant known for its low, mounding form and gray color. It is multiple-branched with aromatic foliage densely covering the plant to the ground. Its species name is Greek and means "small cypress," referring to the dense, scaly quality of the foliage. Growth is moderate to rapid, 12 to 24 inches high and 2-1/2 feet wide. Unclipped plants produce 1/2-inch yellow, buttonlike flowers on slender stems above foliage in summer. Rabbits and other rodents do not seem to like plants. Resistant to heat, drought, wind and poor soil. Native to Spain and South Africa.

Special design features: Undulating gray mound. Handsome all year if clipped. Gray foliage contrast.

Uses: Border or edge. Rock gardens. Formal gardens. Desert or wild gardens where it can be left to grow naturally and become a rangy shrub. Sheared pattern plant.

Disadvantages: Plants lose their compact form if allowed to bloom. They are likely to become woody in a few years even if clipped and may have to be replaced.

Planting and care: Plant from flats or gallon containers any time, or start from cuttings placed in moist sand. Space 12 to 18 inches apart for edging or low clipped border. Space 18 to 24 inches apart for massing. Shear in late spring or early summer as flowers appear.

> **All zones**
>
> **Evergreen**
>
> **Soil:** Tolerant.
> **Sun:** Part, full or reflected sun.
> **Water:** Moderate to occasional.
> **Temperature:** Hardy to cold.
> **Maintenance:** Regular clipping gives this plant a neat appearance. No maintenance is necessary for natural shrub.

Santolina virens is similar to the above, except it has a rounder shape and foliage is bright ferny green. It reaches 2 feet high and spreads 18 inches or more wide. It is very attractive when combined with the gray-foliaged *S. chamecyparissus*, and has the same cultural requirements.

Santolina chamaecyparissus

Santolina virens

SCHINUS molle

Family: *Anacardiaceae*

Bright green, weeping foliage and rough tan bark are the appealing features of this fast growing tree. It can reach 15 feet high in 3 years and may grow to 25 or even 40 feet in favorable situations, with a crown spread of 20 to 35 feet. Because it is susceptible to many problems, it seldom reaches a large size in the desert. Many narrow leaflets are aromatic and sticky with resin and hang from drooping branches. Trunks become large and gnarled with age. Unimportant yellowish flowers in long hanging clusters on females produce sprays of tiny, decorative, pink fruit the size of peppercorns in summer. Pink skin of fruit peels off to reveal hard dark "peppers" only a fraction of an inch in diameter. This tree is attractive in both youth and age, but older trees are often misshapen from breakage and disease. To keep tree as a youthful, multiple-trunk, sapling cluster, cut it back nearly to the ground every few years—it will regrow rapidly. For a large, attractive, shade tree, great care must be taken to treat all problems as they occur. A native of Peru.

Special design features: Fresh-looking weeping foliage. Bright green color contrast. Picturesque. Rustic. Informal.

Uses: Fast shade. Best in dry areas, but will survive lawn conditions. Plants may be set close together to form a tall screen or top-cut and clipped to form hedge or screen at a lower height.

Disadvantages: Tree litters pavement and drips resin. Subject to Texas root rot and root knot nematodes. Trees with diseased roots may be blown over in wind. Heart rot enters trunk and branches through untreated cuts and wounds. It weakens structure so that the brittle branches may blow off in wind. Although it is evergreen, plants can be severely damaged after an Arctic freeze which sometimes occur in these zones. Greedy surface roots can be invasive.

Planting and care: Plant from containers any time except during cold times of year. Prepare planting pit with a soil mix known to discourage Texas root rot. Space 30 to 40 feet apart for roadside planting, 20 feet apart for grove. Space trees 15 feet or less apart for low, informal screen. Stake young plant firmly and head up to make a tree, allowing branches only above head height. Unstaked plants remain low with foliage covering plant to the ground. Treat all wounds and cuts with a dressing to prevent invasion of the fungus which causes heart rot. Treat root problems immediately. Prune only as needed.

Low and middle zones, protected locations in high zone

Evergreen

Soil: Tolerant. Prefers deep improved soils. Needs good drainage.
Sun: Full sun.
Water: Prefers deep periodic irrigation. Tolerates lawn watering with good drainage.
Temperature: Hardy to 20°F (−7°C), but foliage is sometimes damaged above that.
Maintenance: Needs careful maintenance to maintain a healthy tree, especially in residential and garden situations.

Schinus molle

SCHINUS terebinthifolius

Brazilian pepper tree • Christmas-berry tree

Family: *Anacardiaceae*

This shade tree grows at a moderate rate to 15 to 30 feet high. It develops a wide, dense crown with intertwining and twisting branches. Trees naturally branch near the ground and require pruning and training to make a wide umbrellalike crown which can be walked under. Leaves are 2-1/2 inches long, glossy dark green, with several leaflets on rather stiff branchlets. Inconspicuous flowers produce an attractive show of red berrylike fruit in clusters on female plants in late fall or winter. Select trees carefully as individuals vary widely in form and flower. Males are more rangy in form and sparse in foliage. If you definitely want a female tree, buy a plant that has fruit on it. Trees are more attractive with multiple trunks.

Special design features: Spreading canopy. Winter color. Sculptural quality.

Uses: Shade for street, lawn, patio or poolside in warm places. Adapted to well-watered situations such as parks, lawns or golf courses. Weave branches together overhead to create a canopy of shade.

Disadvantages: Subject to Texas root rot, verticillium wilt and winter injury. Wind sometimes breaks branches.

Planting and care: Select plants with care—avoid root-bound plants. Place plants in a well drained planting hole prepared to prevent Texas root rot. Space 20 to 30 feet apart for grove or row, 30 feet or more for street or avenue planting. Stake trunks firmly and cut off lower branches so people can walk under tree. Select scaffold branches carefully so crown will have a balanced form. Thin foliage on larger trees in late summer to avoid wind damage. If branches have become too long, shorten them at that time. Apply fertilizer regularly to maintain growth.

Low zone, marginal in middle zone

Evergreen

Soil: Tolerant. Prefers good drainage.
Sun: Full sun.
Water: Needs deep periodic irrigation. Tree receiving only light, frequent, lawn irrigation may develop nuisance surface roots over a wide area.
Temperature: Young trees may be damaged below 26°F (−5°C), severely at 23°F (−6°C). Older trees are damaged near 20°F (−7°C) and will brown out with wood damage below that temperature.
Maintenance: Careful training, then occasional pruning and clipping.

Schinus terebinthifolius

SIMMONDSIA chinensis

Jojoba • Coffee bush • Goat nut • Deer nut

Family: *Buxaceae*

Jojoba is a shrub of the boxwood family native to the Southwest, Mexico and Baja, California, at elevations between 1,000 and 5,000 feet. It grows naturally in an area of milder winter temperatures. Oil from its nuts produces a wax capable of replacing the oil of the endangered sperm whale. The plant has a long history of uses, indicated by its various common names. Indians and white settlers made a coffee substitute from the nuts. Nuts were also eaten roasted or raw, but are not very appealing because of the bitter tannic acid content. A dense, rounded, sometimes irregular shrub, jojoba grows to 6 feet high with an equal, sometimes wider spread. Oval, gray-green, leathery, almost succulent leaves to 2-1/2 inches long cover plants to the ground, set in upward-pointing pairs along branches. Plants are male and female and produce small, unimportant, yellowish flowers usually in spring. Numerous, 1-inch, acornlike nuts follow on females. Plants bloom any time between December and July, depending on weather. Once established, jojoba can grow at an increasingly fast rate, especially when water is available.

Special design features: Shrubby dense mass of gray-green. Interesting foliage pattern. Informal. Looks great with little or no care.

Uses: Wide screen, hedge or foundation plant. Wild or transitional gardens. Median strips or low maintenance areas.

Disadvantages: Slow to get started. Young plants are particularly sensitive to cold.

Planting and care: Plant from containers in spring, any time in warm regions. Space 3 to 5 feet apart for a loose screen, which will grow together in a few years. Space about 2 feet apart for clipped hedge. Protect young plants from extreme cold. Pruning is not usually necessary.

> **Low and middle zones**
>
> **Evergreen**
>
> **Soil:** Tolerant. Prefers gravelly soil with good drainage.
> **Sun:** Part, full or reflected sun.
> **Water:** Moderate at first and for fast growth, then deep periodic irrigation. Can be neglected completely when it reaches a desired size in areas of 10 inches annual rainfall or more. Taper off irrigation gradually.
> **Temperature:** Young plants are injured by frost at about 20°F (−7°C). Plants are badly damaged or killed at 15°F (−10°C) or below, even mature established plants.
> **Maintenance:** Little to none.

Simmondsia chinensis (hedge form)

SOPHORA secundiflora

Mescal bean • Texas mountain laurel • Frijolito

Family: *Leguminosae*

Mescal bean is admired for its distinctive leaf pattern, wisterialike spring bloom, year-round good looks and lack of bothersome pests and diseases. Tolerant of heat, cold, wind, drought and poor soil, this plant looks attractive in all but the most extreme conditions. Native to Texas, New Mexico and northern Mexico, it is usually grown as a shrub, although it can be slowly trained into a tree 20 to 30 feet high. Upright sculptural branches are covered with silvery bark. They bear 4 to 6-inch leaves with several pairs of oval, glossy, medium green leaflets 1 to 2 inches long. Plant is picturesque at any age. When barely more than a branch it bears fragrant, 8-inch long, violet-blue (rarely white) clusters of pea-shaped flowers early in spring. Decorative, woody, silver-gray pods follow and split open in late summer to reveal bright red beans (*frijolitos*). They are poisonous, but their outer coat is so hard it is believed they pass through digestive systems without harm.

Special design features: Interesting and durable form and foliage. Spectacular spring bloom. Welcome green in even poor conditions. Remains handsome in severe winters.

Uses: Specimen, mass, hedge or row. Espalier. Silhouette against structures. Shrub with spreading branches or upright, short-trunk tree for patio, lawn or median strip. Pool areas. Transitional plant. Natural or wild gardens.

Disadvantages: Poisonous seeds. Bees love flowers. Slow to develop, especially in cool summer areas.

Planting and care: May be grown from seed, but slow. Plant from containers any time. Use a 5-gallon plant for a good start. Space 5 to 6 feet apart for massing. More attractive as specimen. Allow it to grow naturally, or train as bank cover.

Two more *Sophora* species should be mentioned here, although they are not yet widely available. They are both native to the Southwest.

Sophora arizonica, Arizona mescal bean or Arizona mountain laurel, grows on dry rocky hillsides and wash banks of western and central Arizona from 2,000 to 4,000 feet elevation. It is an exceptionally slow growing plant. Flower clusters are violet-blue. Leaflets are deep green, silvery underneath, smaller than *S. secundiflora.*

Sophora gypsophila guadalupensis, Guadalupe mountain laurel or Guadalupe mescal bean, is from the Guadalupe Mountains in western Texas. This low shrub reaches 3 to 4 feet high and spreads 10 to 15 feet wide. Flowers are larger than *S. secundiflora.* Introduced by the Texas A&M Experiment Station at Renner.

> **All zones**
>
> **Evergreen**
>
> **Soil:** Tolerant. Thrives in alkaline soils, but requires good drainage.
> **Sun:** Part to full or reflected sun. Grows fastest with high heat.
> **Water:** Moderate to little. Drought-tolerant when established. Best with moderate irrigation. Accepts periodic deep irrigation.
> **Temperature:** Hardy to cold. Tolerates heat.
> **Maintenance:** None to occasional pruning and training.

Sophora secundiflora

STRELITZIA reginae

African bird of paradise • Bird of paradise • Crane flower

Family: *Strelitziaceae*

(S. parvifolia). The African bird of paradise from South Africa gives a tropical flair to special corners of the garden or patio. Growth is moderate to slow to 3 to 5 feet or more high. Plants are formed of wide, paddlelike, blue-green leaf blades which rise from plant base and radiate outward at top. Flowers that look like tropical birds may bloom any time but usually appear for a long period from fall into winter. Flowers consist of horizontal, boat-shaped bracts to 8 inches long, sometimes edged in purple or red. Orange to yellow petals fan from the top and a dark blue tongue extends sideways toward outer tip of bloom. Cut flowers are long lasting and especially attractive in indoor arrangements.

Strelitzia reginae

Strelitzia reginae

Low zone
Evergreen
Soil: Rich deep soil with much humus and good drainage.
Sun: Full sun in cool areas or during cool periods. Best with filtered to open or part shade.
Water: Moderate to ample.
Temperature: Damaged at 28°F (−2°C). Often killed below 20°F (−7°C).
Maintenance: Periodic feeding and grooming.

Special design features: Exotic tropical appearance. Handsome leafy silhouette. Fantastic flowers last a long time; excellent used in arrangements.
Uses: Sheltered gardens away from hot winds and freezing temperatures. Enclosed patios, entryways, atriums or under overhangs on north or east sides. Containers. Specimens. Tropical groupings. Pool patios.
Disadvantages: Plants are injured from hot sun, winds and frost; recover slowly.

Some plant parts are believed to be poisonous.
Planting and care: Plant from containers when weather warms in spring. Space new plantings 4 to 6 feet apart for grouping. Old clumps may be divided in spring. Because crowded clumps bloom best, don't be in a hurry to divide them. Roots are long and fleshy, and need room to spread. Supply with humidity. Feed frequently and heavily. Groom by removing old blooms and leaves.

SYRINGA persica

Persian lilac

Family: *Oleaceae*

This shrub is a hybrid between a lilac from Afghanistan and a lilac from China. It is the best lilac for warm deserts. Growth is moderate to slow up to 5 to 6 feet high, rarely to 10 feet. It spreads as wide as high if given room. Its medium green leaves are similar to the privet. Small lavender flowers appear in early spring in loose, plumelike, 3-inch clusters. Flowers and plant are much smaller compared to the common lilac often grown in the eastern United States.
Cultivars: 'Alba' is a white-flowering form.
Special design features: Bloom and fragrance are sentimental favorites. Woodsy feeling.
Uses: Place in a sheltered corner where its spring bloom can be appreciated.
Disadvantages: Sometimes gets scale, leaf miner and sunburned leaves in sum-

mer. Close relative of privet (*Ligustrum* species) and subject to the same diseases.
Planting and care: Plant from containers any time. May be planted bare root in winter. Space 5 feet apart for massing. Pinch tips of young plants to shape. Groom by removing spent flowers before seed forms. When removing old flowers, cut them just above where next year's buds are forming, at points where the leaves join the stems. Do not prune plants too heavily or there will be fewer blooms the following year. As plants become old

and woody, cut a few of the oldest stems to the ground each winter to rejuvenate. *Syringa vulgaris,* the eastern or European lilac, is similar to *S. persica,* but with larger more fragrant blooms. It will only bloom in the cooler, high zone where it seldom reaches more than 5 to 6 feet high. A cultivar, 'Lavender Lady', is an exception. It will grow to 20 feet high with leaves to 5 inches long. It produces large clusters of lavender flowers in areas that don't have a long period of winter chill. Cultural requirements are the same as *S. persica.*

Best in higher elevations, but grows and blooms in middle and low zones
Deciduous
Soil: Prefers alkaline soils.
Sun: Part to full sun.
Water: Moderate to ample.
Temperature: Tolerates heat. Likes winter chill, but this lilac, unlike most others, will perform in warm winter areas.
Maintenance: Occasional trimming.

Syringa persica

Syringa persica

TAMARIX aphylla

Family: *Tamaricaceae*

(T. articulata). The athel tree is native to the deserts of North Africa and regions of the Middle East. It tolerates enormous heat, hot winds, drought and soil so salty most plants will not grow in it. Growth is fast to moderate to 30 feet or more high. Spread is 15 to 30 feet wide or more. Cuttings will grow to 10 feet tall in 3 years if grown in deep soil and irrigated. Rough brown bark covers erect trunks. What appears to be long, slender, grayish, blue-green, needlelike leaves are actually jointed branchlets. True leaves are minute scales at the joints. The grayish look of foliage increases with heat and drought in alkaline soils, caused by salts absorbed by tree and secreted by the tiny leaves. It's not very refined or appealing at close range, but its shade is gratefully accepted in conditions where nothing else will grow. An unpruned tree develops a wide spread and deep roots. If cut down it will continue to grow back from the stump. Clusters of pliable needlelike flowers appear in summer as dusty pink to creamy plumes at branch tips.

Special design features: Dense shade or windbreak where little else will grow.

Uses: Low maintenance areas for shade or as tall screen to buffer wind, dust and sun. Roadsides, field edges and river banks. Residential tree in alkaline situations. Plant close together in rows for clipped hedge and keep as low as 3 feet. Replant problem areas such as sanitary landfills or salt flats which are sometimes flooded with brackish water.

Disadvantages: Invasive roots. Constant drop of salty branchlets creates a thatch which discourages other plants from growing under it. Difficult to eradicate once established. To kill plants, drill holes in the stump and fill them with saltpeter. Stumps are hard to remove. Older trees develop an awkward branch structure, have heavy crowns and break in the wind unless thinned. Extremely sensitive to herbicides.

Planting and care: Plants do not keep well in nursery cans because they rapidly develop deep roots. It is best to plant cuttings that are 1/2 to 3/4 inch in diameter and 18 inches long. Bury most of cutting, leaving only 3 inches showing above ground. Keep damp until established. Supply moderate irrigation for a time to encourage size, then taper off gradually. Space cuttings 3 feet apart for a hedge, 10 feet for a windbreak or tall screen, 20 feet or more for group of shade trees.

All zones

Evergreen

Soil: Tolerant of almost any soil.

Sun: Full to reflected.

Water: Once established, this plant accepts very little water, but it grows faster if given occasional irrigation. Tolerant of ample water in boggy situations. Thrives in moderately irrigated situations.

Temperature: Survives to 0°F (−20°C), but damaged at about 15°F (−10°C).

Maintenance: Pruning and raking, depending on location.

Tamarix aphylla

Deciduous tamarisks often have a place in the desert landscape. They would include *T. africana, T. parviflora* and *T. chinensis,* commonly known as salt cedars. These species are shrubs or small trees and are difficult to tell apart. All have attractive feathery flower clusters. Most bloom in spring but *T. chinensis* tends to flower in summer. There are a number of flower selections, from pure white through various shades of pink to deep purple. They are hardy and planted in the Temperate Zone as deciduous flowering shrubs. All are native to the Old World but have escaped cultivation and now grow along irrigation ditches, the Colorado River and many other locations where there is a better than average supply of water. These plants are drought-resistant, even though they are most abundant in such wet areas.

Tamarix aphylla

Tamarix aphylla

TECOMA stans

Family: *Bignoniaceae*

This large shrub delights gardeners with its constant bloom of 2-inch, yellow, bell-shaped flowers that appear during the warm season. In cold winter areas it may die back to its roots each winter and regrow each spring, seldom reaching more than 4 or 5 feet high in a season. Where it is nipped by frost occasionally it may reach 8 feet high. In areas seldom touched by frost it can become a tree to 20 feet. Lance-shaped leaflets to 4 inches long are bright medium green, lush and tropical, densely covering plant. Bright yellow flowers appear in clusters, sometimes followed by slender pods 3 to 8 inches

Tecoma stans

Tecoma stans

long. Native to Florida and Mexico.

Tecoma stans angustata is native to dry stony slopes of desert thermobelts from 3,000 to 5,500 feet elevation in Arizona, New Mexico and Texas. It has less foliage with smaller, narrower leaflets and is rangier in form.

Special design features: Yellow flowers all summer, or all year in the warmest places. Tropical mood.

Uses: Specimen. Containers. Cultivated, wild, natural or desert gardens. Large border or background screen in warm winter areas or as street or patio tree. In cooler winter areas place against south and west walls that have a protective overhang. Plant with evergreens so unattractive winter branches will be masked.

Disadvantages: Abundant seed pods on fertilized plants. Sometimes nipped by cold just when it looks best.

Planting and care: Plant from containers when danger from frost is past. Start easily from seed in porous soil. Space 4 feet apart for massing in colder areas where it will be frosted back, 6 to 8 feet on center for massing in warm areas. Remove dead twigs in spring when plant starts to leaf out. Blossoms appear on new growth, so pinch back branches and remove old blooms and seed pods to encourage them.

Low and middle zones, warm pockets in high zone

Evergreen in warm areas to deciduous in cold winter areas

Soil: Tolerant. Prefers improved soil with good drainage.
Sun: Part to full sun.
Water: Moderate to ample, especially during bloom period.
Temperature: Loves heat. Root-hardy to cold, but top is damaged at 28°F (−2°C) and may die to the ground below that. Fast recovery after cold damage.
Maintenance: Occasional grooming and annual pruning.

TECOMARIA capensis

Family: *Bignoniaceae*

(*Bignonia capensis, Tecoma capensis*). A sprawling, vining plant with luxuriant, glossy, deep green foliage, this South African native bears clusters of brilliant, 2-inch, orange-red, trumpet-shaped flowers in fall and winter. It grows fast, sending out branches up to 12 feet long unless controlled. Tied to a support it can climb 15 to 25 feet high. Pruned carefully, it becomes a shrub 6 to 8 feet high. It thrives in heat and accepts some drought and hot winds.

Special design features: Luxuriant tropical mood. Bright color through cool season when little else is blooming.

Uses: South or west walls under an over-hang in cooler winter areas. Spills over hot banks or planters in areas of reflected heat. Mounding plant in the open in warm locations. Does well as a container plant anywhere, especially if brought indoors during winter to bloom in a sunny window. Plants need to gain a certain size or at least root size before they bloom.

Disadvantages: Can be nipped by cold just when it looks best and has begun to bloom. Although it grows new foliage

rapidly, it may be fall before it recovers to blooming size, and then it can be frozen again.

Planting and care: Plant from containers when frosts are past. Space 6 feet apart for mass planting. Prune and pinch as needed during the growing season to train. Heavy clipping reduces bloom because flowers are at tips of new growth. Keep on dry side to encourage more bloom and hardiness to winter temperatures.

Low zone, protected pockets in middle zone

Evergreen

Soil: Prefers improved garden soil with good drainage.
Sun: Part to full or reflected sun, but it will remain an attractive foliage plant without much bloom with a northern exposure.
Water: Moderate to slightly less than moderate.
Temperature: New growth is damaged at 28°F (−2°C). Loves heat.
Maintenance: Periodic pruning and pinching.

Tecomaria capensis

TEUCRIUM chamaedrys

Germander • Chamaedrys germander

Family: *Labiatae*

This appealing plant is composed of numerous upright stems, densely covered with small, toothed, gray-green leaves set in a distinctive pattern. Growth is slow to moderate to 12 to 18 inches high, spreading by underground rhizomes to 2 to 3 feet, but often less. A member of the mint family, germander produces small, purple, rarely white flowers along its vertical stems in late summer. Native to Europe and southwestern Asia.

Cultivars: 'Prostratum', dwarf germander, is only 4 to 6 inches high and is often preferred. It goes well with stepping stones, in rock gardens, as a lawn extender or as a small area ground cover. Both species and cultivar are tough—tolerant of poor soil, heat, wind, drought and cold.

Teucrium chamaedrys

All zones

Evergreen

Soil: Tolerant, but needs good drainage.
Sun: Part, full or reflected sun.
Water: Moderate with good drainage. Best with periodic irrigation, allowing ground to dry out between waterings.
Temperature: Hardy to cold. Tolerant of heat.
Maintenance: Occasional trimming.

Special design features: Dense green with pleasing leaf pattern. Informal character.
Uses: Both forms can be used as a low mass, ground cover or foreground plant. Edgings, borders, parking strips or spaces adjacent to paved areas. Larger forms can be used as low clipped hedge or natural shrub in rock gardens or in transitional areas. Containers.
Disadvantages: Uneven performance. Slow. Sometimes covers ground unevenly.

Becomes woody in time if not sheared occasionally. Soggy soil causes plant to decline.
Planting and care: Plant any time. Space plants from flats 14 inches on center, 1-gallon plants 18 to 24 inches apart for massing. Fill any bare spots not covered with new plants the following year. Trim lightly in early summer and again after bloom to groom and renew vigor. This also forces side branching, helping create a dense cover.

THEVETIA peruviana

Yellow oleander • Be-still tree • Lucky nut

Family: *Apocynaceae*

(*T. neriifolia*). Yellow oleander is a large, erect, graceful shrub or small tree native to tropical America. It grows fast—6 to 8 feet high as a shrub, or it can be easily trained into a small, spreading tree to 20 feet high in warmer zones. Shiny dark green foliage and yellow or apricot-colored flowers 2 inches across in clusters appear almost any time of year, but most often from June to November. Narrow leaves to 6 inches long densely cover plant. Flowers may be followed by hard, 1-inch, angular fruit, first red then turning black. Plant is root-hardy in cooler areas if roots are well-mulched during winter.

Cultivars: 'Alba' has white flowers.
Special design features: Luxuriant tropical effect. Bloom over a long period.
Uses: Against hot walls. Sheltered patios, atriums and entryways. Specimen. In warm winter areas, it may be used as a canopy for a small patio or as a small street tree. Combine with hardy evergreens where it freezes back in winter.

Disadvantages: All parts of this plant are poisonous. Seeds and flowers can litter.
Planting and care: Plant from containers when frosts are past. Handle plants carefully when planting. Space 6 to 8 feet apart for a continuous mass. Prune to shape or to show trunks. Avoid cultivating at plant base as surface roots are easily damaged. In colder areas apply a mulch several inches deep to protect roots and lower part of trunk from freezing.

Low zone, warm pockets in middle zone

Evergreen

Soil: Improved garden soil with good drainage.
Sun: Part, full or reflected sun.
Water: Ample is preferred, but avoid overwatering young plants. Tolerates moderate or widely spaced irrigation when established.
Temperature: Foliage damaged at 28°F (−2°C), but wood will survive much lower temperatures and new growth comes back quickly in spring. Revels in heat.
Maintenance: Occasional pruning and cleaning up litter.

Thevetia peruviana

TRACHELOSPERMUM asiaticum
Asiatic jasmine • Ground jasmine • Dwarf star jasmine

Family: _Apocynaceae_

(Rhynchospermum asiaticum). Asiatic jasmine trails to as much as 15 feet, with branches rooting as they go. Twiglets bearing shiny 1-1/2-inch leaves rise vertically 6 to 8 inches above the ground. Large established plantings have a dense, even-textured appearance. Small, white, fragrant, starlike flowers sometimes bloom in late spring, but are seldom seen in the desert. This twining, trailing plant needs to be tied up if used as a vine unless grown on a fence. Native to Korea and Japan.

Special design features: Very appealing at close range and looks nice all year. Cool woodland feeling. Tropical appearance when combined with tropical plants.

Uses: Best as dense ground cover in small area or as filler or underplant. Containers. Foreground plant. Attractive in small intimate patios, atriums or entryways. Swimming pool areas.

Disadvantages: May be slow to cover in hot areas and sometimes burns out in spots during summer. Sometimes gets chlorosis in wet situations.

Planting and care: Plant from containers any time. Can be planted from flats, but very slow and uncertain—1-gallon cans are better. Space 18 to 24 inches apart. Mulch to keep roots and new branchlets cool in hot sunny areas, especially new plantings. Avoid planting where it will receive intense or reflected sun. Prune back occasionally in late winter after bloom.

> **All zones**
>
> **Evergreen**
>
> **Soil:** Prefers improved, porous, garden soil.
> **Sun:** Open, filtered or part shade. Full sun in high areas or in middle zone if properly mulched in summer.
> **Water:** Moderate to ample. Established plants tolerate some drought in winter.
> **Temperature:** Tolerant to about 16°F (−10°C), but may be damaged above that if frost follows springlike weather that has resulted in new growth.
> **Maintenance:** Little, some pruning and mulching.

Trachelospermum asiaticum

TRACHELOSPERMUM jasminoides
Star jasmine

Family: _Apocynaceae_

(Rhynchospermum jasminoides). This plant is much like its relative, _T. asiaticum_ but on a larger scale. It is a refined, twining vine with elegant dark green foliage, bearing profusions of fragrant, white, waxy, starlike flowers. It grows at a slow to moderate rate, spreading to 20 feet over the ground. It is also attractive trained up wire or supports. With training, it covers an area 10 by 10 feet at maturity. Shiny, leathery, 2-inch leaves are supported on rich, glossy brown branches. Because star jasmine is slow to develop, it is best to start with 5-gallon size plants.

Special design features: Shrubby sculptural quality. Appealing at close range. Woodsy or tropical effect. Spring fragrance. A choice vine for garlands near entrances or windows where its looks and fragrance can be most enjoyed.

Uses: Walls, fences, trellises or porch posts. A ground cover in middle zone. Container cascade. Small intimate spaces such as patios, atriums or entryways where it looks nice all year and can be enjoyed at close range. Prefers north and east exposures.

Disadvantages: May be very slow to cover or covers unevenly. It sometimes tangles as a ground cover. Older plants in difficult situations begin to look bare and unkempt. Foliage can burn from reflected heat. Occasionally suffers from iron chlorosis. May be damaged by severe cold, especially on south sides where the hot, daytime sun can dehydrate plants.

Planting and care: Plant from containers any time. Space 4 to 5 feet apart for fast cover on a structure, 3 feet apart for ground cover when using 5-gallon plants. Plants will climb if given support. For a ground cover, pinch branch tips to encourage lateral growth, and pin stems down to mulched ground.

> **Low and middle zones, borderline in high zone**
>
> **Evergreen**
>
> **Soil:** Improved garden soil.
> **Sun:** Any exposure in middle zone, except in locations of extreme reflected heat. Some shade in the low zone. In high zone use only as a garland on a warm surface such as a south-facing wall.
> **Water:** Moderate to ample.
> **Temperature:** Damaged by cold at 20°F (−7°C) or below, especially if freeze follows springlike fall weather which has encouraged new growth. Tolerates heat when given water, but may show leaf burn in hot spots.
> **Maintenance:** Regular care.

Trachelospermum jasminoides

TRACHYCARPUS fortunei

Family: *Palmae*

Windmill palm

(Chamaerops fortunei). Windmill palm is a small fan palm reaching only 15 feet high at maturity with a crown to 7 feet in diameter. It grows slowly at first, then at a moderate or even fast rate. Trunk is erect and slender with upward-pointing stubs of old fronds protruding from a dark, fibrous cover which makes the trunk look thicker than it actually is. Lack of fiber at the base makes the plant look top heavy. Crown is formed of small, stiff, dark green fans with toothed bases; the whole frond is about 3 feet long. Clusters of small, dark, unimportant, berrylike fruit occasionally follow creamy white clusters of bloom which are strange in appearance—almost succulent. Native to China and northern Burma.

All zones

Evergreen

Soil: Tolerant. Prefers improved garden soil.
Sun: Open, filtered or part shade to full sun.
Water: Moderate to ample.
Temperature: Hardiest of the palms. Tolerates temperatures to 10°F (−12°C) or below and looks good even after cold spells.
Maintenance: Occasional grooming and feeding.

Special design features: Neat, small-scale palm which stays low for a long period. Older plants have a strong vertical look. Tropical effect. Appears Oriental with right plant combination. Dramatic foliage. Somewhat formal.
Uses: Small-scale areas or narrow spaces. Swimming pool areas, small patios or gardens, atriums or entryways. Specimen, pair or grouping. Overhead tree to smaller plants. Container plant.

Disadvantages: Fronds may be tattered by wind. Susceptible to sunburn in reflected sun of low deserts or in exposed locations.
Planting and care: Plant from containers any time. Field-grown plants or larger specimens are best transplanted spring to midsummer. Cut off old fronds or seed structures to groom. Give regular feeding for best results. Tolerates neglect once established, but it becomes less attractive.

Trachycarpus fortunei

ULMUS parvifolia

Family: *Ulmaceae*

Chinese elm • Chinese evergreen elm

An erect, refined tree from China and Japan, Chinese elm has a spreading canopy of arching branches and weeping branchlets. Fast growing, it can reach a height of 30 feet in 5 years, with an even wider spread if grown in deep moist soil. Glossy deep green leaves 1 inch or more in length glisten in the sun. The slender trunk is smooth and dappled gray and tan as outer layers flake off. Unimportant flowers in late summer produce small decorative fruit. Tree is variable in form and whether it will remain evergreen, so ask about the best selection at your nursery if these features are important to you.
Cultivars: 'Sempervirens' is smaller and more delicate, more nearly evergreen and

better for residential use. 'Drake' ('Brea') is medium-size with an upright and regular form. 'True Green' is said to be nearly evergreen and more uniform in growth than the others.
Special design features: Tree of delicate scale and refined appearance—appealing at close range. Oriental feeling. Weeping form. Fast shade for patio or garden.
Uses: Street, garden or patio tree.
Disadvantages: Subject to Texas root rot. May be slow to develop canopy. Heavy

weeping crown may break in the wind.
Planting and care: Never plant in soil known to be infested with Texas root rot. Prepare soil in planting pit with mix to prevent this disease and treat annually. Space 20 feet apart for canopy, 30 feet or more for row. Stake young tree firmly until trunk is strong enough to hold its crown. Remove lower branches up to the desired height for the crown to branch out, being sure to compensate for hanging branchlets. Shorten extra-long branches.

All zones

Partly evergreen to deciduous in sharp cold

Soil: Prefers deep soil with good drainage.
Sun: Part to full sun.
Water: Best with deep periodic irrigation once established, but tolerates ample water of lawn irrigation.
Temperature: Hardy to cold. Foliage may brown and drop at about 25°F (−4°C).
Maintenance: Occasional pruning and treatment for Texas root rot.

Ulmus parvifolia

ULMUS pumila

Family: *Ulmaceae*

Siberian elm, often miscalled Chinese elm, is native to eastern Siberia and northern China to Turkestan, but grows nearly everywhere. Although it remains a shrub in very difficult circumstances, it grows fast to 50 feet tall with a dense, 30 to 40-foot crown if water and soil depth are adequate. Trunk is erect to slightly bending, covered with deeply furrowed gray bark. Irregular branches support somewhat drooping branchlets and toothed, 2-1/2-inch, lance-shaped leaves.

Special design features: Fast dense shade in summer, sun in winter.

Uses: Shade for difficult areas. Windbreaks, at property edges, along roadsides. Numerous surface roots make it a good bank cover for erosion control.

Disadvantages: Greedy roots. Brittle wood and weak crotches make it subject to wind breakage. Reseeds profusely. Very susceptible to Texas root rot and should only be planted in uninfested soil. Sometimes suffers from slime flux, root knot nematode or mistletoe infestations. Trees at lower elevations may suffer from delayed foliation due to a lack of winter chilling and may show weak growth.

Blossoms may litter pavement. Despite the problems, Siberian elm is still a desirable tree because it is easy to grow in very poor conditions.

Planting and care: Plant from containers any time, bare root in winter. Seedlings may be transplanted in dormant season. New trees are easily started from cuttings. Plant residential trees in pits treated to prevent Texas root rot. Space 10 feet apart for windbreaks or high screens. Space 25 to 30 feet apart for a grouping. Thin crown to prevent wind damage and to shape younger trees.

All zones

Deciduous

Soil: Tolerant. Best with good drainage.
Sun: Full to reflected sun.
Water: Accepts any amount of water. Best with deep periodic irrigation. Grows lush along ditches or in lawns. Tolerates drought, but grows very slowly and parts of tree may die.
Temperature: Tolerates extremes of heat and cold.
Maintenance: Some to none depending on location. Respond quickly to its problems if it is an important tree in your landscape.

Ulmus pumila

VAUQUELINIA californica

Family: *Chamaebatiaria*

Arizona rosewood is a vigorous, dense, erect shrub native to southern Arizona and northern Mexico at the 2,500-5,000 foot level. It grows slowly then moderately to 8 feet high, sometimes up to 20 feet high. Young plants in nursery containers look gawky, but in a year or two they fill out from the base with many branches and become wide, sometimes globe-shaped shrubs. Foliage is dark green and leathery and covers plant to the ground. Serrated leaves are long, slender and grow in an upward-pointing pattern along stems, somewhat reminiscent of oleander (*Nerium* species). Tiny creamy flowers in wide flat clusters appear in summer. This plant is easy to grow and care for, tolerant of adverse conditions of intense sun, poor soil, hot winds and cold.

Special design features: Handsome plant that will grow in tough conditions. Strong vertical shrub. Foliage may take on a bronzy cast in cold weather.

Uses: Specimen. Tall unclipped hedge or screen. Space divider, wind, dust or noise screen. Trains well into patio-size tree. Can be clipped, but loses character. Transitional areas.

Disadvantages: Young plants are slow to get started. Sometimes infested with red spider mites or aphids.

Planting and care: Plant from containers

any time. Space 4 feet apart for clipped hedge, 6 to 8 feet for screen or row planting. Spray for pests as needed. Once established, plants can be almost neglected.

Vauquelinia angustifolia is another desirable rosewood. It is similar to the above, except it has very narrow serrated leaves which give the foliage an unusual, thread-like effect. Its cultural requirements and landscape uses are the same as *Vauquelinia californica*.

Vauquelinia californica

Vauquelinia californica

All zones

Evergreen

Soil: Tolerant. Prefers good drainage.
Sun: Part, full or reflected sun.
Water: Moderate until established, then deep periodic irrigation to encourage growth. Established plants in areas of 12 inches or more annual rainfall can be neglected after second year but grow slowly.
Temperature: Hardy to cold. Young plants with succulent new growth can be damaged by a sudden sharp freeze.
Maintenance: Little to complete neglect.

VERBENA peruviana
Family: *Verbenaceae*

Peruvian verbena

(*V. chamaedrufolia, Glandularia peruviana*). The Peruvian verbena from Argentina and southern Brazil is one of the more colorful low plants grown as an annual or perennial. Fast growing to 6 inches or higher, a single plant can spread to an indefinite width with trailing stems rooting at the nodes. Several plants will weave a dense mat over the ground. Medium to dark green foliage is hairy and finely cut. The best show of flowers is in spring, when plants put on a brilliant show of color. Blooms appear at stem tips as a continuous unfolding of tiny single flowers clustered in small, flat bouquets. Plants stop blooming if weather becomes too cold or if soil becomes dry. In drought they progressively brown but can be revived for a period by irrigation. They continue to bloom in summer heat when given adequate moisture but do better with a little shade, especially in low and middle zones. The species has red flowers with white throats, but there are numerous hybrids with flower shades of pink, salmon, purple and white.

Special design features: Color through late winter and spring. Informal.

Uses: Attractive spilling from containers. Informal bedding plant or edging. Brighten swimming pool patios, parkways or terraces. Rock, natural or wild gardens.

Disadvantages: Eventually dies in spots and needs to be replaced. Sometimes does not cover evenly.

Planting and care: Plant from containers or rooted divisions in spring, spaced 12 to 18 inches apart. Trim occasionally to remove spent flowers and to rejuvenate. Give occasional light feedings. Replace declining plants as they become unsightly.

> **All zones**
>
> **Herbaceous perennial**
>
> **Soil:** Improved garden soil with good drainage.
> **Sun:** Full sun to part shade.
> **Water:** Moderate. Plants in containers require ample.
> **Temperature:** Hardy to cold, but becomes dormant. Tolerates heat with water.
> **Maintenance:** Occasional trimming or replacing plants.

Verbena peruviana

VIBURNUM burkwoodii
Family: *Caprifoliaceae*

Burkwood viburnum

This lustrous green hybrid shrub has sentimental value for many. Young plants are straggly at first, slowly developing the dense, rounded form for which they are valued. Eventually reaching 6 feet high and 4 feet wide, plants are clothed with 3-1/2-inch long leaves that are deep green on top and whitish underneath with brown veins. Spring flowers in 3-1/2-inch wide clusters may produce small, unimportant, blackish berries. Exquisitely fragrant white flowers open from pink buds. Considered a deciduous shrub, leaves may hang on plants but turn purple in warm winter areas.

Special design features: Woodsy feeling. Dense green.

Uses: North sides, shaded gardens, sheltered patios, atriums and entryways. Espaliers on shaded walls, clipped hedges or as a specimen.

Disadvantages: Poor appearance when young and during winter. Sometimes gets iron chlorosis. Foliage may burn from sun, hot winds or soil salts.

Planting and care: Plant from containers any time, but fall is best. Space 2 to 3 feet apart for massing or hedge. Shear as needed. Clipped or heavily pruned plants may not bloom because flowers are produced on wood grown the previous year. Give regular feedings of a balanced fertilizer and extra iron to maintain healthy green plants.

> **Best in high zone, cool shaded pockets of middle zone**
>
> **Deciduous to partly evergreen**
>
> **Soil:** Highly organic garden soil with good drainage. Avoid alkaline conditions.
> **Sun:** Open, filtered or part shade.
> **Water:** Moderate to ample. Occasionally soak soil deeply to leach out salts.
> **Temperature:** Hardy to cold. Dislikes intense heat.
> **Maintenance:** Periodic feeding, soaking and clipping.

Viburnum burkwoodii

VIBURNUM suspensum

Family: *Caprifoliaceae*

This shrub native to Ryukyu Island near Japan grows at a moderate rate to 6 to 8 feet high with an equal spread. A lustrous foliage plant, it provides a welcome splash of green in a shaded garden. Oval leathery leaves to 4 inches long densely cover its rounded form. Tiny pinkish flowers in tightly packed 1-1/2-inch clusters appear in late winter, occasionally followed by small red fruit in summer.

Special design features: Woodsy feeling for shaded places. Attractive at close range. Rich, handsome winter foliage at a time when most plants look barren.

Uses: Specimen, hedge, screen or background plant for north or east sides of buildings. Suitable for shaded patios, gardens and courtyards. Clipped hedge or small patio tree.

Disadvantages: Foliage may sunburn. Sometimes languishes in warm summer areas, but regains its vigor as cool weather approaches. Subject to iron chlorosis.

Planting and care: Plant any time from containers, but best during cooler periods. Space 4 to 5 feet apart for unclipped screen or background planting, 3 to 4 feet for clipped hedge. Clip as needed. Trim and prune after bloom in spring before high heat. Trimmed or sheared plants have little or no bloom. Feed regularly and give extra iron. Leach soil salts with occasional deep soakings.

> **Middle and high zones, marginal in low zone**
>
> **Evergreen**
>
> **Soil:** Highly organic garden soil with good drainage. Avoid alkaline conditions.
> **Sun:** Open, filtered or part shade.
> **Water:** Moderate to ample with occasional deep soakings.
> **Temperature:** Hardy to 10°F (−12°C) or below.
> **Maintenance:** Regular feeding, pinching and soaking.

Viburnum suspensum

VIBURNUM tinus

Family: *Caprifoliaceae*

A shrub from the Mediterranean region, this viburnum accepts more sun than those previously mentioned. Growth is moderate to 6 to 10 feet high with a 3 to 6-foot spread. Leaves are 2 to 3 inches long, rough and dark green on top, lighter underneath. Small pinkish flowers appear in 3-inch clusters in late winter or early spring, and are sometimes followed by clusters of tiny dark berries. There is a warm, rosy glow to the plant caused by a reddish coat on the stems and twigs, especially in winter. Cultivars are planted more often than the species, especially 'Lucidum'.

Cultivars: 'Lucidum', shining laurustinus, has larger leaves and is more resistant to mildew than the species, but less hardy to cold. Best in low and middle zones. 'Dwarf' grows only 3 to 5 feet high and as wide. Good for low screens, hedges or foundation plants. 'Robustum', roundleaf laurustinus, has coarser, rougher leaves than the species and grows dense and erect with pinkish white flowers. It makes a small, narrow tree or medium-size shrub for a narrow space. It is more mildew resistant than the species.

Special design features: Very decorative garden plant. Refined woodsy effect. Handsome form, foliage and flower.

Uses: May be used in the open in middle zone if there is air circulation around it and no reflected heat—avoid west exposure. Specimen, screen, clipped hedge, background plant or espalier.

Disadvantages: Sometimes gets iron chlorosis. The species is subject to mildew, especially in humid areas.

Planting and care: Plant any time from containers, but best during the cool season. If planted in a sunny location it should be set out in spring or fall so it can adjust to summer. Space larger forms 1-1/2 to 2 feet apart for clipped hedge, 2 to 5 feet for loose screen. Space dwarf form 2 to 4 feet on center for massing, 1-1/2 to 2 feet for low clipped hedge. Prune lightly after bloom. Shear as needed.

> **All zones**
>
> **Evergreen**
>
> **Soil:** Improved garden soil with good drainage. Avoid alkaline conditions.
> **Sun:** Filtered, open or part sun. Tolerates full sun in middle and high zones.
> **Water:** Moderate to ample. Give deep soakings occasionally to leach soil salts.
> **Temperature:** The species is hardy to 5°F (−15°C). Cultivars may be damaged by cold at about 15°F (−10°C). All tolerate heat better than the other viburnums.
> **Maintenance:** Periodic feeding, pinching and soaking.

Viburnum tinus 'Lucidum'

VINCA major
Family: *Apocynaceae*

A vigorous, low, trailing plant, periwinkle quickly spreads on long stems that root as they go. Leaves 2 inches long are dark green and grow in opposite pairs along the stems. Plantings become dense mounds in 2 or 3 years. Single blue flowers 2 inches in diameter appear among the foliage in spring. Plant looks amazingly green and lush for the amount of care and water needed. Although it may look wilted if allowed to go too long without irrigation, it restores itself immediately when water becomes available. Native to Europe.

Cultivars: 'Variegata' has leaf margins of yellowish white.

Special design features: Lush, green, mounding ground cover, trailing over planters or climbing fences. Green woodland or jungle feeling.

Uses: Ground cover to extend lawns, vary texture, create patterns. Bank cover for erosion control. Filler plant for bare areas. Containers and planters. Use under trees to hide leaf litter. Naturalizes in woodsy locations of middle and high zone where it survives periods of drought and bounces back quickly when moisture returns.

Disadvantages: Invasive. Greedy roots may overcome less aggressive plants in the competition for moisture. Difficult to eradicate once it has become established. Established plantings tend to take over an area. Sunburns in the hottest locations in midsummer; recovers quickly as the weather cools off.

Planting and care: Plant from containers or flats any time. Bare root divisions may be planted during cool periods. Space 18 to 24 inches apart for fast cover. Cut back occasionally in late winter to renew vigor and keep neat. Plants in shade tolerate more drought.

Vinca minor is a smaller, more delicate and refined version of *Vinca major*. Leaves are small, more pointed at the end and very dark green. This plant grows less rapidly but in time covers as densely, mounding 6 to 12 inches high. It seldom flowers in arid regions but produces 3/4-inch lilac-blue blossoms in more favorable situations. It is an important ground cover plant in the temperate zone and is occasionally planted in the middle and high zones, often as an annual. Its best use is as a ground cover or container plant in a small area, where it can be enjoyed and cared for easily.

All zones

Evergreen

Soil: Tolerates a wide range. Prefers improved garden soil.

Sun: Full, open, filtered or part shade. Full sun in middle and high zones, but may look poor in hottest part of summer. Looks best with afternoon shade in hot regions.

Water: Moderate to ample. Tolerates deep periodic irrigation. Can be allowed to go completely dry, turning brown and shriveling, but recovers miraculously when irrigated.

Temperature: Hardy to cold, but may sustain some foliage damage at 15°F (−10°C). Recovers quickly. Tolerates heat, especially in shade and when given water.

Maintenance: Occasional pruning to control.

Vinca major

VITEX agnus-castus
Family: *Verbenaceae*

Chaste tree is a widely adaptable shrub or small tree with single or multiple trunks and a usually wide-spreading crown. It is native to southern Europe, but has naturalized in warm areas of the United States. Without irrigation it normally remains a shrub, growing no more than 6 feet high. With moderate amounts of water it grows quickly to 15 to 25 feet high with an equal spread. It seems to need heat to develop fully and to bloom well and stays smaller in cool climates. Trunk is often picturesque with gray stringy bark. Leaves are very dark green with 5 to 7 narrow-pointed leaflets fanning out from the center. Numerous flower spikes 7 inches long appear above foliage in early summer and occasionally in early fall. Flowers are usually blue but there are pink and white selections. Tiny, woody, round "peppers" follow bloom.

Cultivars: 'Rosea' has pinkish flowers. 'Alba' has white flowers.

Special design features: Picturesque. Fast shade for summer. Summer bloom.

Uses: Patio or lawn tree. Garden tree as center of interest or along edge of landscape. Transitional or wild gardens as tree or shrub.

Disadvantages: Occasionally gets wood rot. Twiggy winter form is not especially attractive in gardens unless thinned and shaped a bit.

Planting and care: Plant from containers any time or bare root in winter. Prune during dormant season to remove dead wood and to groom and shape.

All zones

Deciduous

Soil: Tolerant of wide range of soil conditions. Grows luxuriantly in rich deep soils, but produces few flowers.

Sun: Part, full or reflected sun.

Water: Moderate to little.

Temperature: Hardy to cold. Revels in heat.

Maintenance: Occasional light trimming.

Vitex agnus-castus

VITIS vinifera

Family: *Vitaceae*

The grape is a tendril-climbing woody vine, with large, deeply lobed leaves and tan stringy bark. Plants produce clusters of small, tasty, round fruit in summer. But this common description doesn't quite express the history and mystique involved with this plant, originating from the Caucasus and grown widely for the wine produced from its fruit. Egyptian tomb paintings show the grape being cultivated in arbors. The Greeks held ribald festivals honoring Bacchus, god of wine. The Jews broke bread and drank wine together in the joyous sharing ceremony of the Kiddush. Christians sanctified wine as the blood of Christ. Places like Bordeaux, France, the Rhine River in Germany and the Napa Valley in California have become famous for the wines they produce. Today there is wide interest in grape growing and wine production. The grape vine is an excellent landscape plant, growing well in arid zones. Once established, a vine can grow rapidly to cover a 20 by 20-foot area and is easily trained to cover arbors or to drape porch posts. The fruit is a decorative and delicious plus. There are many kinds of grapes and the serious grape grower should investigate the varieties and special ways of pruning and caring for them.

Cultivars: 'Thompson's Seedless' from Persia is outstanding for landscape use. It does well up to 4,500 feet elevation. It produces pale, green, sweet, seedless fruit in July. 'Golden Muscat' from Geneva, New York is another fine eating grape. Fruit has seed, is gold tinged with bronze and very sweet. It ripens in late July and August. It does well in high zone, but is not recommended for low deserts because it gets sunburned leaves. 'Black Monukka' is vigorous and productive. Small reddish black seedless grapes ripen in July.

Special design features: Informal garlands of bold leafy form for summer shade and greenery, and winter sun. Interesting classical sculptural effect.

Uses: Trellises, arbors, porches and fences.

Disadvantages: Vine grows rapidly and needs clipping and training often during growing season. Occasionally gets Texas root rot, bacterial crown gall or root knot nematodes. Powdery mildew sometimes causes twigs and leaves to look as if they have been dusted with flour. The most serious threat is the grape leaf skeletonizer—armies of tiny yellow and black-striped caterpillars can sometimes be found on the undersides of leaves. Small numbers can be controlled by picking off infested leaves. For serious attacks, Sevin is an effective spray. Be sure to follow all label directions.

Planting and care: Container-grown plants may be planted any time. Cuttings or bare-root plants may be planted in winter. For landscape purposes, train strong leaders up the support and tie securely. Cut off any unwanted side branches or weave them in and out to form a garland. Plants may be pruned to a basic framework each winter or may be left alone until spring comes. At that time cut off dead wood and less vigorous canes which do not sprout. Head back new growth by pinching off tips to encourage bushiness. When grapes form, there are a number of ways to increase their size. One is to remove a number of clusters so the vine produces fewer but larger fruits. The clusters themselves can be thinned. Long clusters of 'Thompson's Seedless' can be shortened and one or more upper branches removed. Remove about one-third of the cluster. Grapes on landscape plants may be small because pruning for fruit production is considered secondary to shape of vine. Grapes should be covered with paper bags as they grow so that birds and insects do not eat them. Fertilize plants lightly with composted manure in fall. Or give a light sprinkling of ammonium sulfate or complete fertilizer in February: no more than 1 pound to 100 square feet of soil surface. You can skip fertilizing completely, because grapes can often go for many years without a feeding. Grapes can also thrive in poor soil.

All zones

Deciduous

Soil: Tolerant. Prefers porous or gravelly soils with some humus added, having good drainage.

Sun: Part, full or reflected sun, except for 'Golden Muscat', where reflected sun may burn leaves.

Water: Constant soil moisture is necessary in spring and summer for grape production. Otherwise the plant is drought-resistant and tolerates periodic soakings.

Temperature: Hardy to cold, but may suffer some twig damage in cold winters. Late spring frosts can injure or kill grape set and burn leaves. Plant recovers from frost damage quickly in warm weather and loves heat.

Maintenance: Frequent clipping and training and sometimes spraying in summer. Single pruning in winter.

Vitis vinifera

Vitis vinifera

WASHINGTONIA filifera
Family: *Palmae*

California fan palm • Desert fan palm • Petticoat palm

(W. filamentosa, Pritchardia filifera). A large scale fan palm with a wide, heavy trunk, this native of the Southwest grows slowly to as much as 80 feet tall, although it is usually seen at heights of 20 to 40 feet. The dense head spreads 15 feet or more and is composed of stiff, gray-green, fan-like leaves with hairy filaments. Leaves are held well away from the erect trunk on 6-foot tooth-edged leaf bases. Old leaves droop to the trunk and hang, forming a dense, straw-colored thatch or "petticoat." There are two strains of this palm: The California desert type retains its dead leaves all the way down the trunk to the

Washingtonia filifera

ground unless they are removed. The form from native Arizona stands is self-pruning and drops old drying leaves. Long blossoms like streamers emerge from crown in summer and produce small, abundant, white flowers and blue-black fruit on females. Bare trunks of trimmed or self-pruned palms have a brownish fibrous look and often flair at the base. In nature, plants grow in clusters in wet spots.

Special design features: A grand scale palm. Massive. Formal. Ponderous. Strong vertical, eventually becoming a skyline tree. Strong architectural emphasis.

Uses: Too large for the average residence. Boulevards, parks and public spaces for cadence or accent. Groves are impressive and dramatic. Rows make walls in the landscape. A single plant is like an exclamation point. Tiny young plants in containers stay small for a long period. Good transitional plant. Looks attractive rising out of the bare earth.

Disadvantages: Occasionally gets bud rot, which is almost impossible to diagnose until it is too late. Treat palms near an infected palm by saturating their crowns with a Bordeaux mixture. This helps prevent contraction of the disease and may

cure early cases. Slow to develop. Trunks are occasionally infected with a rot near the base. Tall plants are expensive to groom.

Planting and care: May be started from seed, but very slow. Plant from containers after frosts have passed in spring. Transplant field-grown, balled in burlap stock in warm season. Their heavy weight and bulk require an expert to plant them. Space 30 feet or more apart for boulevard or row planting. Place at random distances for natural grove, including several close together in a clump, as they are found in nature. Feed and water generously to encourage fast growth. Established plants tolerate neglect and will grow for years with no care in areas with 10 to 12 inches of annual rainfall. To groom, trim drying fronds and remove fruit garlands.

All zones

Evergreen

Soil: Tolerant. Fastest in rich moist soil.
Sun: Part to full sun.
Water: Moderate to little.
Temperature: Hardy to about 15°F (−10°C). Young plants are more susceptible to cold. Slow to recover from frost damage.
Maintenance: Occasional grooming.

WASHINGTONIA robusta
Family: *Palmae*

Mexican fan palm • Mexican washingtonia • Thread palm

(W. gracilis, W. sonorae, Pritchardia robusta). This fast growing fan palm from Mexico may reach 80 to 100 feet high. The slender trunk may be only 12 to 14 inches in diameter. Its glistening head of fanlike leaves spreads 10 to 12 feet in diameter. Leaves are richer green than *W. filifera* and more festive and refined in appearance. Old leaves become dry and hang down along the trunk as thatch. They can be left or removed, depending on your preference. Young Mexican fan palms may be differentiated from *W. filifera* by a reddish streak along the underside of leaf stalk near the trunk. Trunks shorn of old

Washingtonia robusta

leaf stubs are brown to gray and fairly smooth and may taper from a stout base. It is difficult to determine how fast this palm grows, but it is reasonable to expect a 12 to 15-foot palm at the end of 10 years, perhaps less. Growth depends on the amount of moisture it receives. Long straw-colored streamers in spring develop sprays of tiny white flowers followed by small dark fruit on female trees.

Special design features: Graceful. Festive tropical feeling. Luxuriant vacation mood. Strong vertical. Dramatic, especially in groups. Jungle feeling when used with such plants as bamboo. Eventual skyline silhouette.

Uses: Tall emphasis plant used as specimen, in pairs or clumps. Streets, parks, entrances and public places. Silhouette plant against the sky or tall structures. Two or more palms planted together develop curving trunks as they arch away from each other, or if planted at an angle.

Disadvantages: These plants get *tall*. Do not plant them near power lines or under structures. Mature palms are difficult to groom and require professional tree trimmers. Thatch left on palms may appear ragged and uneven. Foliage on younger palms is subject to frost damage, but plant recovers by mid to late spring after growing a new set of leaves.

Planting and care: Can be started from seed which germinates in 60 days. Container plants are much faster and not too expensive. Plant in spring when danger from frost is past. Transplant palms April through October. Even large plants can be moved successfully at this time. Feed and give ample water for fast growth. Older plants withstand periods of neglect and more frost than young plants. This palm seems to be more resistant to disease than other palms. To groom, remove old dry leaves and flowering parts in late spring. Trunks may be skinned so no leaf bases are left.

Low and middle zones

Evergreen

Soil: Tolerant. Prefers improved garden soil with good drainage.
Sun: Part to full sun.
Water: Moderate for fast growth or even ample. Deep periodic irrigation is satisfactory, but enjoys the ample irrigation of a lawn. Newly transplanted palms greatly benefit from a drip irrigation system.
Temperature: Leaves are damaged in the low 20's°F (−5°C). Plants recover quickly in spring.
Maintenance: Occasional grooming.

WISTERIA floribunda

Family: *Leguminosae*

(W. multijuga). This woody, twining vine is a sentimental favorite, often used in so-called old-fashioned or Victorian gardens. Slightly hardier than Chinese wisteria, *Wisteria sinensis,* it differs by having flower clusters which open over a period of time rather than all at once, thus prolonging the bloom period. Clusters of purple pea-shaped flowers 8 to 18 inches long appear before leaves come out, creating quite a show. Velvety pods 6 inches long may follow. Medium green leaves are formed of 13 or more pointed leaflets to 3 inches long. Trunk, twigs and branches become gray and woody as plant matures. In time, lower trunk can become up to 3 inches thick and rigid. Plants grow at a moderate rate and accept training to almost any shape. A single plant will cover an area of 10 by 10 feet or more. Some plants are trained to become self-supporting trees. Many decorate only a small area of a trellis. Eventual size depends on care and soil conditions. Plants may grow fast at first, then slowly. Select grafted plants because seedlings take years before blooming.

Cultivars: 'Longissima Alba' has white flowers to 2 feet long. 'Rosea' has pink flowers in clusters to 1-1/2 feet long.

Special design features: Oriental effect.

Woodsy or old-fashioned garden mood. Spring color. Refined summer shade and greenery. Sculptural structure or bare branches when dormant.

Uses: Cover for *sturdy* arbors, porches, trellises or other framework. Very attractive on pergolas. May be trained as shrub or small tree.

Disadvantages: Subject to Texas root rot and should not be used in soil known to be infected. Subject to sooty canker which

Wisteria floribunda 'Longissima Alba'

can be prevented by treating cuts and wounds. May get iron chlorosis which can be cured or prevented by feeding plant iron. Pods and seeds are poisonous. Some plants bloom heavily only every other year.

Planting and care: Plant any time from containers. Place in soil mix prepared to prevent Texas root rot. To train a plant on a framework, tie main stem to a strong support at frequent intervals. Trim side shoots and shorten long streamers. Train new shoots in the manner you wish. It is best to train about three shoots to separate vines on supports to prevent excessive intertwining. If a tree form is desired, it is easiest to purchase a plant already trained. Do not fertilize heavily or plants will grow rank and have few flowers.

All zones

Deciduous

Soil: Average garden soil with good drainage.
Sun: Part to full sun.
Water: Moderate to ample.
Temperature: Hardy to cold. Tolerates heat with irrigation.
Maintenance: Considerable early training then only occasional pinching and cutting.

XYLOSMA congestum

Family: *Flacourtiaceae*

(X. racemosum, X. senticosum, Myroxylon senticosum). One of the choice, all-purpose, landscape plants, xylosma always looks nice, is easy to grow and requires very little care. Growth is moderate to fast once it gets started. It may reach 8 feet high as a shrub in 6 or 7 years. Trained as a small garden tree it can reach up to 20 feet high in 15 years. It is easily trained to any shape or kept to any size. Foliage is shiny bright green with a bronzy cast when new. Leaves are very pointed, but are wide at the base with toothed edges. Flowers are green and seldom noticed.

Cultivars: 'Compacta' has a tighter branching habit and can be used for low hedges, borders or screens.

Special design features: One of the most outstanding and agreeable garden plants which looks as nice close-up as at a distance, clipped or unclipped. Versatile, refined and well-behaved.

Uses: Specimen, wide screen, background planting, clipped hedge or wall plant. Topiary. Easily kept narrow for limited spaces. As a single or multiple-trunk tree, it makes one of the loveliest canopies for the

garden or patio. Swimming pool areas.

Disadvantages: May be damaged by spring frost. Slightly susceptible to iron chlorosis and Texas root rot. Sometimes gets spider mites or scale.

Planting and care: Plant from containers any time, but best in spring. Space up to 3 feet apart for clipped hedge, 4 to 6 feet for screen. Do any heavy pruning in late winter or early spring. Clip any time. For a tree form, purchase a plant already trained.

All zones

Evergreen to deciduous in coldest areas

Soil: Tolerant. Prefers improved garden soil with good drainage.
Sun: Filtered or part shade to full sun.
Water: Moderate. Tolerant of some drought when established. Older plants may be given deep periodic irrigation.
Temperature: Hardy to 10°F (−12°C), but may show some damage around 25°F (−4°C). Early spring growth may be nipped by a late frost but plant recovers quickly. Tolerates heat.
Maintenance: Little to regular after initial training as desired.

Xylosma congestum (tree)

YUCCA aloifolia

Family: *Agavaceae*

A bold garden or desert yucca from the southern United States, Mexico and the West Indies, Spanish bayonet grows slowly to 10 feet high. Stiff, smooth-edged, medium to deep green leaves closely set along the stalk have sharp spikes at their tips. Leaves are shorter than most yuccas. Some reach only 12 inches long and stay green on the stalks unless stressed by drought. Plants may be erect, leaning or sprawling, with a single head or several branching out from the central stalk. Summer blooms are clusters of white, purple-tinged, lilylike flowers to 4 inches across. They rise out of plant tips on 2-foot stalks, usually one to each head.

Cultivars: 'Marginata' has yellow margins.

Special design features: Bold foliage for specimen or silhouette. Desert effect. Tropical in combination with tropical plants.

Uses: Accent plant as specimen or in grouping. Containers.

Disadvantages: Very sharp leaf spikes require placement away from walkways or other use areas.

Planting and care: Plant field-grown plants or plants from containers any time. Space 6 to 8 feet apart for grouping. Low leaves may be skinned from plant if you desire a small, palmlike form with a spiky head. If plant grows sideways, be sure to cut off spikes or remove heads before they interfere with walkways. Cut stalks may be placed in the shade for a week to heal and then planted. They will take root and grow. Remove old flower stalks as they become unsightly. All non-native yuccas are subject to the same grub which attacks agaves. Treat with currently recommended insecticide if grubs have infected nearby plants. See page 38.

Yucca aloifolia

> **All zones**
>
> **Evergreen**
>
> **Soil:** Tolerant. Needs good drainage.
> **Sun:** Part, full or reflected sun.
> **Water:** Moderate to little. Tolerates long periods of drought, but lower foliage dries out and leaves become a light green.
> **Temperature:** Hardy to cold. Tolerates heat.
> **Maintenance:** Very little.

YUCCA gloriosa

Family: *Agavaceae*

Spanish dagger is a dramatic yucca from southeast United States and northeast Mexico. It looks at home as an accent plant or grouped with bold foliage plants to produce a tropical effect. Bright to yellowish green leaves are 2-1/2 feet long, 2 inches wide, fleshy and fairly stiff with pointed tips. Growth is moderate to 8 feet tall. It may develop a number of trunks at its base, forming a clump 8 feet wide. When old foliage is skinned off, the resulting slender woody trunks and spiky heads look like miniature palms. Unskinned leaves stay green for several feet down the plant before dying and becoming thatch. Flower spikes are 3 feet tall and produce large creamy to pinkish blooms in clusters. Unlike other yuccas, bloom spikes are short, beginning down among the leaves. Secondary trunks can be removed at the base and healed for a week in shade, then planted elsewhere to make a new plant. Although leaf tips are pointed, they have no sharp spikes and are not dangerous unless at eye level.

Special design features: Bold silhouette. Dramatic vertical form. Tropical or garden effect.

Uses: Accent plants in a bed, containers or garden. Dramatic emphasis at an entrance. May be safely used around swimming pool patios if set away from traffic patterns. Specimen, clump or in combination with other plants.

Disadvantages: May become chlorotic in alkaline soils. May sunburn from reflected heat—avoid western exposure. Yuccas in winter on south exposures are susceptible to sunburn when the sun is low and rays get under leaves. Frost-damaged leaves are unsightly and recover slowly.

Planting and care: Plant from containers any time, but best in spring. Cuttings are best started in spring. Space 6 to 8 feet apart to allow room for a clump to develop. Remove old leaves or allow to remain as thatch according to taste. For an especially dramatic form, remove all leaves but those at the top of trunks. Remove old flower stalks to groom.

Yucca gloriosa

> **Low and middle zones, warmer locations in high zone**
>
> **Evergreen**
>
> **Soil:** Prefers improved garden soil with good drainage.
> **Sun:** Filtered, open or part shade. Full sun.
> **Water:** Moderate. Accepts ample with good drainage. Tolerant of little water but may be unattractive.
> **Temperature:** Damaged at 20 to 24°F (−5°C). Tolerant of heat.
> **Maintenance:** Little to some.

YUCCA recurvifolia
Family: *Agavaceae*

(Y. pendula). Pendulous yucca is native to the southeastern section of the United States and to Mexico. Fast growing to 6 feet high, it develops one or several branches. One plant can spread up to 6 feet wide. Foliage is dark gray-green. Thin leaves to 3 inches wide at the base are 3 feet long and have soft points at their tips and bend downward from the plant about half their length. Spikes 3 to 5 feet long rise vertically above heads in early summer and bear delicate, white, lilylike flowers in loose clusters about 2 feet long. More of a garden plant than a desert yucca, larger plants in a garden setting need grooming to be attractive. They can be cut back every few years and will develop several branches below cut.

Special design features: Bold foliage. Tropical feeling.

Uses: Containers. Garden situations. Accent. Foundation or transitional plant. Attractive in low maintenance situations.

Disadvantages: Sometimes attacked by a grub which eats the roots, killing plant. Grub-damaged plants sometimes resprout from surviving roots. Ungroomed plants become leggy or develop dry leaves at bases which look out of place in a garden setting. Aphids may invade flowers as they begin to open—wash them off with a spray of water.

Planting and care: Plant any time from containers. Branches may be removed, healed in the shade for a few days and then planted. Space plants 4 to 6 feet apart for massing or rows. To groom, cut back plants or trim foliage in late winter. Remove old flower stalks after bloom. To reduce size and to rejuvenate, cut plants back to about 1 foot above ground. New branches will sprout below cut. Prevent grubs by applying diazinon around the base of plant in spring.

All zones

Evergreen

Soil: Tolerant. Needs good drainage.
Sun: Full sun. Open, filtered or part shade. Burns in reflected sun.
Water: Moderate. Tolerates some drought when established or spaced irrigation.
Temperature: Hardy to cold. Accepts heat with water.
Maintenance: Occasional grooming or treatment for grubs.

Yucca recurvifolia

YUCCA species
Family: *Agavaceae*

Desert yuccas are abundantly scattered over the Sonoran and especially the Chihuahuan deserts. Many are transplanted into home landscapes, even though they are not as versatile as yuccas available in nurseries. Native laws protect yuccas and many states require a tag on plants in the home landscape or it is subject to confiscation. Desert yuccas are collected in the wild and are sometimes difficult to establish. Most are so large they need to be propped up until they reroot. Most need some supplemental irrigation in the low zone.

Yucca brevifolia
Joshua tree

This picturesque plant sometimes reaches 40 feet tall. It is very striking, with long arms or branches that often take on irregular shapes. Grayish or dull green leaves are quite narrow and sharp. Clusters of greenish white, lilylike blossoms appear at the end of each branch. A naturally occurring variety, *Yucca brevifolia herbertii*, reaches about 15 feet tall, so is a better plant for most home landscapes. Both are native to the high deserts but hardy in all deserts.

Yucca brevifolia

Yucca elata
Soap tree

This is another desert yucca often seen in home landscape plantings, and is perhaps the most commonly used of desert yuccas described here. It is also native to the high deserts of northern Mexico and the southwestern United States. Form is erect to about 12 feet tall. Leaves are grasslike, with little threads curling along the edges. Old leaves hang down along the trunk, creating a straw-colored thatch. Top leaves are light green with a white edge, giving plants a grayish cast. Plants produce a bold effect and are especially attractive silhouetted against a plain background. Flower spikes produced in late spring are dramatic, topping the thatch of drying leaves. Blossom stalks reach up to 6 feet or more high and display clusters of fragrant snowy white blooms.

Yucca whipplei
Our Lord's candle

This yucca has two chief advantages: It can be raised easily and quickly from seed, and its whorls of sharply spiked leaves on its trunkless form make formidable barriers. This native of California and Baja California, grows slowly or moderately to 3 feet high. Very narrow, gray-green leaves are thick and rigid, to 2 feet long. They often have finely toothed edges and are always sharply armed at the tips. They often grow in clumps and usually have single heads. White, graceful, lilylike flowers sometimes tinted purple appear as 6-1/2-foot clusters at high ends of towering 8 to 12-foot treelike stalks that rise above plants. Flowers are followed by fat, green, stalks—seed capsules which eventually open when dry to scatter many seeds. Plants die after bloom the same as agaves.

There are several varieties found in nature. *Y. w. caespitosa* has dense, compact, secondary rosettes which form in the seeding stage where leaves join. These plants usually create a clump and produce several flower stalks when bloom time arrives. Another variety, *Y. w. percursa*, has more open clumps, formed from underground or rhizomes.

Yucca elata

All zones

Evergreen

Soil: Tolerant. Needs good drainage.
Sun: Part, full or reflected sun.
Water: Moderate to little. Tolerates long periods of drought, but lower foliage dries out and leaves become a light green.
Temperature: Hardy to cold. Tolerates heat.
Maintenance: Very little.

ZIZIPHUS jujuba
Family: *Rhamnaceae*

Chinese date

(*Z. vulgaris*). This decorative tree is often overlooked when selecting plants for landscape purposes. It grows at a moderate to slow rate 20 to 30 feet high and 15 to 20 feet wide. Growth rate and eventual size depend on water supply. Erect trunk or trunks are clothed in interesting, rough, gray bark. Bare winter forms have a unique twig structure of nubby, angular, spiny branches. Glistening green leaves 1 to 2 inches long are slightly curved and shimmer in the breeze. Foliage is dense and fine textured, hanging in weeping cascades from the branches. Small, unimportant, yellow flowers in late spring or early summer produce brown, fleshy, edible "dates" in fall, 1 inch in diameter. They ripen only in hot summer areas and taste something like a blend between a ripe apple and a date. Leaves turn golden in autumn. Basal suckers will form groves unless cut down. This tree is deep rooted and tolerant of periods of drought, heat, cold and alkaline soil conditions. Native to eastern Europe and China.

Cultivars: 'Intermis' has no thorns.
Special design features: Picturesque tree through all seasons. Weeping summer form. Fall color. Naturally forms groves.
Uses: Patio or lawn tree as specimen or grouping. Groves. Looks nice as a silhouette or in combination with other plants. Most effective if planted where its form can be seen and admired and when suckers are removed.
Disadvantages: Plant suckers profusely, sometimes a nuisance. Subject to Texas root rot.
Planting and care: Plant any time from containers or bare root in winter. Prepare planting hole to prevent Texas root rot and do not plant where soil is known to be infected. Space 10 to 12 feet apart for a grove effect or deciduous screen. Prune in late winter to encourage natural weeping form, to shape, to remove dead wood and to contain size. Plants are tolerant of neglect, but grow slowly. They thrive on regular garden care.

All zones

Deciduous

Soil: Widely tolerant. Accepts alkaline or saline soils. Prefers deep, improved, garden soils with ample soil moisture.
Sun: Part shade. Full or reflected sun.
Water: Thrives in lawn situations. Does well in dry areas with deep periodic irrigation. Established plants tolerate periods of neglect.
Temperature: Hardy to cold. Tolerant of heat.
Maintenance: Little to some, depending on situation.

Ziziphus jujuba

Plant Sources

Some of the plant species in this book are new or may be unavailable at your local nursery. The following list of sources may help you in your search.

Adams Company, Inc.
3333 N. Walnut
Tucson, AZ 85712

Arbor Tree and Cactus
1320 W. Newton Drive
Tucson, AZ 85704

Arizona Eldarica Corp.
3111 E. Marilyn
Phoenix, AZ 85032

Boyce Thompson Southwestern Arboretum
P.O. Box AB
Superior, AZ 85273
Arid land plants. Nominal fee required for seed and propagating material.

Clyde Robins Seed Company
P.O. Box 2855
Castro Valley, CA 94546
Wholesale-retail. Seeds only.
80-page retail catalog $2.

Desert Flora
11360 East Edison
Tucson, AZ 85715
Nursery-grown stock.

Desert Products Nursery
2854 E. Grant Road
Tucson, AZ 85716
Trees, shrubs, cacti.

Desert Trees
9559 Camino del Plata
Tucson, AZ 85704
Desert trees and shrubs.

Environmental Seed Producers
P.O. Box 5904
El Monte, CA 91734
Wholesale. Seeds only. Pricelist available.

Gerlach's
1901 North Park
Tucson, AZ 85719
Field-grown cacti and desert plants.

Mountain States Nursery
10020 West Glendale Ave.
Glendale, AZ 85307

Native Plants
360 Wakara Way
University Research Park
Salt Lake City, UT 84108
Wholesale. Catalog. Mail order.

Neel's Nursery
3255 E. Palm Canyon Drive
Palm Springs, CA 92264
Wholesale-retail. Some cacti.

Plants of the Southwest
1570 Pacheco St.
Santa Fe, NM 87501
Wholesale-retail. Free seed catalog.

Sonora
16647 W. Northern Ave.
Waddell, AZ 85355

Stover Seed Company
P.O. Box 21488
1415 E. 6th St.
Los Angeles, CA 90021
Wholesale. Seeds. Catalog.

Acknowledgements

Publishers: Bill and Helen Fisher
Executive Editor: Carl Shipman
Editorial Director: Jonathan Latimer
Editor: Scott Millard
Art Director: Don Burton
Book Design/Illustration: Patrick O'Dell
Major photography: Mary Rose Duffield
Additional photography: Charlie Basham, Ted DiSante, Gill Kenny, Michael MacCaskey, Scott Millard, Muriel Orans
Cover photo: Gill Kenny
Typography: Cindy Coatsworth, Joanne Nociti

We would like to thank the following individuals for their help in preparing this book:
George Brookbank, Extension Agent, Tucson, AZ
Tony L. Burgess, Tucson, AZ
Christopher Duffield, Ph.D, Tucson, AZ
Steve Fazio, Professor of Plant Sciences, University of Arizona
John M. Harlow Jr., Tucson, AZ
Larry Holtzworth, Agronomist, Plant Materials Center, U.S. Soil Conservation Service, Tucson AZ
Eric A. Johnson, Landscape Consultant, Palm Springs, CA
James G. Jerry Lewis, Landscape Architect, El Paso, TX
Michael MacCaskey, horticulturalist, Palo Alto, CA
Laila and Ralph McPheeters, Catalina Heights Nursery, Tucson, AZ
Charles T. Mason Jr., Ph.D, Curator, Herbarium, University of Arizona
Donald F. Post, Ph.D, Soil Scientist, University of Arizona
Charles Sacamano, Ph.D, Extension Horticulturalist, University of Arizona
Carol Shuler, Landscape Architect, Phoenix, AZ
Rubert B. Streets Sr., Ph.D, Plant Pathologist, retired, University of Arizona
Darrell T. Sullivan, Professor of Horticulture, New Mexico State University
Harvey Tate, Extension Horticulturalist, retired, University of Arizona
Lance Walheim, horticulturalist, St. Helena, CA
Jim Wheat, Phoenix, AZ.

Index

174